Remote Sensing and image Processing in Mineralogy

Maged Marghany

Director
Global Geoinformation, Sdn. Bhd.
Kuala Lumpur, Malaysia

CRC Press
Taylor & Francis Group
Boca Raton London New York

CRC Press is an imprint of the
Taylor & Francis Group, an **informa** business

A SCIENCE PUBLISHERS BOOK

First edition published 2022
by CRC Press
6000 Broken Sound Parkway NW, Suite 300,
Boca Raton, FL 33487-2742

ISBN: 978-0-367-89670-6 (hbk)
ISBN: 978-1-032-21481-8 (pbk)
ISBN: 978-1-003-03377-6 (ebk)

DOI: 10.1201/9781003033776

Typeset in Times New Roman
by Radiant Productions

**Visit the Taylor & Francis Web site at
http://www.taylorandfrancis.com**

**and the CRC Press Web site at
http://www.crcpress.com**

Cover image by Maged Marghany.

Dedicated to

My mother Faridah,

COVID-19 victims
and
Richard Feynman and Michio Kaku taught me that a real distinguished
professor invents novelty not exploiting his students and colleagues
to award the highest H-index by false publications.

Preface

◇◇

This book was completed when I was ranked among the top two per cent of scientists in a global list compiled by prestigious Stanford University. Besides, the prestigious Universidade Estadual de Feira de Santana, Universidade Federal da Bahia, and Universidade Federal de Pernambuco, Brazil ranked me as the first global scientist in the field of oil spill detection and mapping during the last fifty years.

To date, various geology and mineral exploration organizations and institutions have been unable to bridge the gap between modern physics, computing and mineral exploration. Conventional geology and mineralogy data collections, even oil and gas exploration model developments, cannot be made to comprehend countless structure geological features, for instance, fault and lineament associations with mineralization zones.

There are many ambiguities exploiting remote sensing image processing in mineral exploration. Remote sensing image processing application in mineral, oil and gas exploration are restricted to specific commercial software such as ERDAS imagine ERDAS, PCI Geomatica, and The Environment for Visualizing Images (ENVI). Numerous remote sensing geology works in a developed country are ill-defined. Since the beginning of remote sensing mineralogy applications, there has been no precise algorithm developing to state the quantity of detected mineralogies. This book, therefore, is devoted to delivering novel algorithms in computing the percentage of any specific minerals in multispectral and hyperspectral satellite data. The author also derived a new algorithm based on quantum mechanics to determine some of the mineral alterations.

Filling the gap between modern physics, quantum speculation, and depletions of remote sensing imaging of geological and mineral features, this book includes technical details allied with the potentiality of various remote sensing sensors and the key techniques exploited to extract the value-added information necessary, including mineral, natural oil seeps, oil and gas exploration from the varieties of optical and microwave measurements. Despite the extensive development of remote sensing technology, most institutions still focus on using conventional image processing algorithms or classic edge detection tools and have isolated or ignored the modern physics behind the measurements.

The scope of this book is to deliver modern and novel image processing tools for mineral, oil and gas explorations rather than demonstrating intensive details of geological maps. In this view, this book is filling a gap between advanced remote sensing technology and mineral, oil and gas explorations. For instance, quantum image processing with a new generation of remote sensing technology such as Sentinel-2, TanDAM-X SAR data and polarimetric radar images are implemented to provide precise mineral, oil and gas explorations.

The understanding of mineralogy, oil and gas occurrences are introduced and overviewed in Chapter 1. It is fundamental to the comprehension of the atom structure elements; for example, oxygen and silicon that are formed in the majority of minerals and rocks.

It is impossible to avoid quantum mechanics when it comes to remote sensing technology. The science of remote sensing is predominantly proven based on quantum mechanics. In this regard, Chapter 2 demonstrates the quantization of remote sensing of mineralogy. Consequently, a novel definition of remote sensing mineralogy is addressed as the mobilizing of the spectra signature of fluctuation of the valence electron energy levels that are different from one element to another, which are based on the interaction of different photon spectra with different level of electron energy in each element. This novel definition can be explained using the quantum mechanics theory, which can be named as quantization of spectral signature of different elements or objects as based on the quantization of the black body radiation.

Chapter 3 introduces specific quantization of the image processing theories. In this regard, some basic quantum image processing such as quantum image storage, quantum image colour transfer and novel enhanced quantum image representation (NEQR) are introduced. These principles of quantum image processing are the cornerstone in dealing with quantum image processing tools in mineralogy. Chapter 4 then implemented quantum entanglement theory to deliver quantum spectral libraries, which is named entanglement spectral signatures. In this sense, Chapter 5 presents a novel formula for mineral identifications, named quantized Marghany's mineral spectral or Marghany Quantum Spectral Algorithms for Mineral Identifications (MQSA). In Chapter 6, a novel approach for automatic detection of the lineament from multispectral satellite data such as Landsat-8 is presented. In this chapter, an automatic detection tool for lineaments based on quantum fuzzy B-spline and quantum edge detection is demonstrated in such heavy vegetation covers as the subtropical zone. Consequently, Chapter 7 presents the novel formula for computing clay saturation in optical remote sensing data. Such a formula is titled "quantized Marghany clay saturation algorithm".

Oil and gas explorations based on identifying marine nature oil seeps require accurate detection tools in radar images. In this regard, Chapter 8 presents precise quantum image processing for oil seeps' automatic detection exploiting quantum immune fast spectral clustering. This chapter is a pioneer study for natural oil seeps exploration along the East coast of Malaysia, Terengganu, Dungun in synthetic aperture radar (SAR). Consequently, Chapter 9 introduces a novel technique for interferometry synthetic aperture radar processing. The modification of Differential-InSAR (DInSAR) is performed using quantum mechanics. In this context, the quantum Hebbian learning algorithm (qHob) is exploited for weighting the matrix of the Hopfield network to reconstruct accurate phase unwrapping. The new technique, which is named quantum Differential-InSAR (QDInSAR), can detect the subsidence activity along with the oil and gas company. Subsequently, Chapter 10 presents advanced quantum computing technique based on the quantum machine learning algorithm and quantum artificial neural network (QANN) for mapping gold, iron, and copper minings. Lastly, Chapter 11 presents a novel technique for 4-D automatic detection of copper mineralization or open-pit mining. This technique is based on four-dimensional hologram interferometry that is known as Marghany's 4-D phase unwrapping technique, which is a high-quality promise technique for 4-D reconstruction of copper mineralization and other objects in SAR complex data.

I wish to convey my appreciation to Dr. Samy Ismail El Mahady, Mr. Mohamed El Hakami-Head of the Remote Sensing Division at Geological Department of Saudi Geological Survey (SGS) for offering geological consultancy, and editorial project manager Raju Primlani who afforded the opportunity to publish this book. Without their intense commitment, this book would not have become such a precious piece of novel knowledge.

Prof. Maged Marghany

Director
Global Geoinformation, Sdn. Bhd.
Kuala Lumpur, Malaysia

Contents

CHAPTER 1
Principles of Mineralogy, Oil and Gas

◇◇

It is well recognized that minerals are solidified by atoms. To comprehend, describe, and predict the manner of minerals, and rocks, which are clotted of minerals, or accumulation of thick mineral deposits over longest geological periods. Therefore, some elementary facts about atoms must be understood and how they behave. In this view, a full understanding of atom behaviors leads to mineral quantization theory. This chapter is devoted to building up the fundamental mineral quantization theory. In fact, majority of the research minerology work cannot explain probably how photon of remote sensing sensors interact with atom structure of every element. This forms complicated restriction in algorithm development to identify exact minerals in remote sensing data. In this regard, this chapter introduces the principle of minerals, rocks, and oil and gas explorations.

1.1 What is a Mineral?

A mineral is a naturally occurring inorganic solid with chemical composition and crystalline structures: these are the two most vital properties of a mineral. They somewhat discriminate mineral from all others. Before discussing them, however, let us briefly consider the other properties of minerals described by this definition [1].

Mostly, organic substances are naturally formed of carbon that is chemically bonded to hydrogen or other elements. Most of the Earth's organic material, therefore, is created by plants and animals, despite the fact that organic elements can be manufactured in laboratories through chemical activities and by industrial technology [2]. On the contrary, inorganic compounds do not comprise carbon-hydrogen bonds and commonly are not fabricated by living organisms. In these regards, entire minerals are inorganic and most form autonomously of life. Exclusion is the calcite that creates limestone, which is generally comprised of the shells of dead corals, clams, and similar marine organisms. In turn, shells are formed of the mineral calcite or an analogous mineral termed aragonite. Even though they are produced by organisms and comprise of carbon, calcite and aragonite are real minerals [1, 2].

1.2 What is the Relationship between Atoms, Elements, Minerals and Rocks?

Now an important question arises: what is the basic correlation between atoms, elements, minerals and rocks? Minerals are composed of chemical elements. A chemical element is a substance that is made up of merely one sort of atom. In this regard, oxygen, hydrogen, iron, aluminium, gold and copper are all chemical elements. Basically, mineral formations are based on combination or reaction between different element atoms such as oxygen and silicon. Consequently, minerals are uniform and arise from simply one compound, while rocks develop from sequences of minerals [3].

Generally, a mineral is an element or chemical compound that is crystalline and has been materialized as a consequence of geological processes. This signifies that the calcite in the shell of a clam is not counted as a mineral. However, calcite can be considered as a mineral as soon as that clamshell experiences burial, diagenesis, or other geological processes. Typically, substances comparable to coal, pearl, opal, or obsidian that do not accept the definition of a mineral are termed as mineraloids [1–3].

Rocks are composed of minerals that have a specific chemical composition. To understand mineral chemistry, it is essential to examine the fundamental unit of all matter, the atom.

1.3 Atom Structure

Minerals are made of atoms. Atoms comprise subatomic particles—protons, neutrons, and electrons. In this context, every mineral, rock, oil and gas is composed of neutral or ionized atoms. However, atoms do not have well-identified boundaries, and there are dissimilar approaches to outline their dimensions that deliver dissimilar but approximate values. Every atom is made up of three sorts of smaller particles, referred to as protons (which are positively charged), neutrons (which have no charge) and electrons (which are negatively charged). The protons and neutrons are heavier and remain in the centre of the atom. They are known as the nucleus [1, 3]. They are surrounded by a cloud of electrons which are very light in weight and are attracted to the nucleus' positive charge. This enchantment is known as the electromagnetic force (Figure 1.1).

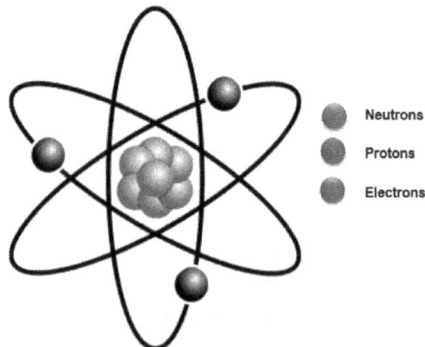

Figure 1.1. Structure of atom.

Electron

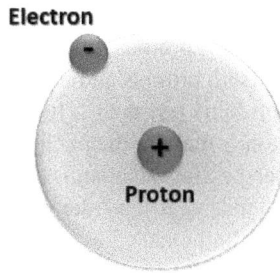

Figure 1.2. Hydrogen atom structure.

Protons and neutrons are formed of even tinier particles termed quarks. Electrons, therefore, are basic or essential particles; they cannot be fragmented into tiniest parts. Subsequently, the number of protons, neutrons and electrons an atom has is the keystone to identify what element it is. Hydrogen, for instance, has one proton, no neutrons and one electron (Figure 1.2).

In the periodic table, the hydrogen atom is the meekest atom of all the elements. It is made up of a proton, which is restricted to the nucleus and an electron that spins around the nucleus in a circular motion. The Bohr theory of the hydrogen atom, consequently, was the first fruitful speculation that prophesied the construction and energetics of the hydrogen atom. This theory, indeed, presumed the quantization of the angular momentum.

Electrons orbit, or spin around, the nucleus. They are known as the atom's electron cloud (Figure 1.3). They are attracted closer to the nucleus due to the electromagnetic force. Electrons have a negative charge and the nucleus constantly has a positive charge, so they attract each other. Around the nucleus, electrons are found in distinctive layers or orbitals. These are known as electron shells. In most atoms, the first shell has two electrons, and all after that have eight. Exceptions are rare; however, they do manifest and are challenging to predict. The further away the electron is from the nucleus, the weaker the pull of the nucleus on it. This is why greater atoms, with extra electrons, react more effortlessly with other atoms. The electromagnetism of the nucleus is now not robustly adequate to hold onto their electrons and atoms lose electrons to the strong attraction of smaller atoms [1, 4].

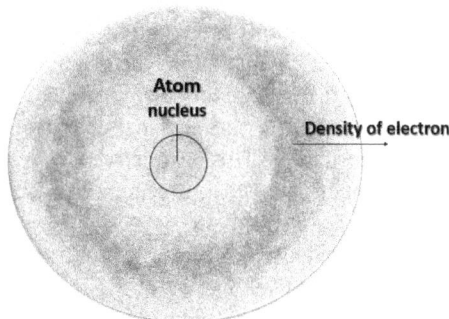

Figure 1.3. Electron cloud model of the atom.

A cloud of negatively charged electrons are surroundings the nucleus, the number of electrons matching the number of protons, thus equalizing the positive charge of the protons for a neutral atom. Both protons and neutrons have a mass number of 1. The mass of an electron, therefore, is less than 1/1000th that of a proton or neutron, denoting furthermost of the atom's mass in the nucleus [1, 6].

According to the above perspective, an atom is the smallest unit of any chemical element. They are creating chunks that generate every chemical element and are extensively too tiny to realize with the naked eye. Imagine a trivial piece of copper, for instance. Yet the tiniest quantity of copper is created of billions and billions of copper atoms (Figure 1.4). There are 103 types of atoms, and because each chemical element is made up of only one kind of atom there are 103 chemical elements [2, 5].

Figure 1.4. Billions and billions of copper atoms stack together to form a piece of copper.

1.4 Minerals in Periodic Table

A significant question now arises: what are the minerals in the periodic table? These compounds form the periodic table, which in turn consist of elements; for example, carbon, hydrogen, oxygen, nitrogen, phosphorus, calcium, iron, zinc, magnesium, manganese, and so on are the key elements in creating minerals, sediments and rocks. In other words, every mineral species is composed of a precise array of chemical compounds that form it what it is. Therefore, modification of the elements in any chemical reaction turns the mineral species into diversities of different sorts of crystallizations The discovery of the chemical elements was due largely to chemical experiments on minerals by alchemists and scientists. This section emphasizes the association of minerals to the Periodic Table of Chemical Elements [1, 7].

Minerals are composed of elements, which are atoms that have a precise number of protons in the nucleus. This number of protons, therefore, is known as the Atomic Number for the element. An oxygen atom, for instance, has 8 protons and an iron atom has 26 protons. An element cannot be split chemically into a more naive form and maintains exclusive chemical and physical characteristics. In this regard, each element acts uniquely in nature. This distinctiveness led scientists to elaborate a periodic table of the elements, a level arrangement of all known elements itemized in order of their atomic number.

In 1869, Dmitri Mendeleev pioneered the first arrangement of elements into a periodic table using the elements known at the time. In the periodic table, every element has a chemical symbol, name, atomic number, and atomic mass (Figure 1.5). The chemical symbol, therefore, is an acronym for the element, regularly originated

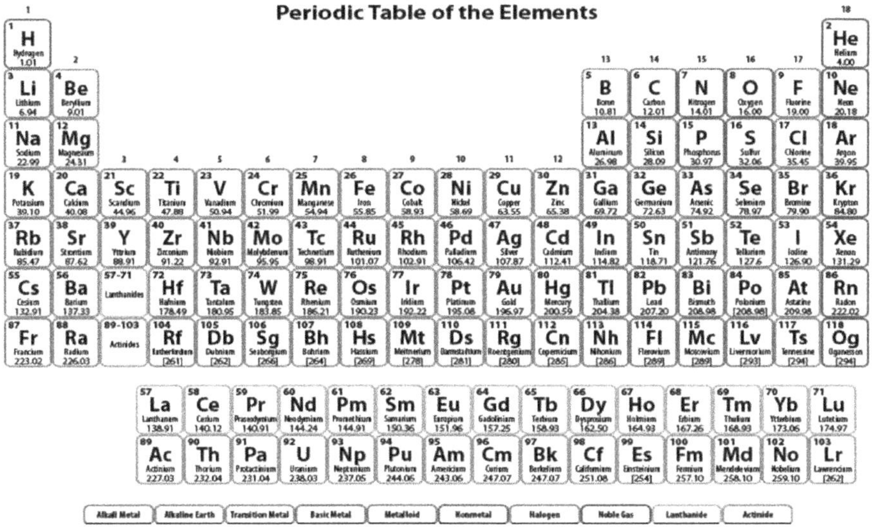

Periodic Table of the Elements

1 H Hydrogen 1.01																	2 He Helium 4.00
3 Li Lithium 6.94	4 Be Beryllium 9.01											5 B Boron 10.81	6 C Carbon 12.01	7 N Nitrogen 14.01	8 O Oxygen 16.00	9 F Fluorine 19.00	10 Ne Neon 20.18
11 Na Sodium 22.99	12 Mg Magnesium 24.31											13 Al Aluminium 26.98	14 Si Silicon 28.09	15 P Phosphorus 30.97	16 S Sulfur 32.06	17 Cl Chlorine 35.45	18 Ar Argon 39.95
19 K 39.10	20 Ca 40.08	21 Sc 44.96	22 Ti 47.88	23 V 50.94	24 Cr 51.99	25 Mn 54.94	26 Fe 55.85	27 Co 58.93	28 Ni 58.69	29 Cu 63.55	30 Zn 65.38	31 Ga 69.72	32 Ge 72.63	33 As 74.92	34 Se 78.97	35 Br 79.90	36 Kr 84.80
37 Rb 85.47	38 Sr 87.62	39 Y 88.91	40 Zr 91.22	41 Nb 92.91	42 Mo 95.95	43 Tc 98.91	44 Ru 101.07	45 Rh 102.91	46 Pd 106.42	47 Ag 107.87	48 Cd 112.41	49 In 114.82	50 Sn 118.71	51 Sb 121.76	52 Te 127.6	53 I 126.90	54 Xe 131.29
55 Cs 132.91	56 Ba 137.33	57-71 Lanthanides	72 Hf 178.49	73 Ta 180.95	74 W 183.85	75 Re 186.21	76 Os 190.23	77 Ir 190.22	78 Pt 195.08	79 Au 196.97	80 Hg 200.59	81 Tl 204.38	82 Pb 207.20	83 Bi 208.98	84 Po [208.981]	85 At 209.98	86 Rn 222.02
87 Fr 223.02	88 Ra 226.03	89-103 Actinides	104 Rf [261]	105 Db [262]	106 Sg [266]	107 Bh [264]	108 Hs [269]	109 Mt [278]	110 Ds [281]	111 Rg [280]	112 Cn [285]	113 Nh [286]	114 Fl [289]	115 Mc [289]	116 Lv [293]	117 Ts [294]	118 Og [294]

57 La 138.91	58 Ce 140.12	59 Pr 140.91	60 Nd 144.24	61 Pm 144.91	62 Sm 150.36	63 Eu 151.96	64 Gd 157.25	65 Tb 158.93	66 Dy 162.50	67 Ho 164.93	68 Er 167.26	69 Tm 168.93	70 Yb 173.05	71 Lu 174.97
89 Ac 227.03	90 Th 232.04	91 Pa 231.04	92 U 238.03	93 Np 237.05	94 Pu 244.06	95 Am 243.06	96 Cm 247.07	97 Bk 247.07	98 Cf 251.08	99 Es [254]	100 Fm 257.10	101 Md 258.10	102 No 259.10	103 Lr [262]

Alkali Metal · Alkaline Earth · Transition Metal · Basic Metal · Metalloid · Nonmetal · Halogen · Noble Gas · Lanthanide · Actinide

Figure 1.5. Elements periodic table.

from a Latin or Greek name for the substance. In this view, the atomic number is the number of protons in the nucleus. Consequently, the atomic mass is the number of protons and neutrons in the nucleus, individually with a mass number of one. On the other hands, the atomic mass is efficiently the number of protons plus neutrons since the mass of electrons is accordingly considerably less than the protons and neutrons [7].

Additionally, the natural element atomic masses signify a typical mass of the atoms containing that mineral substance and is generally not an entire number as realized on the periodic table, signifying that a mineral occurs in nature with atoms holding diverse numbers of neutrons. The contradictory number of neutrons influences the mass of a mineral in nature and the atomic mass number represents this average. This gives rise to the perception of 'isotope'. Isotopes, in this view, are arrays of an element with an identical number of protons but dissimilar numbers of neutrons. According to this view, there are generally numerous isotopes for a specific element. In this understanding, 98.9% of carbon atoms, for instance, have 6 protons and 6 neutrons, which is termed carbon-12 (^{12}C). On the contrary, limited carbon atoms, carbon-13 (^{13}C), have 6 protons and 7 neutrons. Moreover, a trace quantity of carbon atoms, carbon-14 (^{14}C), has 6 protons and 8 neutrons. As well as the 118 well-identified elements, the heaviest are flitting human creations recognized merely in high energy atom accelerators, and they decay precipitously. The heaviest naturally occurring element is uranium, atomic number 92. To sum up, Table 1.1 shows the eight most abundant elements in Earth's continental crust, which originate in the most common rock-forming minerals [1, 6, 8].

Atoms move faster when they are in gas form (because they are free to move) than they do in liquid form and solid matter. In solid materials, the atoms are tightly packed next to each other so they vibrate, but are not able to move (there is no room) as atoms in liquids do. However, something is surprising through the electron

Table 1.1. The eight most abundant elements in the Earth's continental crust.

Elements	Symbols	Percentage (%)
Oxygen	O	47%
Silicon	Si	28%
Aluminium	Al	8%
Iron	Fe	5%
Calcium	Ca	4%
Sodium	Na	3%
Potassium	K	3%
Magnesium	Mg	2%

organization of the group of noble gases. Excluding helium, which merely has two electrons, all members of the noble gases, no matter how many electrons they have, constantly have eight electrons each in the last or outer orbit or shell [6, 8].

The main characteristics of last elements in the Periodic Table such as alkali metal group (hydrogen, lithium, sodium, potassium, rubidium, caesium, and francium) is that they have similar chemical reaction and several of their characteristics are matched. In this regard, these alkali metal elements have a similar electron oddity. In this view, the outer, last orbit, or shell of every element has merely one electron. The alkali metal elements, moreover, are among the furthermost chemically energetic of the recognized elements, which is contradictory to the noble gases [6–8].

The outer shell electrons would be vital to the chemical activity of the elements. In this regard, those elements whose outer orbit has less than four electrons in it are chemically active and are termed metals. They are assembled to the left of the Table. Therefore, elements whose outer orbital have more than four electrons in their outer shells are assembled to the right side of the Table and are deliberated as non-metals. According to this perspective, valence electrons are termed as outer orbital electrons when they participate in chemical activities [7, 9].

The Table correspondingly assembles elements along with their characteristics. In this sense, elements that reveal such characteristics as malleable, capable to conduct electricity, and ductile are assembled on the left side of the chart as metals. Gold is the most malleable metal. In contrast, ductility is the ability of a solid material to deform under tensile stress. In other words, certain metals like zinc, arsenic, mercury, antimony are non-malleable in nature; which are incapable of being shaped or altered. In this perspective, the elements that behave like metals have fewer than four electrons in their outer orbit.

A critical question now arises: what about those elements whose outer orbit comprise precisely four electrons? It should be no wonder that they occasionally perform similar to a metal and infrequently behave as a non-metal. In this understanding, Carbon is an excellent example, which has specific metal properties. In this regard, Carbon is unique to the metalloid elements. Therefore, metalloid elements in the periodic table; which generally border the stair-step line. The metalloids are located close to that line [8].

1.5 Chemical Bonding

Most substances on Earth are composites comprising numerous elements. Chemical bonding designates how these atoms bond to create compounds; for instance, sodium and chlorine combining to form NaCl, common table salt. Compounds that are organized together by chemical bonds are termed molecules. Water, for example, is a compound of hydrogen and oxygen, in which two hydrogen atoms are covalently bonded with one oxygen making the water molecule. In this view, oxygen is produced when one oxygen atom covalently bonds with another oxygen atom to produce the molecule O_2. In this understanding, subscript 2 in the chemical formula specifies that the molecule comprises dual atoms of oxygen [6–8].

Most minerals, therefore, are similar compounds of more than one element. Consequently, the conventional mineral calcite has the chemical formula $CaCO_3CaCO_3$ signifying that the molecule contains one calcium, one carbon, and three oxygen atoms. In calcite, one carbon and three oxygen atoms are arranged together by covalent bonds to form a molecular ion, named carbonate, which has a negative charge. In this sense, calcium as an ion has a positive charge of plus two [7]. According to this perspective, the dual oppositely charged ions magnetize each other and merge to create the mineral calcite, $CaCO_3CaCO_3$. The name of the chemical compound is calcium carbonate, where calcium is CaCa and carbonate refers to the molecular ion CO_3^{-2}.

Moreover, the mineral olivine has the chemical formula $(Mg, Fe)_2SiO_4$, in which one silicon and four oxygen atoms are merged with dual atoms of either magnesium or iron. The comma between iron (Fe) and magnesium (Mg) means that the dual elements can dominate a similar position in the crystal structure and substitute for one another [1, 8].

1.6 Valence and Charge

The electrons circumnavigate the atom's nucleus that are positioned in shells representing unique energy levels. The outermost shell is referred to as the valence shell. Electrons in the valence shell are concerned with chemical bonding. In 1913, Niels Bohr proposed an easy model of the atom that states atoms are extra steady when their outermost shell is full. Atoms of most elements, as a result, tend to acquire or lose electrons so the outermost or valence shell is full. In Bohr's model, the innermost shell can have a maximum of two electrons and the second and 1/3 shells can have a maximum of eight electrons. When the innermost shell is the valence shell, as in the case of hydrogen and helium, it obeys the octet rule when it is full with dual electrons. For elements in higher rows, the octet rule of eight electrons in the valence shell applies. Figure 1.6 reveals that a molecule of carbon dioxide (CO_2) is made up of one carbon atom and two oxygen atoms [7, 9].

The rows in the periodic table donate the elements in order of atomic number and the columns form elements with similar physical characteristics; for instance, the identical number of electrons in their valence shells. Columns, therefore, are

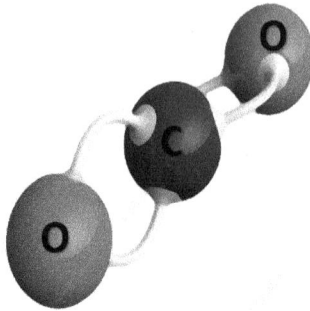

Figure 1.6. The carbon dioxide molecule.

regularly categorized from left to right with Roman numerals I to VIII, and Arabic numerals 1 through 18. The elements, consequently, in columns I and II have 1 and 2 electrons in their particular valence shells and the elements in columns VI and VII have 6 and 7 electrons in their particular valence shells (Figure 1.5).

Moreover, in row 3 and column I, sodium (Na) has 11 protons in the nucleus and 11 electrons in three shells—2 electrons in the inner shell, 8 electrons in the second shell, and 1 electron in the valence shell. To sustain a complete outer shell of 8 electrons per the octet law, sodium has 11 electrons. The first two fills the innermost energy level. The second energy level is also full, holding eight electrons and one electron remaining in the outer energy level. It is the number of outer electrons that give an element its chemical properties. With 11 positively charged protons in the nucleus and 10 negatively charged electrons in two shells, sodium when generating chemical bonds is an ion with an overall net charge of +1 [6, 8].

In column I, all elements have a single electron in their valence shell and a valence of 1. These further columns I elements similarly dexterously reveal this single valence electron and thus develop ions with a +1 charge. Subsequently, elements in column II easily allocate 2 electrons and end up as ions with a charge of +2. Note that elements in columns I and II, which freely end their valence electrons, regularly generate bonds with elements in columns VI and VII, which willingly occupy these electrons. In columns 3 through 15, elements are frequently elaborate in covalent bonding. Consequently, the last column 18 (VIII) comprises the noble gases. In this view, these elements are chemically inactive since the valence shell is occupied with 8 electrons; accordingly, they do not obtain or lose electrons. In this view, the noble gas helium, for instance, which has 2 valence electrons in the first shell as its valence shell, is hence complete. Consequently, in column VIII, all elements hold complete valence shells and do not create bonds with other elements [1, 7].

In this understanding, an atom with a net positive or negative charge as a consequence of gaining or losing electrons is termed an ion. Generally, on the left side of the table, the elements lose electrons and turn into positive ions, termed cations since they are captivated to the cathode in an electrical device. However, the elements on the right side incline to obtain electrons, which are called anions. In fact, in an electrical device, these elements are attracted to the anode. The elements in the middle of the periodic table, columns 3 through 15, do not steadily obey the octet rule, which are known as transition elements. In this view, iron is considered a common case, which has a +2 or +3 charge relying on the oxidation status of the element. Moreover, oxidized Fe^{+3} conveys a +3 charge and diminished Fe^{+2} is +2. These two diverse oxidation phases of iron regularly convey dramatic colours to rocks containing their minerals—the oxidized form creating red colours and the diminished form creating green. In other words, a transfer of electrons between the microbe and the mineral, in this case, carries approximately a fluctuation in the chemical state of the iron (its 'reduction'), which also initiates the mineral to dissolve. Interactions of this sort are now known to play important roles in the release and movement of metals and other elements, including pollutants such as arsenic, at the Earth's surface [8].

1.7 Ionic Bonding

Ionic bonds, additionally named electron-transfer bonds, are created by the electrostatic attraction between atoms having contradictory charges. In this circumstance, atoms of dual contradictory charges attract each other electrostatically and create an ionic bond, in which the positive ion conveys its electron (or electrons) to the negative ion, which engage them. Beyond these transfers, both atoms accordingly attain a complete valence shell. In this view, one atom of sodium (Na^{+1}) and one atom of chlorine (Cl^{-1}), for instance, create an ionic bond to form the compound sodium chloride (NaCl), which is known as the mineral halite. An alternative example is calcium (Ca^{+2}) and chlorine (Cl^{-1}) combining to form the compound calcium chloride ($CaCl_2$). In this demonstration, subscript 2 specifies that dual atoms of chlorine are ionically bonded to one atom of calcium [6, 8, 10].

1.8 Covalent Bonding

Ionic bonds are commonly made up between a metal and a nonmetal. Another type, known as a covalent or electron-sharing bond, frequently takes place between nonmetals. Covalent bonds share electrons between ions in which the electron pairs are shared between them. For instance, oxygen (atomic number 8) has 8 electrons—2 in the inner shell and 6 in the valence shell. Gases such as oxygen frequently structure diatomic molecules employing sharing valence electrons. In the case of oxygen, two atoms connect and allocate 2 electrons to fill their valence shells to turn out to be the frequent oxygen molecule we breathe (O_2). Methane (CH_4) is another covalently bonded gas. The carbon atom requires 4 electrons and each hydrogen requires 1. Each hydrogen shares its electron with the carbon to shape a molecule [7, 10] as shown in Figure 1.7.

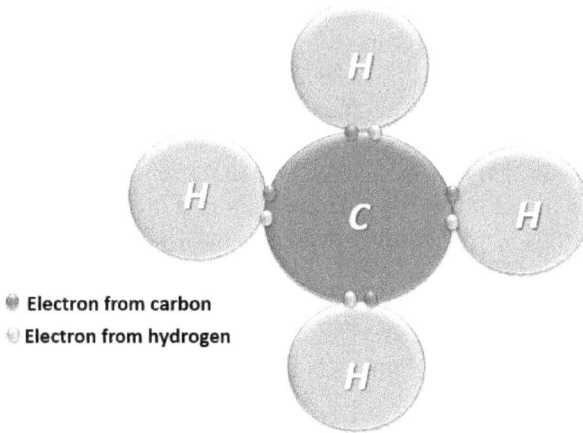

Figure 1.7. Methane molecule.

1.9 Natural Crystallization of Minerals

The main question normally arises: why are most of the minerals initially form as tiny crystal? In accordance with the atomic structure explained in previous sections, minerals grow as layer after layer of atoms are eventually added to their surface. In this scenario, a crystal is somewhat substance whose atoms are organized in a systematic, periodically constant pattern, which shows that all minerals are naturally crystalline. For instance, the mineral halite (common table salt) has the composition NaCl: one sodium ion (Na^+) for every chlorine ion (Cl^-). Figure 1.8 demonstrates the ions in halite, which indicates the crystalline structure of halite. In this view, the sodium and chlorine ions interchange in tidy rows and columns interconnecting at perpendicular angles.

Following crystal structure, the minerals' discrimination is based on the various crystal forms. Indeed, the mineral molecule structures are arranged accurately as a function of different crystal forms. In this regard, a very simple outline of six systems of the crystal is the fundamental approach of mineral identifications. This can assist in describing the mineral structure, symmetry, and shape.

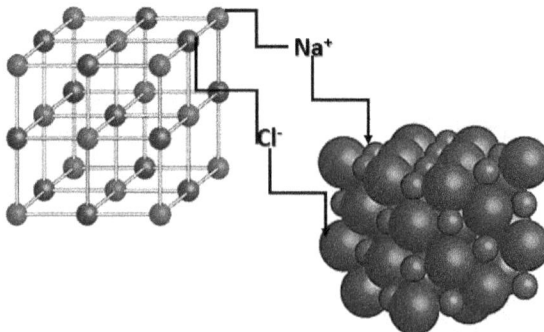

Figure 1.8. The crystalline structure of halite.

The speculation of symmetry defines the periodic repetition of structural features. Two archetypal sorts of symmetry exist. These encompass translational symmetry and feature symmetry. Translational symmetry designates the periodic repetition of a motif throughout a size or through a region or volume. Point symmetry, alternatively, labels the periodic repetition of a motif around asymmetry feature. In this regard, all point symmetry operations are reflection, rotation, inversion, and root inversion [11–13].

Foremost, a reflection arises when a motif on one side of a plane circulates through the middle of a crystal. In this regard, a reflection is the mirror image of a motif, which looks on the other side of the plane. The motif is expressed to be reflected across the mirror plane, which splits the crystal. On the other hand, in a rotational symmetry, the line of points that stay in the same place constitute a symmetry axis; in a reflection the points that remain unchanged make up a plane of symmetry. In the circumstance of a crystal possessing inversion symmetry, formerly each line drawn across the crystal's centre would interlink dual matching features on reverse margins of the crystal. Lastly, root inversion is a composite symmetry operation, which is formed by acting a revolution tailed by an inversion.

The reflection, rotation, inversion, and root inversion symmetry operations perhaps combine in a multiplicity of different approaches. In this view, there are thirty-two probable unique amalgamations of symmetry operations. Consequently, minerals acquiring the diverse amalgamations are sorted as affiliates of thirty-two crystal categories; each category set matches an exclusive set of symmetry operations. Each of the crystal categories is termed consistent with the variation of a crystal form, which it presents. Each crystal category, therefore, is assembled as one of the six diverse crystal systems consistent with the property symmetry operation it maintains. In this sense, each crystal category is a member of one of the six crystal systems, which involve the isometric, hexagonal, tetragonal, orthorhombic, monoclinic, and triclinic systems [10, 12, 14].

1.9.1 Isometric

The isometric crystal system is correspondingly identified as the cubic system (Figure 1.9). The crystallographic axes exploited in this system are of equivalent length and are jointly perpendicular, striking at right angles to one another. In these regards, entire crystals of the isometric system have four 3-fold axes of symmetry, each of which continues diagonally from corner to corner across the cubic's central unit cell. Crystals of the isometric system, therefore, may also reveal up to three separate 4-fold symmetry axes. These axes, subsequently, if exist, continue from the middle of every face across the origin to the middle of the opposite face and correspond to the crystallographic axes. Besides, crystals of the isometric system perhaps have six 2-fold axes of symmetry, which develop from the middle of every edge of the crystal across the origin to the middle of the contradictory edge. According to this view, minerals of this system perhaps reveal up to nine different mirror planes. For instance, halite, magnetite, and garnet are crystallized in the isometric system. Minerals of this system, therefore, tend to generate crystals of equidimensional or equant habit [11, 14].

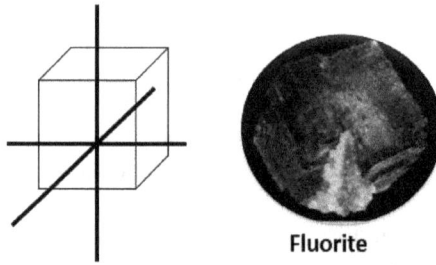

Fluorite

Figure 1.9. Isometric crystal system.

1.9.2 Hexagonal

Minerals of the hexagonal crystal system are denoted by three crystallographic axes, which crisscross at 120° and a fourth, which is upright to the other three (Figure 1.10). Therefore, this fourth axis is regularly depicted perpendicularly. In this sense, the hexagonal crystal system is alienated into the hexagonal and rhombohedral or trigonal partitions. Under this view, entire crystals of the hexagonal partition have a single 6-fold axis of rotation. Also, the crystals of the hexagonal partition perhaps have up to six 2-fold axes of rotation. In this regard, they perhaps reveal a middle of inversion symmetry and equal to seven mirror planes. Crystals of the trigonal division, moreover, entirely have a single 3-fold axis of revolution rather than the 6-fold axis of the hexagonal partition. Therefore, crystals of this division perhaps have up to three 2-fold axes of revolution and perhaps reveal a middle of inversion and up to three mirror planes. Apatite, beryl, and high quartz are such mineral species, which crystallize in the hexagonal division. Moreover, these minerals tend to form hexagonal prisms and pyramids. On the other hand, calcite, dolomite, low quartz, and tourmaline are the mineral species, which crystallize in the rhombohedral division and tend to create rhombohedra and triangular prisms [10, 13, 15].

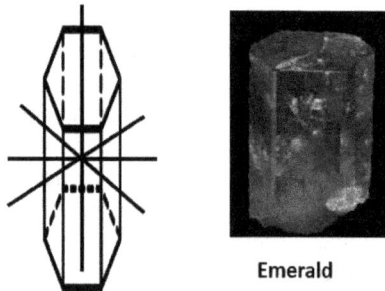

Emerald

Figure 1.10. Hexagonal crystal system.

1.9.3 Tetragonal

Minerals of the tetragonal crystal system are represented by three jointly perpendicular axes (Figure 1.11). In this regard, the dual horizontal axes are of equivalent length, while the upright axis is of dissimilar length and can perhaps be either shorter

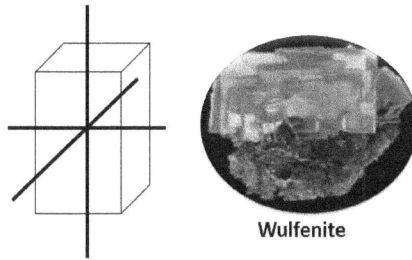

Wulfenite

Figure 1.11. Tetragonal crystal system.

or longer than the other two. Minerals of this system would have a single 4-fold symmetry axis. In this view, they perhaps have up to four 2-fold axes of rotation, a middle of inversion, and up to five mirror planes. For instance, zircon and cassiterite are mineral species, which crystallize in the tetragonal crystal system and are inclined to generate the short crystals of prismatic habit [10, 12, 14].

1.9.4 Orthorhombic

Olivine and barite are mineral spices, which belong to the orthorhombic system (Figure 1.12) and tend to be of prismatic, tabular, or acicular habit. In this regard, these orthorhombic crystal systems are shown in three jointly upright axes, every one of which is of a dissimilar length than the others. Moreover, orthorhombic systems uniformly have three 2-fold rotation axes and/or three mirror planes. Therefore, the holomorphic category reveals three 2-fold symmetry axes and three mirror planes besides a middle of inversion. Subsequently, other categories perhaps display three 2-fold axes of rotation or one 2-fold rotation axis and dual mirror planes [12–15].

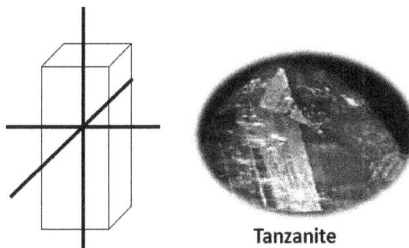

Tanzanite

Figure 1.12. Orthorhombic crystal system.

1.9.5 Monoclinic

Minerals such as pyroxene, amphibole, orthoclase, azurite, and malachite, adhere to the monoclinic crystal system and tend to form long prisms. Therefore, the monoclinic crystal system is signified with three unequal axes (Figure 1.13). Dual of these axes are leant toward each other at an oblique angle; these are regularly represented steeply. In this view, the third axis is upright to the other two. The dual upright axes, consequently, do not interconnect one another at perpendicular angles,

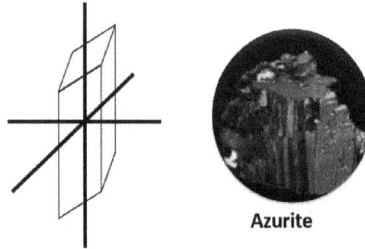

Figure 1.13. Monoclinic crystal system.

even though both are upright to the horizontal axis. A single 2-fold rotation axis and/ or a single mirror plane, therefore, can be shown through monoclinic crystals. The holomorphic categories, consequently, have the single 2-fold rotation axis, a mirror plane, and the middle of symmetry. Moreover, different categories demonstrate an appropriate 2-fold rotation axis or proper mirror plane [13–16].

1.9.6 Triclinic

Plagioclase and axinite are the minerals species devoted to the triclinic category and tend to be of tabular habit. In this regard, triclinic crystals are referred to as three unequal axes (Figure 1.14), all of which crisscross at oblique angles. However, none of the axes is upright to any other axis. Perhaps this sort of crystal has only a 1-fold symmetry axis, which is equal to having no symmetry at all. In other words, the triclinic category has no mirror planes. Therefore, the holomorphic category reveals a middle of inversion symmetry [1, 8, 10, 16].

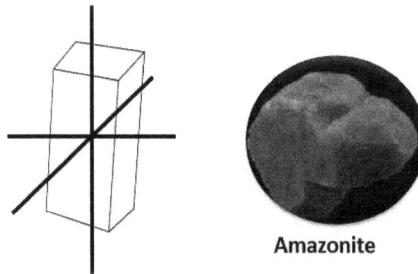

Figure 1.14. Triclinic crystal system.

1.9.7 Trigonal or Rhombohedral

A geometric system of trigonal or rhombohedral is based on a diagonally-stretched cube. For instance, Rhodochrosite crystallizes in the trigonal system (Figure 1.15). In this view, Rhodochrosite is the tendency of crystalline materials to split along definite crystallographic structural planes. These planes of relative weakness are a result of the regular locations of atoms and ions in the crystal, which create smooth repeating surfaces that are visible both in the microscope and to the naked eye [14–17].

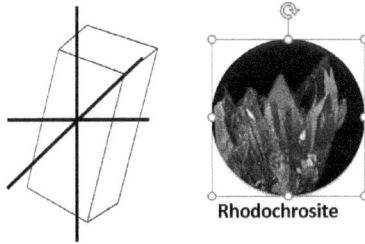

Rhodochrosite

Figure 1.15. Trigonal or Rhombohedral crystal system.

1.10 Occurrence and Formation

In the circumstances of temperature and pressure fluctuations through Earth's internal layers. In this understanding, minerals form when atoms bond together in a crystalline arrangement. Three foremost approaches this occurs in nature are (i) precipitation directly from an aqueous (water) solution with a temperature change; (ii) crystallization from a magma with a temperature change; and (iii) biological precipitation by the action of organisms.

In this regard, solutions contain ions or molecules, recognized as solutes, dissolved in a medium or solvent. Naturally, this solvent is generally water. Therefore, numerous minerals can be liquified; for instance, halite or table salt, which has the composition of sodium chloride, NaCl. According to this view, the Na^{+1} and Cl^{-1} ions disconnected and dissolve into the solution.

Precipitation, consequently, is the reverse process, in which ions in the solution hold together to create solid minerals. Precipitation, therefore, is reliant on the dilution of ions in solution and other factors; for example, temperature and pressure. The point at which a solvent cannot hold any more solute is termed saturation. In this context, precipitation can arise when the temperature of the solution drops when the solute evaporates, or with varying chemical circumstances in the solution [10–17].

Naturally, fluctuations in environmental circumstances perhaps instigate the minerals dissolved in water to create bonds and raise into crystals or cement grains of sediment together. For instance, in Utah, deposits of tufa created from mineral-rich springs occurred in the ice age, Lake Bonneville.

Atom vibrations arise owing to the energy of the highest temperature such as magma. In this circumstance, chemical bonds are broken and the crystals melt discharging the ions into the melt. Consequently, magma is molten rock with spontaneously stirring ions. When magma is emplaced at depth or erupts out onto the surface (then named lava), it begins to cool and mineral crystals can produce.

Numerous organisms create bones, shells, and body casings by mining ions from water and causing minerals biologically. The most popular mineral caused by organisms is calcite, or calcium carbonate ($CaCO_3$), which is regularly precipitated by organisms as a polymorph named aragonite. In this regard, polymorphs are crystals with a similar chemical formula but dissimilar crystal structures. According to this perspective, marine invertebrates, for example corals and clams, precipitate aragonite or calcite for their shells and structures. Upon death, their hard fragments amass on the ocean floor; for example, sediments and ultimately may convert into the sedimentary rock limestone. Nevertheless, limestone can be created inorganically as the massive mainstream is shaped through upon decomposition process.

Radiolaria are the alternative example in marine organisms, which are zooplankton that precipitates silica for their microscopic external shells. When the organisms perish, the shells amass on the ocean floor and can create the sedimentary rock chert. Moreover, most of the world biologic deposits are a function of bone, which is created generally from a sort of apatite, a mineral in the phosphate set. In this sense, the apatite originates in bones and comprises calcium and water in its structure and is known as hydroxycarbonate apatite, $Ca_5(PO_4)_3(OH)Ca_5(PO_4)_3(OH)$. According to the previous perspective, such substances are not precisely minerals till the organism dies and these hard parts convert into fossils [6, 10, 19].

1.11 How are Minerals Categorized?

Minerals are scientifically categorised into dissimilar chemical sets based on their chemical composition. Corresponding classes are grounded on the configuration of the predominant anion or anion complex. For instance, minerals with sulfur, S^{2-}, as an anion are clustered under sulfides; those with $(SiO_4)^{4-}$ as an anionic cluster are silicates. The comprehensive sorting subdivisions exploited in entirely mineralogical literature are grounded on the anionic unit in the chemical formula as specified in the succeeding (Table 1.2):

Table 1.2. Mineral chemical symbols.

Elements	Chemical symbol
Native elements	Au, Cu, Pt
Sulfides	ZnS, FeS_2, $CuFeS_2$
Sulfosalts	Cu_3AsS_4, $Cu_{12}Sb_4S_{13}$
Oxides	Al_2O_3, Fe_3O_4, TiO_2
Hydroxides	$Mg(OH)_2$, $FeO(OH)$
Halides	$NaCl$, KCl, CaF_2
Carbonates	$CaCO_3$, $CaMg(CO_3)_2$
Phosphates	$Ca_5(PO_4)_3(F, Cl, OH)$
Sulfates	$CaSO_4 \cdot 2H_2O$
Tungstates	$CaWO_4$
Silicates	Mg_2SiO_4, Al_2SiO_5

Exactly 4150 minerals are recognized, of which 1140 are silicates; 624, sulfides and sulfosalts; 458, phosphates; 411, oxides and hydroxides; 234, carbonates; and 90, native elements [7, 11, 14, 16, 20].

1.11.1 Silicate Minerals

A tetrahedron possesses a pyramid-like geometry with four sides and four corners (Figure 1.16). Minerals, therefore, are classified based on their structure and composition, which are made up of a molecular ion known as the silicon-oxygen tetrahedron (Figure 1.17). Silicate minerals, therefore, build up the massive class of minerals on Earth, containing the massive joint of the Earth's mantle and crust. Out

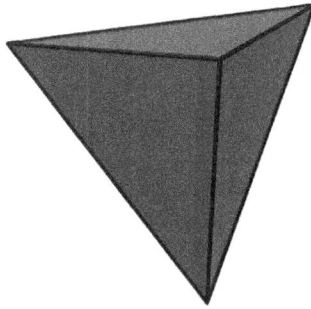

Figure 1.16. Tetrahedron geometry.

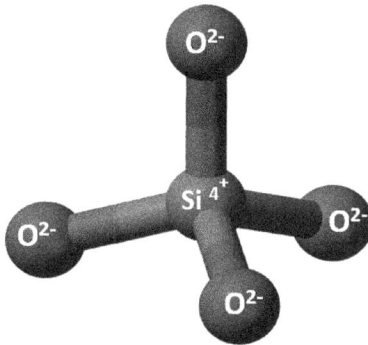

Figure 1.17. Silicon-oxygen tetrahedron.

of the approximately four thousand identified minerals on Earth, the majority are rare. There are only a few that form major of existing rocks in nature, which tending to be confronted by surface-dwelling creatures identical to us, which are commonly known as the rock-forming minerals.

Moreover, the silicon-oxygen tetrahedron (SiO_4) contains a single silicon atom at the centre and four oxygen atoms that are located at the four corners of the tetrahedron (Figure 1.17). Each oxygen ion has a –2 charge and the silicon ion has a +4 charge. The silicon ion, therefore, designates one of its four valence electrons with each of the four oxygen ions in a covalent bond to generate asymmetrical geometric four-sided pyramid form. On the other hand, just half of the oxygen's valence electrons are portioned, offering the silicon-oxygen tetrahedron an ionic charge of –4. This silicon-oxygen tetrahedron, consequently, creates bonds with several other amalgamations of ions to create the massive set of silicate minerals or other positively charged ions such as Al_{+3}, $Fe_{+2,+3}$, Mg_{+2}, K_{+1}, Na_{+1}, and Ca_{+2}.

Furthermore, silica-oxygen tetrahedra can amalgamate with other tetrahedra in numerous diverse configurations as a function of the original magma chemistry. In this circumstance, tetrahedra can be separated, devoted to chains, sheets, or three-dimensional structures. These combinations, therefore, create the chemical structure in which positively charged ions can be injected for an exclusive chemical composition producing silicate mineral classes [12, 17, 19, 21].

1.11.2 The Dark Ferromagnesian Silicates

The mantle rock such as peridotite and basalt, its primary element is olivine, which is characteristically green when not weathered. The chemical formula of olivine is $(Fe, Mg)_2SiO_4$. Olivine, therefore, is denoted as a mineral set owing to the competence of iron and magnesium to replace each other. In this sense, iron and magnesium in the olivine class specify a solid solution generating a compositional series within the mineral set, which can make up crystals of iron only. In this regard, the crystal structure of olivine is formed from autonomous silica tetrahedra. Minerals with autonomous tetrahedral structures, consequently, are known as neosilicates or orthosilicates. Therefore, olivine, and other familiar neosilicate minerals contain garnet, topaz, kyanite, and zircon.

In this regard, other two similar compositions of tetrahedra are similar in structure to the neosilicates and rank toward the succeeding cluster of minerals (Figure 1.18), the pyroxenes. In a discrepancy on autonomous tetrahedra termed sorosilicates, some minerals reveal one oxygen between two tetrahedra and contain minerals such as pistachio-green epidote, a gemstone. Moreover, the cyclosilicates consist of tetrahedral rings, and involve gemstones like beryl, emerald, aquamarine, and tourmaline [2, 9, 11, 15, 21].

In the olivine chain of minerals, the iron and magnesium ions in the solid solution are approximately the equivalent in size and charge. In this regard, either atom can suit into a similar position in the maturing crystals. In the case of the cooling magma, the mineral crystals remain to cultivate till they freeze into igneous rock. The relative amounts of iron and magnesium, in the maternal magma regulate the minerals are in the formed chain. On the other hand, such rarer elements like manganese (Mn) has similar properties to iron or magnesium, which can be relieved into the olivine crystalline structure in tiny quantities. Such ionic substitutions in mineral crystals, consequently, give rise to the huge diversity of minerals and are frequently accountable for variances in colour and other characteristics within a cluster or class of minerals. In this regard, olivine has a pure iron end-member, which is also termed fayalite and pure magnesium is known as forsterite [10–18].

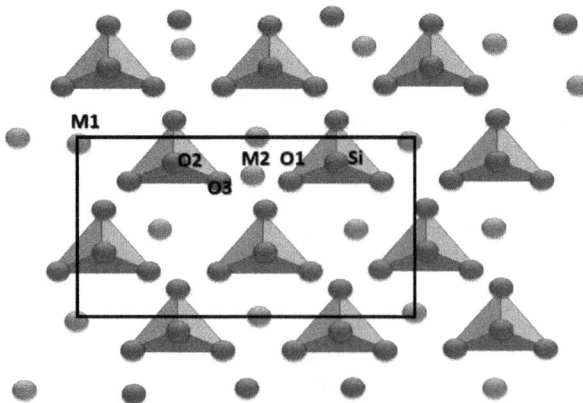

Figure 1.18. Tetrahedral structure of olivine.

Chemically, olivine is typically silica, iron, and magnesium and, therefore, is clustered among the dark-coloured ferromagnesian. Iron is equivalent to Ferro, and magnesium equals magnesian or mafic minerals. Subsequently, the contraction of their chemical symbols are Fe and Ma, respectively. In other words, Ferro represents iron and magnesian means magnesium. According to this view, mafic minerals likewise belong to dark-coloured ferromagnesian minerals [15, 18, 22].

1.11.3 Pyroxene Family

Pyroxene is an alternative cluster of dark ferromagnesian minerals, naturally black (Figure 1.19) or dark green. Therefore, the pyroxene's member family retains a complex chemical composition that comprises iron, magnesium, aluminium, and the alternative elements merged in polymerized silica tetrahedra. In this regard, polymers are single chains (Figure 1.20), sheets, or three-dimensional structures, and are shaped by multiple tetrahedra covalently bonded through their corner oxygen atoms. Pyroxenes are commonly found in mafic igneous rocks such as peridotite, basalt, and gabbro, as well as metamorphic rocks like eclogite and blue-schist.

Figure 1.19. Pyroxene crystal deposit.

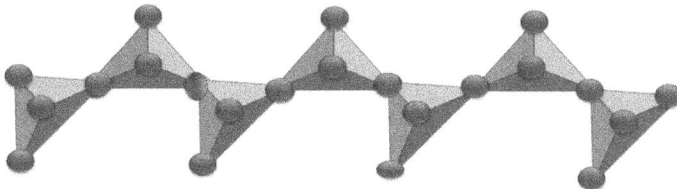

Figure 1.20. Single chain.

Pyroxenes are made up of long, single chains of polymerized silica tetrahedra in which tetrahedra contribute dual corner oxygens. The silica chains, therefore, are merged into the crystal structures by metal cations. A general member of the pyroxene family, consequently, is augite, itself comprising numerous solid solution sequences with a complex chemical formula (Ca, Na)(Mg, Fe, Al, Ti)(Si, Al)$_2$O$_6$ that gives rise to several discrete mineral designations.

This single-chain crystalline structure, subsequently, bonds with numerous elements, which can similarly spontaneously swap for each other. The comprehensive chemical configuration for pyroxene is $XZ(Al, Si)_2O_6$. In this view, X signifies the ions Na, Ca, Mg, or Fe, and Z embodies Mg, Fe, or Al. These ions, therefore, possess comparable ionic sizes, which permits countless probable replacements among them. Even though the cations perhaps spontaneously relieve each other in the crystal, they hold dissimilar ionic charges that would balance out in the ending crystalline structure. For instance, Na possesses a charge of $+1$ but Ca carries a charge of $+2$. If a Na^+ ion replaces the Ca^{+2} ion, it forms a dissimilar charge that would be balanced by other ionic replacements somewhere else in the crystal. In this regard, that ionic size is more significant than ionic charge in forming the chains of crystals [17–23].

1.11.4 Amphibole Minerals

Let us imagine double pyroxene chains that interlink together by allocating the third oxygen on each tetrahedron. In this view, amphiboles regularly originate in igneous and metamorphic rocks and stereotypically possess a long-bladed crystal habit. The furthermost usual amphibole, hornblende (Figure 1.21), is generally black; nevertheless, they are produced in a diversity of colours relying on their chemical composition. According to this view, amphibole minerals are made up of polymerized dual silica chains (Figure 1.22) and they are correspondingly denoted as inosilicates. The metamorphic rock, amphibolite, is mainly comprised of amphibole minerals.

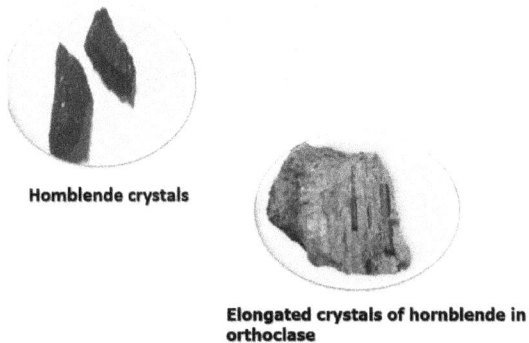

Homblende crystals

Elongated crystals of hornblende in orthoclase

Figure 1.21. Hornblende crystals and elongated crystals of hornblende in orthoclase.

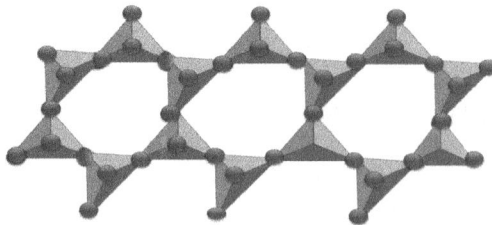

Figure 1.22. Double chain structure.

Amphiboles are comprised of iron, magnesium, aluminium, and additional cations bonded with silica tetrahedra. For instance, gabbro, basalt, and diorite are dark ferromagnesian minerals that form the black specks in granite. In this regard, they have a very complex chemical formula and are commonly inscribed as $(RSi_4O_{11})_2$, where R signifies numerous dissimilar cations. In other words, they can also be inscribed more precisely as $AX_2Z_5((Si, Al, Ti)_8O_{22})(OH, F, Cl, O)_2$, where A perhaps be Ca, Na, K, Pb, or blank; X equates Li, Na, Mg, Fe, Mn, or Ca; and Z is Li, Na, Mg, Fe, Mn, Zn, Co, Ni, Al, Cr, Mn, V, Ti, or Zr. The substitutions, therefore, form an extensive diversity of colours; for example, green, black, colourless, white, yellow, blue, or brown. Moreover, amphibole crystals can similarly comprise hydroxide ions (OH^-), which befalls from an interface between the development minerals and water dissolved in the magma [1, 6, 9, 14, 19, 21, 23].

1.11.5 Sheet Silicates

Sheet silicates are made up of tetrahedra which allocate completely three of their bottom corner oxygens, thus creating sheets of tetrahedra with their top corners accessible for relating with other atoms. In this regard, micas and clays are general sorts of sheet silicates, as well known as phyllosilicates. Moreover, mica minerals generally originate in igneous and metamorphic rocks, while clay minerals more frequently initiate in sedimentary rocks. For instance, micas are presented as the dark-coloured biotite and light-coloured muscovite, which frequently exist in granite and metamorphic rock (i.e., schist); respectively.

Chemically, sheet silicates typically comprise silicon and oxygen in a 2:5 ratio (Si_4O_{10}). Mostly, micas contain silica, aluminium, and potassium. Besides, biotite mica includes more iron and magnesium and is believed to be a ferromagnesian silicate mineral. Muscovite micas, therefore, belong to the felsic silicate minerals. Felsic is a contraction created from feldspar, the prevailing mineral in felsic rocks [17–24].

Clay minerals arise in sediments created by the weathering of rocks and are an alternative cluster of silicate minerals with a tetrahedral sheet structure. Clay minerals, therefore, make up a complex cluster and are a vital component of countless sedimentary rocks. Additional sheet silicates comprise serpentine and chlorite occurring in metamorphic rocks.

Clay minerals, consequently, are made up of hydrous aluminium silicates. One sort of clay, kaolinite, comprises a structure corresponding to an open-faced sandwich, which begins with an individual layer of silicon-oxygen tetrahedra and ends up by a layer of aluminium as the array in an octahedral arrangement with the top oxygens of the sheets [17–22].

1.11.6 Framework Silicates

In the continental crust, quartz and feldspar are the two extreme abundant minerals (Figure 1.23). Feldspar itself is the single most abundant mineral in the Earth's crust. There are two sorts of feldspar, one comprising potassium and abundant in felsic

Figure 1.23. Quartz crystals.

rocks of the continental crust, and the other with sodium and calcium abundant in the mafic rocks of oceanic crust. Feldspars are mostly silica with aluminium, potassium, sodium, and calcium. Orthoclase feldspar ($KAlSi_3O_8$), also called potassium feldspar or K-spar, is made of silica, aluminium, and potassium. Organized with quartz, these minerals are classified as framework silicates [22–26].

Quartz is created of pure silica, SiO_2, with the tetrahedra organized in a three-dimensional framework. Impurities, in this regard, containing atoms within this framework gives rise to numerous diversities of quartz, among which are gemstones similar to amethyst, rose quartz, and citrine. Quartz and orthoclase feldspar are felsic minerals. Felsic, therefore, is the compositional term used to identify the continental igneous minerals and rocks that encompass plenty of orthoclase feldspar. Alternative feldspar is plagioclase with the chemical formula $(Ca, Na)AlSi_3O_8$. Thus, the solid solution (Ca, Na) signifies sequences of minerals, and one end of the sequences with calcium $CaAl_2Si_2O_8$ is termed as anorthite. Moreover, the other end signifies sodium $NaAlSi_3O_8$ and is termed albite. In this view, the mineral accommodates the substitution of Ca^{++} and Na^+.

Consequently, aluminium, which puts up with a similar ionic size to silicon to replace silicon inside the tetrahedra. Since the potassium ions are so abundant and higher than sodium and calcium ions, which are similar in size, the incapability of the crystal lattice to hold both potassium and sodium/calcium gives rise to the two clusters of feldspar: orthoclase and plagioclase, respectively. It can be said that framework silicates are known as tectosilicates and comprise the alkali metal-rich feldspathoids and zeolites [17, 19, 24].

1.12 Non-Silicate Minerals

In comparison with silicate minerals, the crystal structure of non-silicate minerals does not involve silica-oxygen tetrahedra (Table 1.3). In this understanding, countless non-silicate minerals are economically imperative and deliver metallic resources; for example, copper, lead, and iron. They similarly comprise valued non-metallic products; for instance, salt, construction materials, and fertilizer.

Table 1.3. Mineral class chemical formulas.

Mineral classes	Chemical formula	Example
Native elements	Au, Ag, Cu	gold, silver, copper
Carbonate	$CaCO_3$, $CaMg(CO_3)_2$	calcite, dolomite
Oxides	Fe_2O_3, Fe_3O_4, a mixture of aluminium	hematite, magnetite, bauxite
Halides	oxides	halite, sylvite
Sulfides	NaCl, KCl	galena, chalcopyrite, cinnabar
Sulphates	PbS, $CuFeS_2$, HgS	gypsum, epsom salts
Phosphates	$CaSO_4 \cdot 2H_2O$, $MgSO_4 \cdot 7H_2O$	apatite
	$Ca_5(PO_4)_3(F,Cl,OH)$	

1.12.1 Carbonate

Calcite ($CaCO_3$) (Figure 1.24) and dolomite ($CaMg(CO_3)_2$) are the two most commonly forming carbonate minerals, and regularly form in sedimentary rocks; for example, limestone and dolostone rocks, respectively. Some carbonate rocks, such as calcite and dolomite, are created through evaporation and precipitation. Nevertheless, most carbonate-rich rocks, for instance limestone, are formed by the lithification of fossilized marine organisms (Figure 1.25). These organisms, involving numerous microscopic organisms, have shells or exoskeletons containing calcium carbonate ($CaCO_3$). When these organisms die, their remnants amass on the seafloor, in which they remain and the soft body parts decompose and dissolve away. The calcium carbonate hard parts turn out to be contained in the sediments, ultimately developing

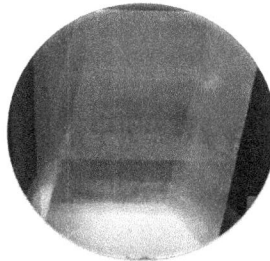

Figure 1.24. Calcite crystal.

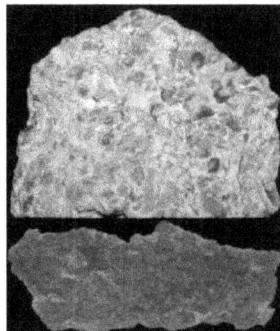

Figure 1.25. Limestone is full of small fossils.

the sedimentary rock known as limestone. While limestone perhaps comprises huge, easy to observe fossils, most limestones encompass the remains of microscopic creatures and thus originate from biological processes [14, 20, 25].

1.12.2 Oxides

Succeeding carbonates, oxides, halides, and sulfides are the most common non-silicate minerals. In this regard, oxides involve metal ions covalently bonded with oxygen. In this view, rust is the most familiar oxide, which is a combination of iron oxides (Fe_2O_3) and hydrated oxides. Hydrated oxides are, therefore, created when the iron is revealed to oxygen and water. Iron oxides, consequently, are vital for forming metallic iron. Iron oxide or ore manufactures carbon dioxide (CO_2) and metallic iron when it is smelted [2, 18, 20, 26].

The red sandstone cliffs in Zion National Park and throughout Southern Utah is usually due to the presence of iron oxides. In this sense, the red rocks involve white or colourless grains of quartz covered with iron oxide, which act as cementing agents maintaining the grains jointly.

Alternative iron oxides involve limonite, magnetite, and hematite. In this view, hematite arises in countless dissimilar crystal forms. The enormous structure reveals no external form. In this perspective, botryoidal hematite demonstrates massive concentric blobs. Specular hematite, therefore, seems like a mass of shiny metallic crystals. Oolitic hematite, subsequently, appears as a mass of dull red fish eggs. In this regard, these dissimilar kinds of hematite are polymorphs and completely enclose the similar formula, Fe_2O_3. Consequently, the alternative general oxide minerals are (i) ice (H_2O), an oxide of hydrogen; (ii) bauxite ($Al_2H_2O_4$), hydrated oxides of aluminium, an ore for producing metallic aluminium; and (iii) corundum (Al_2O_3), which includes ruby and sapphire gemstones [7, 10, 16, 27].

1.12.3 Halides

The halides comprise halogens in column VII, regularly bound to fluorine or chlorine, ionically merged with sodium or another positive ion. These contain halite or sodium chloride (NaCl), common table salt, sylvite or potassium chloride (KCl), and fluorite or calcium fluoride (CaF_2) [20–25].

1.12.4 Sulfides

Sulfides are important metal ores, in which metals are bound to sulfur. Significant cases comprise galena (lead sulfide), sphalerite (zinc sulfide), pyrite (iron sulfide, sometimes called "fool's gold"), and chalcopyrite (iron-copper sulfide). Sulfides, therefore, are well-identified for being significant ore minerals. Galena is, therefore, the foremost foundation of lead. Moreover, sphalerite, consequently, is the foremost basis of zinc, while chalcopyrite is the cornerstone of copper ore mineral mined in porphyry deposits. The main causes of nickel, antimony, molybdenum, arsenic, and mercury formations are sulfides [1, 20, 26].

Sulfate minerals encompass a metal ion; for example, calcium, bonded to a sulfate ion. In this view, the sulfate ion is a mixture of sulfur and oxygen (SO_4^{-2}).

In this view, the sulfate mineral gypsum ($CaSO_4 \cdot 2H_2O$) is exploited in construction materials; for instance, plaster and drywall. Subsequently, gypsum is frequently created from evaporating water and typically comprises water molecules in its crystalline structure. In this regard, the $2H_2O$ in the formula specifies the completed H_2O of water molecules. This is dissimilar from minerals such as amphibole, which encompasses a hydroxide ion (OH^-) that results from water but loses a hydrogen ion (H^+). Calcium sulfate, therefore, without water is a dissimilar mineral to gypsum named anhydrite ($CaSO_4$) [12, 15, 21, 27].

1.12.5 Phosphate Minerals

Phosphate minerals encompass a tetrahedral phosphate unit (PO_4^{-3}) commingled with numerous anions and cations. In some situations, arsenic or vanadium can replace phosphorus. The great well-known phosphate mineral is apatite, $Ca_5(PO_4)_3(F, Cl, OH)$, variations of which originate in teeth and bones. Moreover, the gemstone turquoise [$CuAl_6(PO_4)_4(OH)_8 \cdot 4H_2O$] is a copper-rich phosphate mineral that, similar to gypsum, contains water molecules.

1.12.6 Native Element Minerals

Lastly, native element minerals, typically metals, form in nature in a pure or almost pure state. Gold is an instance of a native element mineral; it is no longer very reactive and rarely bonds with alternative elements, so it is generally located in a remoted zone or pure state. The non-metallic and poorly-reactive mineral carbon is regularly discovered as a native element, such as graphite and diamonds. Mildly reactive metals identical to silver, copper, platinum, mercury, and sulfur once in a while take place as native component minerals. Reactive metals, for example iron, lead, and aluminium, almost continually bond to different elements and are rarely located in a native state [13, 17, 27].

1.13 Oil and Gas Formation

Sedimentary rocks are the most significant sort of rock to the petroleum industry since most oil and gas are developed in sedimentary rocks. In this understanding, sedimentary rock is the main source of all petroleum accumulation. In this regard, hydrocarbons form from the source rock in a sedimentary basin owing to organic evolution. Therefore, oil and gas are created when huge extents of organic (plants and animals) debris are endlessly suppressed in deltaic, lake and ocean environment. In the subsiding sedimentary basin, subsequently, this organic debris is entombed swiftly. As a result, sediments containing the organic debris sink deeply into the earth as their weight increases under the highest pressure and continuous sedimentation. In these circumstances, organic debris naturally decays in the presence of oxygen. Yet, the sediment shields the organic matter by producing an oxygen-free environment through the deepest earth. It can be said that the organic matter are accumulated rather than be annihilated by bacteria. In this view, sediments and the organic debris are heated up while they are as buried beneath younger sediments due to the highest temperature of Earth internal layers. Over millions of years, consequently, chemical reaction arises,

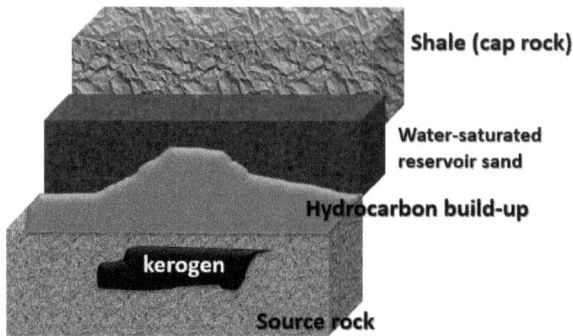

Figure 1.26. Formation of kerogen.

altering the organic debris into kerogen (Figure 1.26) owing to continuous increments in heat and pressure through Earth's internal layers [1, 5, 13, 28].

Kerogen, is key stone of geological hydrocarbon explorations since it is the substance that creates coal, oil, and gas. In this regard, the generation of coal, oil, and gas is a function of over thousands of years of temperature and pressure increments, assisting the decomposition of the organic debris. In this understanding, the generation of hydrocarbon seems to be very slow at temperature below 150 F, while the generation process reaches its maximum at a temperature range between 225 and 350 F. Subsequently, at maximum temperature, the source rock becomes hotter, chains of hydrogen and carbon atoms ultimately secede and create heavy oil (Figure 1.27).

As well as the temperature continues to increase, the heavy hydrocarbon is rehabilitated to lighter oil or gas. Gas can perhaps similarly be generated immediately from the decomposition of kerogen from the woody fragment of plants (Figure 1.28). This can be depicted as an oil and gas reservoir.

The oil and gas are generated by these developments perhaps in somewhat amalgamation and are permanently sorted out with water. In this view, a pocket of oil or gas is now created. Relying on the dimensions of the reservoir, the hydrocarbon generated could convert into a lucrative oil and gas field. Finally, the kerogen is then

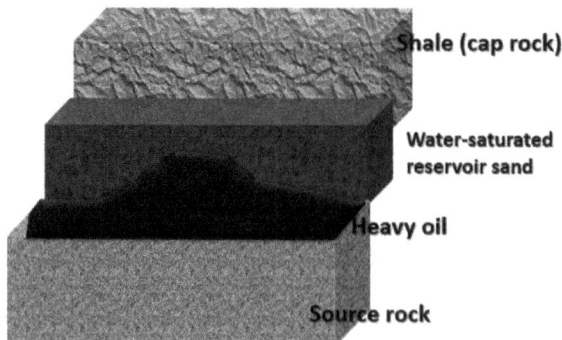

Figure 1.27. Concept of heavy oil formation.

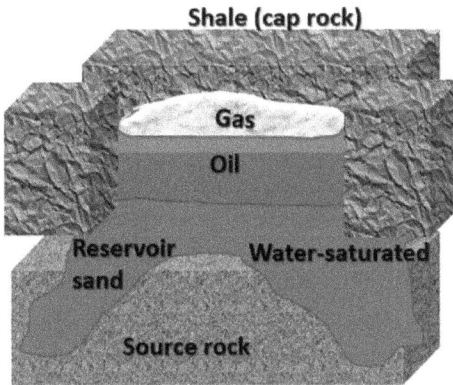

Figure 1.28. Oil and gas reservoir.

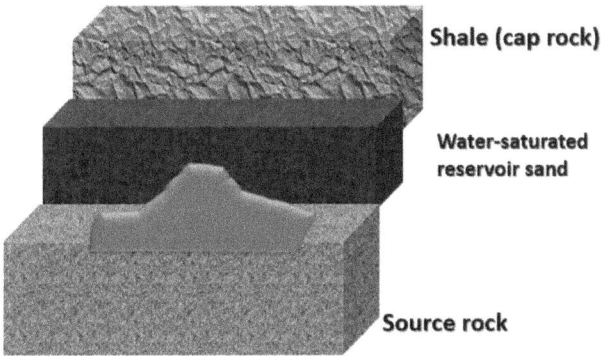

Figure 1.29. Intellectualized carbonization of kerogen.

carbonised, and hydrocarbons are no longer generated at temperatures above 500 F (Figure 1.29).

The hydrocarbon generated within the source rocks then is drifted into sandstone, i.e., the pores of the permeable rocks. In the permeable rock, it becomes lighter than the water. Consequently, these hydrocarbons drift, that is, they migrate up through the rock until disallowed moving further up by an impermeable rock (shale), where they coalesce into larger volumes [1, 29–32].

This chapter has demonstrated some aspects of the physical properties of the minerals and rocks. Besides, their fundamental occurrences and formation are also addressed. Consequently, the list of the sort of native and nonnative elements are also explained. The end of this chapter addressed the formation of oil and gas; for instance, the formation of Kerogen. The principles of minerals, oil and gas can assist us to understand the mechanism of remote sensing for distinguishing between different minerals. The next chapter will address the quantization of minerals and their interaction with remote sensing electromagnetic spectral and signals.

References

[1] Longwell CR, Flint RF. Introduction to physical geology. Soil Science. 1962 Apr 1; 93(4): 294.

[2] Tarbuck EJ, Lutgens FK, Tasa D, Linneman S. Earth: An Introduction to Physical Geology. Upper Saddle River: Pearson/Prentice Hall; 2005.

[3] Skinner BJ, Porter SC, Park JJ, Levin HL. Dynamic Earth: An Introduction to Physical Geology. Chichester, West Sussex: John Wiley & Sons. Inc.; 2013.

[4] Duff PM, Duff D (eds.). Holmes' Principles of Physical Geology. Taylor & Francis; 1993.

[5] Chernicoff S. Geology: An Introduction to Physical Geology. Houghton Mifflin Harcourt (HMH); 1999.

[6] Davidson JP, Davis PM. Exploring Earth: An Introduction to Physical Geology. Prentice Hall; 1997.

[7] Levi P, Rosenthal R. The Periodic Table. New York: Schocken Books; 1984 Dec.

[8] Scerri E. The Periodic Table: Its Story and Its Significance. Oxford University Press; 2019 Oct 21.

[9] Lengler R, Eppler MJ. Towards a periodic table of visualization methods for management. InIASTED Proceedings of the Conference on Graphics and Visualization in Engineering (GVE 2007), Clearwater, Florida, USA 2007 Jan 3.

[10] Wedepohl KH. The composition of the continental crust. Geochimica et cosmochimica Acta. 1995 Apr 1; 59(7): 1217–32.

[11] Evans AM. Ore Geology and Industrial Minerals: An Introduction. John Wiley & Sons; 2009 Jul 10.

[12] Putnis A. An Introduction to Mineral Sciences. Cambridge University Press; 1992 Oct 22.

[13] Barnes-Svarney PL, Svarney TE. The Handy Geology Answer Book. Slovenia: Visible Ink Press; 2004.

[14] Raines GL, Sawatzky DL, Connors KA. Great Basin geoscience data base. US Geological Survey; 1996.

[15] Taylor EW. Correlation of the Mohs's scale of hardness with the Vickers's hardness numbers. Mineralogical Magazine and Journal of the Mineralogical Society. 1949 Sep; 28(206): 718–21.

[16] Marinari S, Masciandaro G, Ceccanti B, Grego S. Influence of organic and mineral fertilisers on soil biological and physical properties. Bioresource Technology. 2000 Mar 1; 72(1): 9–17.

[17] Jackson SA, Cartwright AG, Lewis D. The morphology of bone mineral crystals. Calcified Tissue Research. 1978 Dec 1; 25(1): 217–22.

[18] Ford DJ. The challenges of observing geologically: Third graders' descriptions of rock and mineral properties. Science Education. 2005 Mar; 89(2): 276–95.

[19] Le Maitre RW, Streckeisen A, Zanettin B, Le Bas MJ, Bonin B, Bateman P, Bellieni G, Dudek A, Efremova S, Keller J, Lameyre J. Igneous rocks. A classification and glossary of terms. 2002; 2.

[20] Cox KG (ed.). The Interpretation of Igneous Rocks. Springer Science & Business Media; 2013 Apr 9.

[21] Streckeisen A, Zanettin B. Igneous rocks: IUGS classification and glossary: recommendations of the International Union of Geological Sciences, Subcommission on the Systematics of Igneous Rock. Le Maitre RW, editor. University of Cambridge; 2004.

[22] O'connor JT. A classification for quartz-rich igneous rocks. Geological Survey Professional Paper. 1965; 525: 79.

[23] MacKenzie WS, Donaldson CH, Guilford C. Atlas of Igneous Rocks and Their Textures. London: Longman; 1982 Dec 27.

[24] Middlemost EA. Magmas and Magmatic Rocks: An Introduction to Igneous Petrology. Longman Group Ltd., London; 1985.

[25] Murase T, McBIRNEY AR. Properties of some common igneous rocks and their melts at high temperatures. Geological Society of America Bulletin. 1973 Nov 1; 84(11): 3563–92.

[26] Pettijohn FJ. Sedimentary Rocks. New York: Harper & Row; 1975 Jan.

[27] Mackenzie FT, Garrels RM. Evolution of Sedimentary Rocks. New York: Norton; 1971.

[28] Tucker ME (ed.). Sedimentary Petrology: An Introduction to the Origin of Sedimentary Rocks. John Wiley & Sons; 2009 Apr 1.

[29] Tucker ME. Sedimentary Rocks in the Field. John Wiley & Sons; 2003 Jul 25.

[30] Buerger MJ. The role of temperature in mineralogy. American Mineralogist: Journal of Earth and Planetary Materials. 1948 Apr 1; 33(3-4): 101–21.

[31] Osborn EF. Reaction series for subalkaline igneous rocks based on different oxygen pressure conditions. American Mineralogist: Journal of Earth and Planetary Materials. 1962 Apr 1; 47(3-4_Part_1): 211–26.

[32] Young DA. Norman Levi Bowen (1887–1956) and igneous rock diversity. Geological Society, London, Special Publications. 2002 Jan 1; 192(1): 99–111.

CHAPTER 2

Quantization of Minerals and their Interactions with Remote Sensing Photons

∞∞∞

In the previous chapter, we demonstrated that atoms of elements, for example oxygen and silicon, are the keystone in the majority of mineral formations. Minerals are homogeneous and makeup of only one compound, but rocks form from combinations of minerals.

Therefore, atoms, molecules, and the essential electron and proton charges are all examples of physical entities that are quantized, that is, they exist merely in convinced discrete values and do not possess all conceivable value. Quantized is the opposite of continuous. For instance, a fraction of an atom is impossible to be obtained, or part of an electron's charge, or 14-1/3 cents. Rather, the entirety is made up of integral multiples of these substructures.

In this view, quantum mechanics is implemented in this chapter for full comprehension of the behaviour of mineral atoms' interaction with photons. Image processing algorithms are mainly based on a perfect understanding of the quantization of both minerals and photon interactions.

2.1 Quantization in the Atom

Nowadays, two models of atomic structure are in use: (i) the Bohr model and the (ii) the quantum mechanical model. Therefore, the quantum mechanical model is grounded in mathematics. Even though it is extremely complex to comprehend than the Bohr model, it can be exploited to clarify inspections initiated on complex atoms. Consequently, the quantum mechanical model is established on quantum speculation, which articulates matter also has properties allied with waves. Consistent with quantum theory, it is impossible to distinguish the precise position and momentum of an electron at an equal time using the Uncertainty Principle. In this regard, the quantum mechanical model of the atom exercises the complex forms of orbitals, which is rarely termed as electron clouds, and volumes of space that are occupied by electrons. Accordingly, this model is built on probability rather than certainty.

In this understanding, four numbers, known as quantum numbers, were acquainted with atoms to designate the physical characteristics of electrons and their orbitals: (i) principal quantum number, (ii) angular momentum quantum number, (iii) magnetic quantum number, (iv) spin quantum number.

2.1.1 Principal Quantum Number

The principal quantum number n designates the mean distance of the orbital from the nucleus, and the energy level of the electron in an atom. For instance, it can enclose a positive integer, i.e., entire number values: 1, 2, 3, 4, and so on. In this perspective, the higher the value of n, the maximum the energy and the bigger the orbital. In this view, chemists have mostly termed the orbitals electron shells. According to this view, electron shells can be split into subshells, where every subshell comprises a dissimilar sort of orbital. For instance, Shell 1 comprises merely one subshell, 1s, which encompasses its single s-orbital. Subsequently, shell 2 comprehends dual subshells, 2s and 2p, which correspondingly comprise its s-orbital and its three p-orbitals. Figure 2.1 shows the s-orbital of shell 1 and the s-orbital and p-orbitals of shell 2. The p-orbitals, for instance, have matching forms in three dissimilar directions [1].

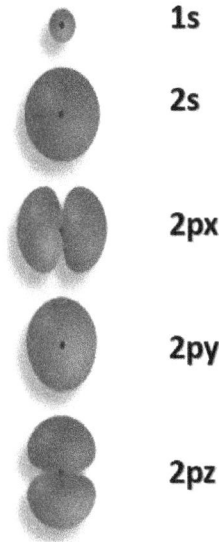

1s

2s

2px

2py

2pz

Figure 2.1. Orbital atom shells.

2.1.2 Angular Momentum Quantum

Therefore, the angular momentum quantum number l designates the form of the orbital. However, the orbital shape is restricted by the principal quantum number n. In other words, the angular momentum quantum number l can include positive integer values from 0 to $n - 1$. For instance, if the value of n is 3, three values are permitted for l: 0, 1, and 2. Therefore, the value of l describes the profile of the orbital, and the value of n defines the size.

The significant question is: why do the orbitals make a dumbbell shape instead of something like a circle? In this perspective, there are, for instance, d and f orbitals along with the *p* and *s* orbitals. In this circumstance, if *n* = 3 there will be five 3*d* orbitals, five 4*d* orbitals, etc. because every shell has five possible values for the *n* numbers. Reliance on the *d* orbital will modify the shape of the "dumbbell" (orbital cloud). A *d*-orbital of four reveals a "four-leaf clover" shape of the dumbbell, with each of the leaves laying orbital cloud on the planes of the axis.

Therefore, the most exciting thing about orbitals is the shape that they form. Instead of just spinning around a centre point, they turn into ellipse shapes forming a dumbbell owing to them higher energy and various wavelengths. On the contrary, *d*-orbitals only has a spherical shape. Indeed, they do not possess approximately as great of energy as the other orbitals.

The *p* orbitals make up a shape practically similar to a rounded bow tie. In this view, p-orbitals have a two of the dumbbell shapes coupled by their slightest close at the origin of the axis. They merely possess sufficient energy to form that single shape, and there is just two of them; subsequently, merely dual shapes are created. The *d* orbitals form a rounded four-leaf clover shape (Figure 2.2). In this circumstance, four orbitals are formed, creating four shapes that join in the origin of the axis.

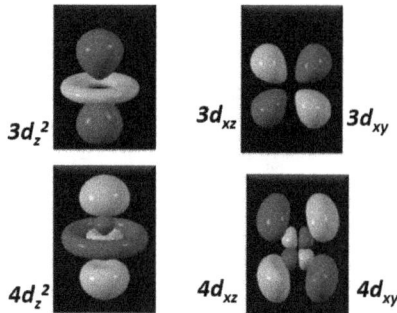

Figure 2.2. Shapes of atomic orbitals.

2.1.3 *Magnetic Quantum Number*

The magnetic quantum number m_1 defines how the numerous orbitals are oriented in space. The value of this number relies on the value of *I*. The values permitted are integers from –*I* to 0 to +*I*. For instance, if the value of *I* = 1 (*p* orbital), three values for this number are stated as –1, 0, and +1. This signifies that there are three dissimilar *p* subshells for a particular orbital. In this regard, the subshells possess similar energy but dissimilar orientations in space.

2.1.4 *Spin Quantum Number*

The fourth and last quantum number is the spin quantum number *S*, which designates the direction the electron is spinning in a magnetic field—either clockwise or counterclockwise. In other words, there are only two values that are considered: +1/2 or –1/2. In this understanding, there can be merely dual electrons, one with a spin of +1/2 and another with a spin of −1/2 for every subshell. For instance, quasi-one-dimensional spin systems, such as fedotovite $K_2Cu_3O(SO_4)_3$, have an unfamiliar magnetic behaviour at an identical low temperature, in which the ground state is a one-dimensional chain in a triplet configuration with $S = 1$ spin. In this regard, the $S = 1$ arises since there is an even number of S = 1/2 on the magnetic Cu^{2+} ions at the ends of the spin chain. Therefore, quasi-one-dimensional chains, for example fedotovite, do not have a single spin but a bunch of spins creating a cluster. In this sense, one cluster of atoms then insipidly intermingles with the neighbouring cluster of atoms [3].

2.2 Quantum Mechanics of Bonding

The significant question is: what is the role of quantum mechanics in understanding minerals? There is no doubt that all minerals involve electrons. A full theory of the chemical bond requests to revert to the origins of the behaviour of electrons in molecules. Specifically, the role of the electron pair and the quantitative explanation of bonding must be grounded on the Schrödinger equation and the Pauli exclusion principle.

In this understanding, the behaviour of electrons in molecules and atoms is designated by quantum mechanics; classical (Newtonian) mechanics cannot be implemented since the de Broglie wavelengths $\lambda = \dfrac{h}{mv}$ of the electrons are comparable with molecular (and atomic) dimensions. Consequently, the applicable quantum-mechanical concepts are explained by:

i) Electrons are categorized by their entire distributions, which is termed as the wave functions or orbitals. In this view, an electron perhaps always considers to be with appropriate probability at entire points of its distribution rather than by instantaneous positions and velocities. In this circumstance, an electron does not vary with time.

ii) The kinetic energy of an electron diminishes as the volume dominated by the bulk of its spreading grows; consequently, delocalization reduces its kinetic energy.

iii) The potential energy of interaction between an electron and other charges is as computed by classical physics. In this view, using the appropriate distribution (wave function) for the electron: an electron distribution is, consequently, attracted by nuclei and its potential energy, which diminishes as well as the average electron-nuclear distance shrinks.

iv) A minimum-energy electron spreading signifies the greatest conciliation between concentration near the nuclei (to diminish potential energy) and delocalization (to decrease kinetic energy).

According to the above perspective, a bond would make up between dual atoms when the electron variation of the shared atoms (molecular orbital) produces pointedly minor energy than the separate-atom spreading (atomic orbitals). For instance, a covalent bond, in which the sharing of electron pairs. These electron pairs are known as shared pairs or bonding pairs, and the stable balance of attractive and repulsive forces between atoms, when they share electrons, is known as covalent bonding.

A molecular formula, consequently, is exercised in the chemistry field to designate how many atoms of each element are existing in a molecule [2]. The ball-and-stick model, for example, is one of the well-recognized molecular formulae, where the balls depict atoms and stick describe bonds (Figure 2.3).

A molecular formula, yet, does not exhibit the construction of the molecule. Scientists regularly implement physical formulas to demonstrate the number and array of atoms in any composite [4]. Contrariwise, the quantum theory of atoms (QTAIM) can sculpt molecular and condensed matter electronic structures. In practice, the electron density distribution function is articulated by essential objects of molecular structure—atoms and bonds. An electron density distribution of a molecule is a probability distribution, which terms the standard behaviour in which the electronic charge is scattered through definite space in the nuclei. Consistent with QTAIM, the molecular structure has been revealed to employ the immobile features of the electron density together with the gradient tracks of the electron density, which create and dismiss at these points [5].

In chemistry principle, as per valence bond theory (Chapter 1), the carbon atom requires to perceive the "4-hydrogen rule", which affirms that the maximum diversity of atoms reachable to bond with carbon is equivalent to the number of electrons that are involved into the outer shell of carbon. In watchwords of shells, carbon requires of a partial outer shell, which comprises 4 electrons and subsequently holds 4 electrons accessible for covalent or dative bonding [6, 7]. In this understanding, the quantum array of the hydrocarbon (HC) molecule, for instance, is substantiated. Therefore, hydrocarbon (HC) molecules of numerous forms and elements are shaped on account of the interaction of the valency electron orbits of carbon (C) and hydrogen (H) atoms, the elementary particles of which are quantum matters and contains chemical and physical information of the HC [1, 8].

In this understanding, the numbers that specify quantized physical quantities, for instance, angular momentum, energy, and quantum spin are known as quantum number. For instance, the quantum energy of the Hydrogen atom is denoted by a set of numbers beginning with $n = 1$ of the ground state, where n is a principal quantum number [7–9].

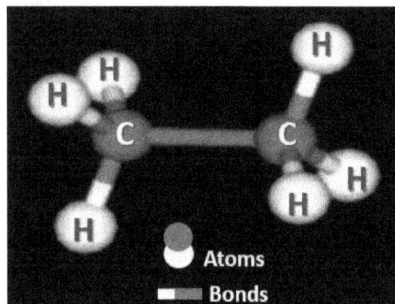

Figure 2.3. Ball-and-stick model of the molecular formula.

2.3 Quantum Mechanics of Mineral Atomics

Quantum mechanics can investigate a reliable quantification of various mineral atoms as a function of its absorption spectra. Let us assume that minerals contain number of carbon atoms (n) with edge length b and d represent the average spacing between carbons $\sim 1.4 A^{\circ}$. The mathematical description of any mineral atom as a function of its compound chain can be revealed as [10]:

$$\eta(x) = \begin{cases} 0, & \text{for } 0 \le x \le L_n \\ \infty, & \text{Otherwise} \end{cases} \tag{2.1}$$

Equation 2.1 represents the mineral compound chain in the one-dimensional 1-D infinite potential well $\eta(x)$ where the Fermi level L is defined as:

$$L_n = (n-1)d + 2b \tag{2.1.1}$$

The Fermi energy is an insight in quantum mechanics regularly denoting the energy alteration between the highest and lowest (Figure 2.4), which is occupied single-particle states in a quantum system of non-interacting fermions at absolute zero temperature. Specifically, the levels of potential energy E_k are formulated as:

$$E_k = \frac{\hbar^2 k^2}{2m(nd)^2}. \tag{2.2}$$

where $\hbar$ is h-bar, which is a modified form of Planck's constant and equals $\dfrac{h}{2\pi}$. In this regard, if electrons are pointed in $n+1$, the ground state energy E_0 of the mineral atom is calculated using:

$$E_0 = \frac{\hbar^2 \pi^2}{24m(nd)^2}(n^3 + 3n^2 + 5n + 3). \tag{2.3}$$

Equation 2.3 demonstrates that the energy of the first excited state can be computed by projecting an electron from the Fermi level to the next excited state as:

$$E_1 = \frac{\hbar^2 \pi^2}{24m(nd)^2}(n^3 + 3n^2 + 17n + 27). \tag{2.4}$$

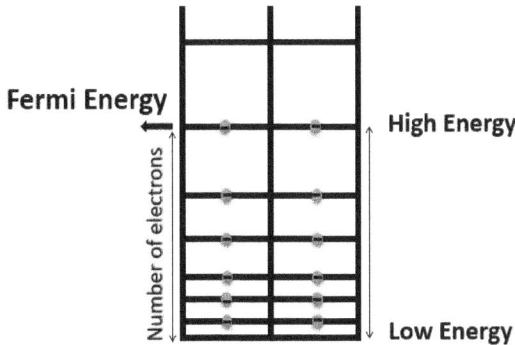

Figure 2.4. The Fermi energy concept.

The wavelength of light absorbed in a ground state-to-first excited state transition is a function of the Fermi level. In this view, the difference between E_0 and E_1 is proportional inversely with the absorbed wavelength λ_n and is obtained by [11]:

$$\Delta E = \frac{hc}{\lambda_n} \tag{2.5}$$

In this view, the wavelength of light absorbed λ_n by mineral atoms is computed by:

$$\lambda_n^N = \frac{\lambda_n^0}{1+(-1)^{\frac{n+3}{2}}\gamma\dfrac{n}{n+2}} \tag{2.6}$$

where λ_0 is the wavelength of absorbed light at the first Femi level, and γ is given by:

$$\gamma \equiv \frac{4m\alpha\eta_0 d}{\hbar^2\pi^2} \tag{2.6.1}$$

where α is level of absorbed energy Equation 2.6 explains the level of the light wave spectra absorption by mineral atom. In this understanding, it was documented that minerals' molecular ions with different numbers of element atoms absorb a different wavelength of electromagnetic spectra (Figure 2.5). In this view, the presence of the nitrogen atoms in the centre of atom spectra illustrates the shifting of the red or blue wavelength on whether $\dfrac{n+1}{2}$, where $n + 1$ is the number of electrons, is even or odd, respectively [17–19]. Moreover, Figure 2.5 reveals that there is series of dark lines owing principally to the absorption of specific frequencies of light by cooler atoms in the outer atmosphere of the sun. By comparing these lines with the spectra of mineral elements measured on Earth, it can be noticed that the sun encompasses huge quantities of hydrogen, iron, and carbon, along with smaller amounts of other elements such as Mg, and Na.

According to the above perspective, an electrical discharge electrifies neutral atoms to a higher energy phase, and light is radiated when the atoms decay to the ground state. For instance, mercury has most of the emission lines, which are below 450 nm and generates blue light. On the other hand, sodium has the most intense emission lines at 589 nm, which forms an intense yellow light.

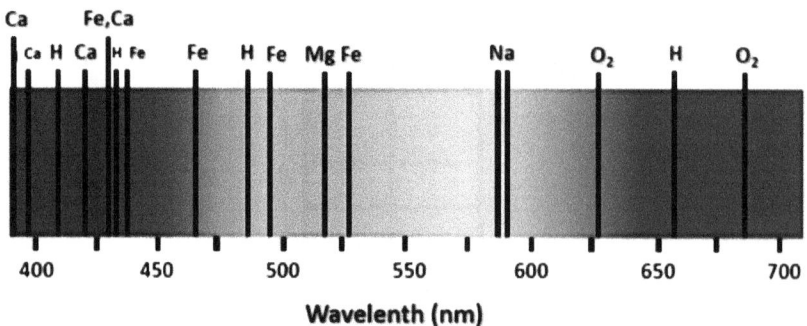

Figure 2.5. Light spectrum absorption with different mineral elements.

2.4 Energy Variations Based on Schrödinger Wavefunction

Let us assume that $\psi(x, t)$ is the wavefunction, in which the energy variation can be expressed as

$$\psi(x, t) = \psi(x)e^{-iEt/\hbar} \tag{2.7}$$

Equation 2.7 demonstrates the energy variations based on the time-independent Schrödinger equation $\psi(x)$, which is computed as:

$$\frac{d^2\psi}{dx^2} = \frac{2m}{\hbar^2}[\eta(x) - E]\psi. \tag{2.8}$$

In particular, the solution, $\psi(x)$, to equation 2.8 must be finite; otherwise, the probability density $|\psi|^2$ would develop immeasurable. Similarly, the solution must be incessant, else, the present probability would measure infinite. Equation 2.8, therefore, reveals that E is merely exploited to acquire confident discrete quantities. In this understanding, the energy operator would have the discrete eigenvalues, which is a conventional feature of bounded solutions. In mathematical form, the solutions would satisfy $|\psi| \to 0$ as $|x| \to \infty$. In this understanding, despite the fact the energy eigenvalues are perhaps discrete for tiny quantities of energy, they regularly develop incessantly at extraordinary sufficient energies since the mineral structures can no longer occur as a bound state. In the case of a more realistic harmonic oscillator potential, for instance a diatomic molecule, the energy eigenvalues become denser and denser as it approximates the dissociation energy. This can be demonstrated within the formation of igneous rocks when magma cools and crystallizes. The energy levels, therefore, post dissociation energy can take the continuous values associated with free particles. In this regard, the stationary solutions of Equation 2.8 can lead to the linear superposition of typical *time-dependent* solution as:

$$\psi(x,t) = \sum_{n=0,\infty} \int_0^a \psi_n(x)\psi(x,0)dx.[e^{-iE_nt/\hbar}] \tag{2.9}$$

Equation 2.9 demonstrates the typical concept of quantum mechanics within endless of n = 1,2,3,…….. In this regard, the infrared spectrum (IR) can distinguish between different mineral elements. For instance, IR spectroscopy smoothly recognizes the carbonyl group C=O of organic compounds: of amides, or esters, or ketones, of acids as a robust piercing absorption at approximately 1900–1700 cm^{-1}. This reveals that the chemical bonds are not inelastic, steady sticks; rather, they are elastic and are accomplished in both stretching and bending. They are continuous in mobility: the bonds vibrate, and they can absorb light of energy equivalent to this vibration. This absorption leads to turn into an 'electrified' vibrational state. It can be said that the energy of the electron in hydrogen is quantized since it is in a specific orbit. Since the energies of the electron can have merely convinced quantities, the variations in energies would have only specific values. This explanation can guide us for a general definition of quantization. In this view, quantities that have convinced specific values are termed quantized.

In a dissimilar crystal field, the energy level for a similar ion would be different, which is termed as the crystal field effect. In the case of transition elements, for instance Ni, Gr, Cu, Co, Mn, etc., it is the 3d shell electrons that primarily reveal the energy levels in which they are not shielded (outer shell orbital). In this circumstance, the external field of the crystal regulated electron energy levels, and the electrons perhaps adopt new energy values relying on the crystal fields. In such circumstances, the new energy levels, the transition between them, and subsequently, their absorption spectra, are revealed mainly by the valence state of the ion (e.g., Fe^{2+} or Fe^3) and by its co-ordination number, position equilibrium and, to a partial extent, the sort of ligand formed (e.g., metal-oxygen) and the degree of lattice distortion. For instance, ferrous ions, when situated in dissimilar crystal fields, yield absorption peaks at dissimilar wavelengths.

2.5 What is Quantum Influences?

The main question that arises is: what is the quantum effects? Quantum impacts turn out to be dominant when mineral atom sizes are equivalent to the electron wavelength. This stereotypically arises when a mineral's dimensions are on the nanometer scale with new singularities associated with the mineral's electronic and magnetic characteristics evolving in minor realms. The sighting of these characteristics has delivered novel fields of exploration forward to develop and rule the physical singularities confirmed by such minerals. In this view, quantum effects initially occur at the boundary of the Earth's inner core where a phase change leading to a reverse of geomagnetic field polarity proceeds. This leads to another significant question: what are those minerals that exhibit quantum characteristics?

Plentiful copper minerals present an assembly of quantum characteristics counting dimerized quantum spin dimers. In this regard, dimers are principally antiferromagnetically coupled atoms. In other words, the magnetic moment of every contiguous copper atom is disparate. This permits a quantum critical point, which is the point at which a quantum phase shift arises, to be approached under the function of a magnetic field. In this understanding, such minerals comprise the copper carbonates malachite [$Cu_2CO_3(OH)_2$] and callaghanite [$Cu_2Mg_2(CO_3)(OH)_6 \cdot 2H_2O$], and the copper arsenates urusovite [$CuAl(AsO_4)O$] and clinoclase [$Cu_3(AsO_4)(OH)_3$]. In this sense, naturally arising minerals such as copper are ruled with quantum spin chains that impact both ferromagnetism. In this view, the magnetic domains of contiguous atoms are affiliated and become antiferromaganetism, wherever they are contradicted. In these circumstances, interactions between these chains can impact spin-spiral magnetic order tolerating suppression of long-range magnetic order. Therefore, copper arsenates, sulfates, and molybdates entirely demonstrate this kind of quantum behaviour. Such characteristics and their control on the behaviour of minerals exhibiting them to have potential functions in low-temperature processes; for example, the liquefaction of the hydrogen. Consequently, quantum properties are also shown in the transition metal sulfide minerals, which are revealed in iron sulfides crystallizing in the tetragonal crystal system. In this circumstance, superconductivity, a quantum mechanical phenomena associated with electrical resistivity, practically entirely disappears beneath a convinced critical temperature. Awkwardly, the occurrence of other metals in naturally ensuing tetragonal iron sulfides, e.g., mackinawite [$(Fe,Ni)_{1+x}S$], would overwhelm these characteristics and it is improbable an abstract phase tetragonal iron sulfide would be revealed in the geological environment. It can be said that an enormous band of minerals reveal quantum characteristics that impact their magnetic and electrical characteristics. Yet in the field of quantum magnetism, there is an excess of other phenomena kinds. Accordingly, the complex exchanges between electrons and the countless metallic element exhibit in the crystal structure present chemists an enormous chance to discriminate the countless factors that escort to such characteristics and exploit them in advanced-tech usages. Recognizing the temperatures at which minerals and their electrons display quantum characteristics, for instance, magnetism and superconductivity can assist in the growth of advanced-operation computers as predictable processors naturally depend on cooling systems to optimize operating.

Let us note the role of quantum linked particles which accumulation in high-pressured kalicinite ($KHCO_3$), the mineral in which hydrogen bonds show the capacity for quantum entanglement. For instance, the changes of hydrogen bonds' density or its unit-cell volume would obey the fluctuation of its atomic bonds, which perhaps due to adding a specific amount of energy such as a shock wave that swelling extreme energy in the Earth's internal structure layers. Consequently, the structural phase transition of hydrogen bonds occurs instantly with other minerals containing hydrogen and oxygen (nitrogen, fluorine) over a large volume. In this view, this phenomenon is well-known as cooperativity [14].

2.6 Quantization of Minerals from Point View of Wavefunction

The main question that arises is how minerals can be imagined in remote sensing data? The conventional answer, which is constrained by the interaction of electromagnetic radiation with minerals and the level of mineral absorptions under the circumstance of the blackbody radiation concept, is extremely consumed. Is there any alternative theory under the quantum that can deliver a new approach of explanation? How to establish a general concept for mineral quantization?

Let us assume that mineral particles as a quantum bound state. The mineral particles are assumed as N_P, which can be described as completely as possible by an acceptable, square-integrable function ψ^2. In this regard, $\psi(q_1, q_2,......, q_N, \omega_1, \omega_2,...., \omega_N, t)$, where the q_N are spatial coordinates, ω_N are the spin coordinates and t is the time coordinate [12].

In this view, many mineral particles must be spinning and their spinning causes electrons to oscillate through their orbits. Therefore, if ψ is normalized, then $\psi^*\psi d\tau$ represents the probability that the electron space-spin coordinates, which are located on a part of minerals volume-space $d\tau (\equiv d\tau_1 d\tau_2 d\tau_3.....d\tau_N)$ at time t.

In this understanding, the wavefunction ψ must describe the interaction of mineral particles with each other and their surrounding environment as one dynamic system, which is a function of a time-dependent state. The mathematical wavefunction can be described as $\psi(x_1, y_1, z_1, \omega_1, x_2, y_2, z_2, \omega_2, t)$. In fact, the first particle mineral wave function is $\psi_s(x_1, y_1, z_1, t)$ while the second one's wavefunction is $\psi(y_2, z_2, \omega_2, t)$. In this circumstance, the ω_s spin coordinates would each be some combination of spin functions of first mineral particles α_s and the other mineral particles β_w, respectively. Consequently, a spin-free density function is produced by the integration of $\psi^*\psi$ over the spin coordinates of numerous of mineral particles. This can be mathematically expressed as:

$$\rho(x_1, y_1, z_1, x_2, y_2, z_2, t) \equiv \rho(v_s, v_w, t) \tag{2.10}$$

Equations 1.10 shows that the electron velocity v_s is a function of their velocity probability changes in dv_s between the initial orbit, and location (x_1, y_1, z_1) to induce shifting of the electron to the next orbit based on gaining a certain amount of energy $(x_1 + dx_1, y_1 + dy_1, z_1 + dz_1)$ and the last orbit that holds electron energy in dv_w [13]. In quantum mechanics, therefore, it indicates that entirely allied particles of minerals are designated by a single wavefunction and they have a precise coherence.

In this understanding, a new density function $\rho'(v_s, t)$ is produced to describe the probability of determining the electron behaviour in various dynamic states of mineral particles due to time-energy fluctuations of their propagations. In this view, electron energy cannot be separated from the mineral particles. In remote sensing technology, this explains why remote sensing sensors can imagine mineral with the surrounding environment as the different quantum state, which enable to discriminate any mineral particle from each other owing to its different dynamic interaction with an electromagnetic wave. This indicates the significant role of quantum mechanics based on wave function to understand the mechanics of mineral imaging in remote sensing data. This also can be used as a new approach to quantize mineral imaging mechanism in remote sensing data as density function variations, which are based on mineral particle interactions with electromagnetic spectra.

2.7 Antiferromagnetic Spin-frustrated Layers of Minerals

The recent superior study done by Baraban [16] proves the atomic spin behaviour through minerals due to different sorts of atom magnetic moment. The closest interactions of the atomic spins allied to specific spots in the lattice regulate macroscopic magnetic characteristics of numerous mineral phases. Magnetic arrays, therefore, have a vital impact on the physical characteristics of materials such as stacked triangular antiferromagnets or high-temperature superconductors. In this view, the interface between electron spins in materials expresses the nature of the condensed phase. Nevertheless, investigations of the organization of the magnetic moments in mineral crystals are rather difficult and correlated to common space, e.g., neutron scattering, or spectroscopic Mössbauer measurements, or macroscopic quantities, such as heat capacity. In this regard, the most common phenomena, which determine the properties of magnetic crystals are paramagnetism, ferromagnetism, antiferromagnetism, and ferrimagnetism in which they are distinguished from each other by the fluctuation of the atom magnetic moments.

In a paramagnetic material, atoms have permanent magnetic moments, whose routes are, nevertheless, randomized owing to the thermal fluctuations. In this scenario, the overall magnetic moment of paramagnet, therefore, turns to be zero. Since an external magnetic field is operated, the magnetic moments being to ally along the field lines, causing the steady growth of the magnetization of the material. This is known as magnetic moment per unit volume. In this view, the conventional speculation of paramagnetism, which deliberates the arrangement of magnetic moments in paramagnets, is also known as Langevin's theory. Consistent with Langevin's theory, weak magnetic field allies with the linear growth of the entire magnetic moment. In this circumstance, the magnetization achieves its saturation value once total magnetic moments in a material are associated.

In the circumstance of ferromagnetism, the mineral atoms have a parallel order of magnetic moments (Figure 2.6). In this perspective, the ferromagnetic state is detected beneath the critical Curie temperature (Figure 2.7).

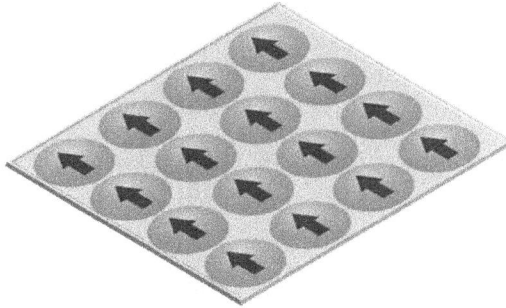

Figure 2.6. Ferromagnetic ordering in a crystal.

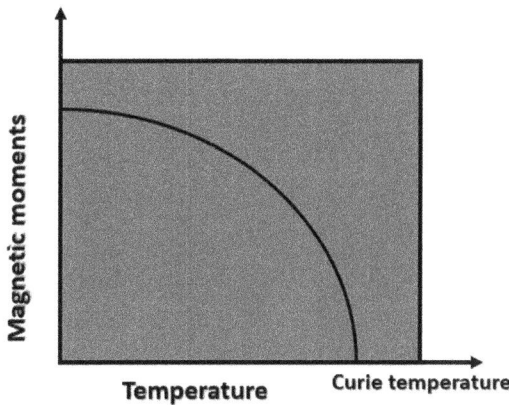

Figure 2.7. Curie law.

Consistent with Curie law, at higher temperatures than critical Curie temperature, the phase transition from the ferromagnetic state to the paramagnetic state arises. Consequently, entirely metallic ferromagnets ally to the set of the phase transition metals, whose atoms possess a partial 3d– (Fe, Co, Ni) or 4f–shell (Gd, Tb). In this understanding, the electron shells of these elements turn to have nonzero magnetic moments owing to the occurrence of spin and orbital moments. This circumstance, therefore, is essential for the growth of the magnetic ordering in crystals. Sequentially, the parallel configuration of the magnetic moments in ferromagnets delivers more energy for the entire system owing to the interchange interface of the electrons from the outer shells. In this regard, why and how does ferromagnetism behave above and below the Curie? The moments are allied beneath the Curie temperature (at ferromagnetic) but then above the Currie temperature, it turns into paramagnetic. This is anticipated since beneath the Curie temperature the spins are of a similar magnitude with the order. But then circulating the Curie temperature indicates the moments can become randomly aligned, resulting in collapsing down, forming the material paramagnetic.

The quantum mechanical phenomena are the most accurate approach to designate the ferromagnetism since merely certain angles of magnetic movement are permitted. Through the conventional viewpoint, entire angles are apportioned since the Langevin theory delivering this approach is tremendously improbable. Consequently, the normalized power-law length with a gamma of 0.5 is a precise depiction of the ferromagnetism phenomena. According to the above perspective, what are the differences between ferromagnetism and paramagnetism? The ferromagnetic mineral has harmonized electron spins steering in a similar direction, while paramagnets have spun in entire directions. This, subsequently, induces the ferromagnets to have robust attractive or revolting forces when applies to a permanent magnet. On the other hand, ferromagnets possess poor attractions to robust permanent magnets.

The significant question is: what is meant by antiferromagnetic spin? The specific sort of magnetism in solid, for instance, manganese oxide (MnO) is termed as antiferromagnetism (Figure 2.8). In this view, manganese ions (Mn^{2+}) operate as minor magnets that freely ally themselves at moderately low temperatures into reverse, or antiparallel arrays across the minerals so that it shows practically no gross exterior magnetism. In this regard, antiferromagnetic minerals involve specific metals and alloys as well as some ionic solids. Therefore, from magnetic atoms or ions, the magnetism is pointed in a single direction, which is terminated by the set of magnetic atoms or ions. On the contrary, ions or magnetic ions are affiliated in the inverse direction.

The magnetic moments, therefore, are produced from the spin of the atomic valence electrons. In other words, the magnetic moments of every orbiting electron are entirely randomly oriented exclusively of a magnetic field that operates on the minerals. In this circumstance, the magnetic moments of the electrons ally and act antiparallel to every adjacent moment [15]. In this context, the Néel temperature associated with heating disrupts the antiparallel coupling of atomic magnets. In this view, magnetic ordering temperature is known as the Néel temperature that is the temperature above which an antiferromagnetic fabric turns into paramagnetic. In

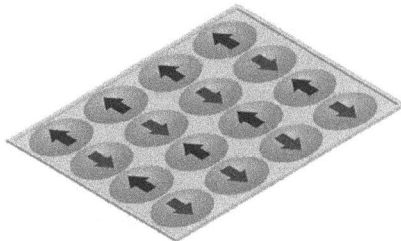

Figure 2.8. Antiferromagnetic arrangement of atomic spins in crystals.

this circumstance, the thermal energy turns into huge kinetic energy to smash the microscopic magnetic ordering inside the material.

According to the above perspective, the antiparallel orientation is the consequence of the marvellous exchange of spin energy within the minerals. A case of this antiferromagnetic phenomenon is manganese II oxide. In this regard, MnO is ionic with linear chains of Mn^{2+} and O^{2-} ions. The oxygen ion, consequently, has a complete set of valence electrons in the orbitals directly influencing the spin of the neighbouring Mn^{2+} ions. Figure 2.9 reveals the marvellous exchange mechanism example. Since both the Mn and O ions have complete electron shells, hybridization occurs by the contribution of the O^{2-} electrons to the open orbitals of the Mn^{2+} ions. The Mn^{2+} orbitals containing up-spin electrons receive one down-spin electron from the O^{2-} p orbital leaving one up-spin O^{2-} electron. The O^{2-} ion is then able to donate its up-spin to the next Mn^{2+} ion in the chain completing the bonds. This contribution, nevertheless, just arises if the subsequent Mn^{2+} ion has its d elections in the down-spin orientation. Since all unpaired electrons must affiliate with parallel spins within an orbital, all of the Mn^{2+} electron spins must be reversed. In other words, the Mn^{2+} ions are allied with inverse spins within the crystalline structure.

Some antiferromagnetic materials have Néel temperatures at, or even several hundred degrees above, room temperature, but usually, these temperatures are lower. The Néel temperature for manganese oxide, for example, is 122 K, for $MnSO_4$ is 12 K, and 25 K for $NiCO_3$; however, CoO has the highest Néel temperature of 290 K.

As well as in the case of antiferromagnetism, the ferrimagnetic state of materials is characterized by the creation of limited ferromagnetic sublattices with antiparallel orientations of the atomic magnetic moments. On the contrary, the magnetic array in antiferromagnets with sublattices comprise dissimilar sorts of atoms with dissimilar magnetic moments (Figure 2.10). Accordingly, the magnetisation of the ferrimagnet equates to the vector sum of the magnetisations of entire sublattices.

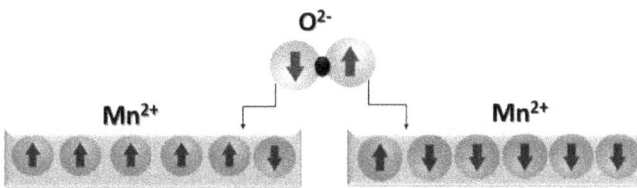

Figure 2.9. Schematic of marvellous exchange of MnO.

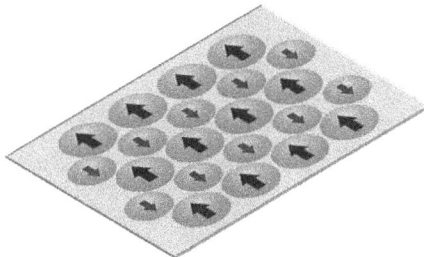

Figure 2.10. Ferrimagnetic ordering.

The ferrimagnetic order in a crystal is examined beneath the Curie temperature, as in ferromagnets. At temperatures higher than the Curie temperature, a phase transition of the second-order occurs, which swaps the ferrimagnet to the paramagnetic state.

Lodestone, or magnetite (Fe_3O_4), corresponds to a set of substances identified as ferrites. Ferrites and some other sets of magnetic substances discovered more recently possess many of the properties of ferromagnetic materials, including spontaneous magnetization and remanence. Unlike ferromagnetic metals, they have low electric conductivity, however. In alternating magnetic fields, this greatly reduces the energy loss resulting from eddy currents. Since these losses rise with the frequency of the alternating field, such substances are of much importance in the electronics industry [18].

One of the well-known example of ferrimagnetic material is nickel ferrite, $NiO \cdot Fe_2O_3$. The Fe^{3+} ions are consistently circulated across both A and B sublattices, and accordingly, their magnetic moments contradict, while Ni^{2+} ions only assemble on B sites. In this view, the electron configuration of nickel is $3d^8 4s^2$ (Figure 2.11), from which dual electrons are occupied from $4s^2$ to form Ni^{2+}, delivering an electron configuration of $3d^8$. Therefore, the electron spins of the ion are organized consistent with Hund's Rule. In this understanding, every orbital in a subshell is singly occupied with one electron before any one orbital is doubly occupied, and all electrons in singly occupied orbitals have the same spin. In these circumstances, all five states are occupied with one spin-up electron, parting three electrons with down spin to be combined. Therefore, adding electrons in different orbital levels forms two unpaired spin-up electrons. These dual unpaired electrons create a net magnetic moment on Ni^{2+} ions; since these completely sit on a similar sublattice, they are parallel and form a net magnetic moment in the crystal [17].

Therefore, what is the net magnetic moment in iron oxide $FeO \cdot Fe_2O_3$? Fe_2O_3 consists of Fe^{3+} and O^{2-} and has a net magnetic moment of zero. FeO consists of Fe^{2+} and O^{2-} ions; taking two electrons from the 4s orbital of Fe leaves Fe^{2+} with 6 electrons in its 3d orbital, and, according to Hund's rule, each Fe^{2+} ion will have 4 unpaired electrons (in its high-spin state).

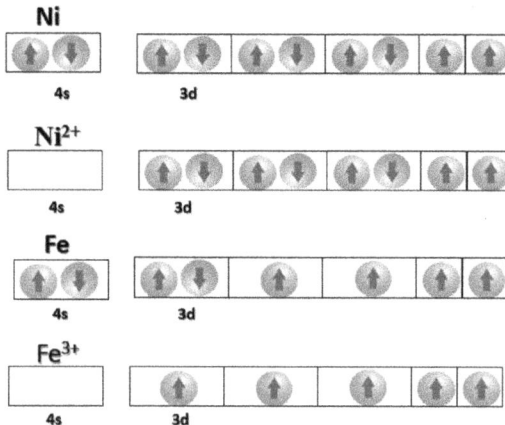

Figure 2.11. The electron configurations of Ni^{2+} and Fe^{3+}.

The critical question is: what are the major similarities and differences between ferrimagnetism and antiferromagnetism? The major similarity between ferri- and antiferromagnetism is that they both experience superexchange behaviour between ions that causes their magnetic moments to be antiparallel. The major difference between the two is that the net magnetic moment of the former is non-zero while in the latter it is zero. Antiferromagnetic materials thus do not exhibit spontaneous magnetization, while ferrimagnetic materials do.

2.8 General Quantization of Mineral Remote Sensing Imagines

Mineral, its characteristics and the modification it experiences, to the eyes of a physicist, meaningfully relies on the electronic structure of the atoms present. According to early debates in Chapter 1 and previous paragraphs, electrons, nevertheless, are very tiny and satisfy an equation of motion that varies from the one that is used. Electrons demonstrate wave properties and their position cannot be known precisely without determining their energy (Section 2.4).

Quantum mechanics represents the philosophies that rule electrons and atoms. These particles follow the equations of motion verbalized by quantum speculation. Understanding electrons and atoms at this level will enable one to interpret or explain various properties of minerals owing to their interaction with the electromagnetic spectrum. In this view, quantum mechanics instigates the uncertainty principle. Consequently, it maintains the equation of motion that comprises wave functions (Section 2.6).

These wavefunctions have a physical consequence; for example, their square correlates precisely to the probability of obtaining a particle at an exact position. In this understanding, each wavefunction is categorized by a (or a set of) quantum number, which relates to a precise amount of energy, which can assist to distinguish between different minerals in optical remote sensing data. Through the knowledge of these wavefunctions, one can determine any mineral property.

The keystone speculations of quantization of minerals in remote sensing data involve Max Planck—quantization of energy, the photoelectric effect, particle-wave duality, modern atomic theory, the uncertainty principle, the wave equation, the wave function, and the exclusion principle.

2.8.1 Plank Quanta

Plank initiated the concept that the radiation from a hot body is not given off as an endless stream, similar to water flowing from a tap, but rather like a leaking tap, in tiny packets—quanta—which he primarily termed 'energy elements'. He presumed that the energy E of these packets was inversely proportional to their wavelength λ, the energy of the packets with the shortest wavelengths being the highest. The relation between the wavelength of these quanta and their energy became known as the Plank relation, which is mathematically expressed as:

$$E = \hbar\lambda^{-1} \tag{2.11}$$

Furthermore, each quantum could only have certain energy values, but no value between these. This fact would not be noticeable to exploit in our huge-scale world but it somewhat turns out to be vitally important at an atomic level. It was, therefore, in the study of heat radiation and not in the realms of nuclear physics that the quantum theory was born. Therefore, Planck construed an equation for the energy variation of black body radiation grounded on Plank's quantum theory which tailored the experimental results exactly.

$$E_\lambda = \frac{8\pi hc\lambda^{-5}}{\left[e^{-a(T\lambda)^{-1}} - 1\right]} \tag{2.12}$$

where E_λ represents the energy density, T indicates the absolute temperature of the body, c is the speed of electromagnetic radiation in free space, a presents a constant equal to ch/k where k is Boltzmann's constant (a = 1.44×10^{-2} m K), and h is Planck's constant which is equivalent to:

$6.62607015 \times 10^{-34}$ Joule second.

When Planck applied this concept to the computation of the energy of the wavelengths of light emitted by hot bodies, he found that his formula fitted laboratory measurements exactly.

In this understanding,

$$h = \begin{pmatrix} 6.626 \times 10^{-34} & \text{joule seconds} \\ 4.136 \times 10^{-15} & \text{eV seconds} \end{pmatrix} \tag{2.12.1}$$

Equation 2.11.1 reveals that the electron volt (eV) is an appropriate unit of energy correlated to the typical metric unit (joules, or J) by the standard correlation 1 eV = 1.602×10^{-19} J. Consistent with quantum physics, stream of radiated black-body or light is formed of squillions of tiny packets of light, termed photons, spilling through the air. Yet, the critical question of what exactly is a photon is not answered.

A photon is the tiniest discrete quantity or quantum of electromagnetic energy, which is the elementary unit of fully light spectra. Photons are always in motion and, in a vacuum, travel at a constant speed of light of 2.998×10^8 ms^{-1} (Equation 2.11). In this regard, the relationship between the wavelength (λ) and frequency (v) for electromagnetic radiation is formulated as:

$$\lambda v = c \tag{2.13}$$

Equation 2.13 can be used to reveal the relationship between energy and wavelength as:

$$E = hc\lambda^{-1} \tag{2.14}$$

Rearranging Equation 2.14 leads to inversion relationship between λ and E, as is expressed by:

$$\lambda = hc(E)^{-1} \tag{2.15}$$

Equation 2.15 demonstrates that photon energy E is determined by the wavelength of electromagnetic radiation: the shorter the wavelength, the higher the energy. Even though the photons propagate at the speed of light, they have zero rest mass, so the rules of special relativity are not ruined. In this understanding, the elementary characteristics of photons are signified as follows. Photons travel at the speed of light in space. They stream continuously with zero mass and rest energy at the speed of light. In this regard, they only exist as moving particles with no electrical charges, which is known as spin-1 particles, which makes them bosons. In other words, a particle after a full rotation is again back in the same state. According to Equations 2.14 and 2.15, they carry energy and momentum which are dependent on the frequency. In this regard, they can have interactions with other particles such as electrons; for example, the Compton effect. Moreover, the photon is smashed or formed by numerous natural processes; for example, when radiation is absorbed or discharged.

2.8.2 Requantization of Photoelectric Effect

In precise circumstances, light can be depleted to impulse electrons, freeing them from the surface of a solid. In fact, Einstein believed that light is a particle (photon) and the stream of photons is a wave. In this understanding, this process is well-known as the photoelectric effect or photoelectric emission or photoemission in which a material that can exhibit this phenomenon is said to be photoemissive, and the ejected electrons are also termed photoelectrons. However, there is nothing that would distinguish them from other electrons. All electrons are identical to one another in mass, charge, spin, and magnetic moment.

In 1887, the photoelectric effect was originally discovered by Heinrich Hertz during experiments with a spark gap generator, which was the earliest device that could be known as a radio. In these experiments, sparks formed between dual tiny metal spheres in a transmitter cause sparks that jump between dual dissimilar metal spheres in a receiver.

The photoelectric effect, therefore, is characterized by the ejection of electrons from a metal surface when light is shined on the surface (Figure 2.12). Consequently, the electrons derive from within the metal; they are expelled from the material owing to the huge amount of energy they gain from the absorption of light.

What Lenard observed was that the intensity of the incident light had no impact on the utmost kinetic energy of the photoelectrons. Those ejected from exposure to very bright light had identical energy as those ejected from exposure to a very dim

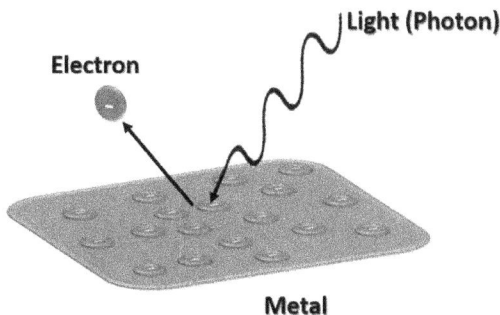

Figure 2.12. Photoelectric effect.

light of equal frequency. In keeping with the law of conservation of energy, however, massive electrons had been ejected with the aid of a bright source than a dim source.

Later in 1914, the American physicist Robert Millikan established the concept of the threshold frequency. In this context, the light was observed with frequencies below a certain cut-off value that would not eject photoelectrons from the metal surface no matter how bright the source was. In this view, these consequences were entirely unanticipated. Assuming the possibility of emitting an electron from an atom by interaction with a photon, especially an electron emitted from a solid surface by the action of light. Classical physics, however, would predict that a more intense beam of light would eject electrons with greater energy than a less intense beam no matter what the frequency. On the contrary, these consequences perhaps are not completely that normal.

In this understanding, most elements have ultraviolet threshold frequencies and only a few recedes down with the minimum frequency of radiation to be green or yellow. The materials with the lowest threshold frequencies are all semiconductors. Some have threshold frequencies in the infrared region of the spectrum. In this scenario, the red spectrum *does not* eject photoelectrons even if it is very bright. On the contrary, both green and blue spectra do eject photoelectrons even if they are very dim. However, the blue spectrum ejects photoelectrons with more energy than the green spectrum (Figure 2.13).

Along with the above perspective, the frequency of the incident radiation and the material on the surface are the keystone factors ruling the maximum kinetic energy of photoelectrons. Figure 2.14 reveals that electron energy grows with frequency in a direct linear relationship above the threshold. In this view, the entire three curves have a similar slope, which equals *Planck's constant*. The energy-frequency relationship, therefore, is constant for all materials. On the contrary, photoemission does not arise beneath the threshold frequency. In this regard, every curve has a dissimilar intercept on the energy axis, which reveals that the threshold frequency is a function of the material.

The mathematical formula that is used to describe the photoelectric effect which relates to the maximum kinetic energy $E_{K_{max}}$ of the photoelectrons to the frequency

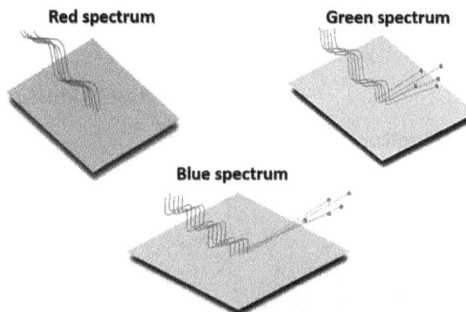

Figure 2.13. Variation of photoelectrons energy due to different light spectrum, i.e., red, green and blue spectra.

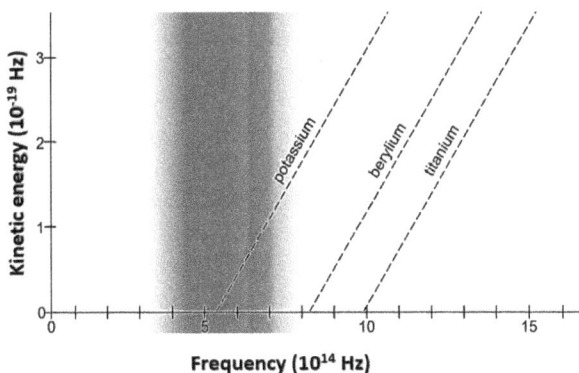

Figure 2.14. The energy-frequency relationship.

of the absorbed photons v and the threshold frequency v_0 of the photoemissive surface is

$$E_{K_{max}} = h(v - v_0) \qquad (2.16)$$

Let us assume that the work function is φ of the surface for the photon absorbed energy E then Equation 2.16 can be rewritten as:

$$E_{K_{max}} = E - \varphi \qquad (2.17)$$

Equation 2.14 can be expressed in the terms of the work function of the surface as:

$$\varphi = hv_0 = hc\lambda_0^{-1} \qquad (2.18)$$

where v_0 and λ_0 are threshold frequency and wavelength, respectively. Consequently, the maximum kinetic energy $E_{K_{max}}$ of the photoelectrons with charge e can be determined from the stopping potential P_0 as:

$$E_{K_{max}} = eP_0 \qquad (2.19)$$

The unit of energy will differ as well as the charge (e) unit change. For instance, $E_{K_{max}}$ is estimated in joules when charge (e) is given in coulombs, the energy will be calculated in joules. Besides, when charge (e) is specified in elementary charges, the energy will be designed in electron volts. However, these results have a lot of constants. The one that would be used is the one that's most appropriate for the specific problem (Table 2.1).

Lastly, the rate R at which photoelectrons with charge e are emitted from a photoemissive surface can be formulated as a function of the photoelectric current (I):

$$R = Ie^{-1} \qquad (2.20)$$

Table 2.1. Variation values of Planck's constant.

Planck's constants	Acceptable non SI units	SI units
h	4.14×10^{-15} eV s	6.63×10^{-34} J s
hc	1240 eV nm	1.99×10^{-25} J m

2.8.3 The Uncertainty Principle

Besides light, numerous other particles such as electrons behave both like waves and particles. In other words, the particles have wave-particle duality. One identical weird detail of quantum speculation, the Heisenberg uncertainty principle, states that the momentum (or velocity) and position of a particle cannot be observed precisely at once. This puzzling dilemma, therefore, arises since the simple performance of observing either the momentum or position of particle. In quantum mechanics, a quantum particle can deliver some quantity of energy to the particle, which interferes with the accuracy of the measurement of the other quantity! In other words, in quantum mechanics, the uncertainty principle (also known as Heisenberg's uncertainty principle) is any of a variety of mathematical inequalities asserting a fundamental limit to the accuracy with which the values for certain pairs of physical quantities of a particle, such as position, x, and momentum, p, can be predicted from initial conditions.

$$\Delta x \Delta p \geq 0.5\hbar \tag{2.21}$$

Equation 2.21 reveals that the product of the uncertainties in x (position) and p (momentum) of a single, non-relativistic particle is greater than or equal to $\hbar = \dfrac{h}{2\pi}$ over two where $\hbar$ is Planck's constant, 1.05×10^{-34} Joule seconds. This denotes that there is an essential boundary to the knowledge of any particle's position and velocity—we can never recognize both of them for unquestionability. This is the origin of particles possessing a probabilistic nature on the quantum scale.

2.8.4 Photovoltaic Effect

The photovoltaic effect is the physical phenomenon responsible for the creation of an electrical potential difference (voltage) in a material when exposed to light, which is very similar to the photoelectric effect. In this understanding, it is the practical physical phenomenon for the formation of an electrical potential difference (voltage) in a material when exposed to light. Therefore, the photovoltaic effect in semiconductors allows the running of solar cells as current-creating devices. In comparison to the photoelectric effects, the photovoltaic effect involves photons from a light source knocking electrons only out of their atomic orbitals, but keeping them in the material; this allows them to flow freely through the material. On the contrary, the photoelectric effect involves light photons knocking electrons out of material completely.

At the atomic level, electrons start to behave more like waves and particles. As the famous double-slit experiment shows, these waves can interfere, constructively or destructively, just like normal waves. The function that describes the amplitude of an electron is called the wave function, which is a function of position r and time t, and is denoted by the Greek letter ψ.

From ψ, the formulations for the physical properties of the electron such as energy, momentum, and position can be obtained; intrinsically, the wave function is articulated to designate the state of the electron. In this view, the function $\psi(r, t)$ is governed by the potential energy function $E_p(r, t)$. Therefore, $\psi(r, t)$ and $E_p(r, t)$

functions are correlated through the Schrödinger equation, which can be expressed by a second-order differential equation as:

$$i\hbar \frac{\partial}{\partial t}\psi(r,t) = \frac{-\hbar}{2m}\nabla^2\psi(r,t) + E_p(r,t)\psi(r,t) \tag{2.22}$$

The Schrödinger equation exhibits that, for an electron orbiting a nucleus, the states of the electron are quantized, meaning there are definite states that the electron can and cannot be in. For instance, an electron can only have an energy of x, $x+1$, $x+2$, etc., but not $x+0.5$. The consequence of this is the present energy levels (or orbitals), states of definite energy that electrons can exist in. Providing sufficient energy, an electron can "jump" up into the succeeding highest energy level. However, an electron must obtain there all in one jump. In this circumstance, an electron cannot jump halfway there twice. There are four distinct types of energy levels, denoted as s, p, d, and f, with many possible states in each.

This can assist to understand how electromagnetic radiation that is reflected from the minerals can distinguish easily between different minerals. When the electromagnetic radiation strikes the minerals, their atoms fill up with electron and fill from the state closest to the nucleus outwards. In this view, if a mineral atom has several electrons, the outermost valence electrons are the only ones that change the state into dissimilar orbitals, since the inner electrons are "stuck" in between the nucleus and the valence electrons and hence have nowhere to go. The valence electrons cannot fill states that are already taken by the inner electrons, and so they can only take on higher states. The valence electron energy levels are varied from one element to another one. The reflected energy is received by a remote sensing sensor is just the spectra signature of fluctuation of the substance's valence electron energy levels due to the interaction with electromagnetic radiation.

If the wave function of one of the photons is determined, it is possible to predict the form of the second from the positions of brightening variations across pixels on remote sensing data.

2.8.5 De Broglie's Wavelength

In 1924, Louis-Victor de Broglie articulated the de Broglie theory, declaring that all matter, not just light, has a wave-like nature; he correlated wavelength (λ), and momentum (P_{EM}): $\lambda = \hbar p^{-1}$. He stated that the momentum of a photon is given by $P_{EM} = Ec^{-1}$ and the wavelength (in a vacuum) by $\lambda = cv^{-1}$, where c is the speed of light in vacuum.

The relationship between the wavelength (λ) and momentum ($m*v$) for DeBroglie's "particle-wave" is determined from:

$$\lambda = \frac{\hbar}{m*v} \tag{2.23}$$

From the above relationships, we can calculate the relationship between energy (E) and momentum ($m*v$) as:

$$\frac{\hbar}{m*v} = \hbar\frac{c}{E} \tag{2.24}$$

Simplify, and solve for *E*:

$$E = mvc \tag{2.25}$$

The highest velocity (*v*) attainable by matter is the speed of light (*c*); therefore, the maximum energy would seem to be:

$$E = mcc \tag{2.26}$$

or

$$E = mc^2 \tag{2.27}$$

Equation 2.27 is the De Broglie equation where photons behave like matter particles and extend to all particles (Figure 2.15). Consequently, Equation 2.25 relates the wave-like behaviour of matter to its momentum.

De Broglie's concept illustrated just how far quantum physics of electromagnetic beam was instigated to comprehend old postulations. Electrons were components of matter-of substance-while the photons were insubstantial electromagnetic beaming. Nonetheless, they behave in some circumstances as if they are waves and in others as particles. In this understanding, this postulation, thereby, fitted well with Bohr's model of the atom. Here, an electron could occupy only specific orbits around the nucleus, as if it were running on tracks, and jumped between these orbits in a quantum leap as it gained or lost energy in the form of a photon. But what determined which orbits were acceptable and which were not? Bohar had revealed that electrons could be in only specific orbitals, with leaps in between. Orbital could only present if they could retain integer numbers of wavelengths (Figure 2.16). In this view, the electron is in the form of a standing wave that is occupied within the atom, with no loss of energy.

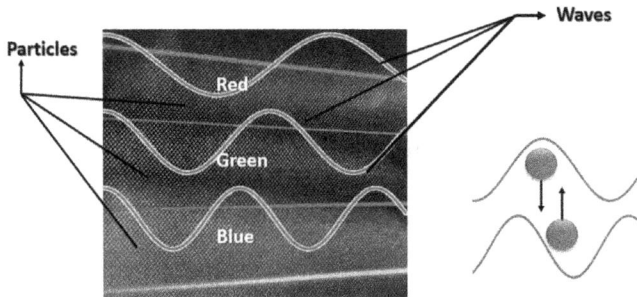

Figure 2.15. Photon likes wave-particle behaviour.

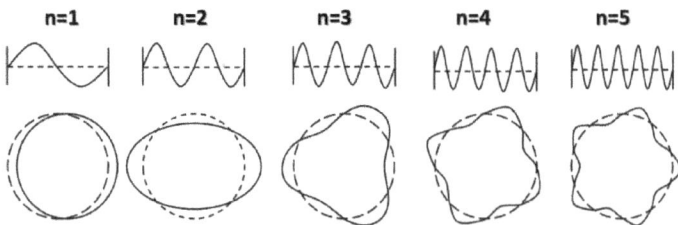

Figure 2.16. De Broglie's wavelength concept.

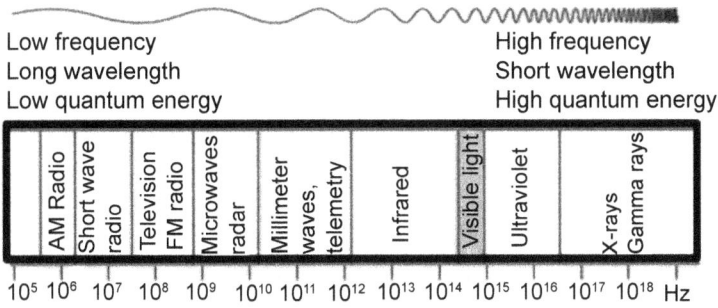

Figure 2.17. Electromagnetic spectra.

Therefore, Figure 2.17 summarizes the concepts of EM energy spectra. For instance, visible-light wavelengths correspond to a wavelength range from 0.38–0.75 μm of 2–3 eV.

In this understanding, to image an object, the object must be larger than the wavelength of the electromagnetic radiation used. In the case of visible light microscopes, this gives a maximum resolution of ~ 400 nm. Table 2.2 summarizes the brief of the electromagnetic spectra characteristics. In addition, Tables 2.3 and 2.4 deliver the units of both wavelength and frequency, respectively.

Table 2.2 shows that the longer wavelength is radio waves (> 1 m), while the Gamma-ray is the shortest wavelength (< 10 pm). The highest frequency is presented by Exahertz (EHz) which equals 10^{18} Hz, while the lowest frequency is presented by Kilohertz (kHz), which equals 10^{3} Hz. According to Table 2.2, the Gamma-ray has the highest frequency of > 30 EHz (Table 2.4).

Table 2.2. Summary of the essentials of electromagnetic spectra.

Electromagnetic spectra	Wavelength	Frequency
Gamma ray	< 10 pm	> 30 EHz
X-rays (Hard) X-rays (Soft)	1 pm–100 pm 100 pm–10,000 pm	300,000–3000 (PHz) 3000–30 (PHz)
Ultraviolet (UV)	0.30 μm–0.38 μm	750–30,000 (THz)
Visible Spectrum	0.4 μm–0.7 μm	379–769 (THz)
Infrared (IR) Spectrum	0.7 μm–100 μm	0.3–430 (THZ)
Microwave Region	1 mm–1 m	1–110 (GHz)
Radio Waves	(> 1 m)	< 3 Hz–3000 (GHz)

Table 2.3. Units of wavelength.

Units and symbols	Values
Millimeter (mm)	10^{-6} m
Micrometer (μm)	10^{-6} m
Nanometer (nm)	10^{-9} m
Picometer (pm)	10^{-12} m

Table 2.4. Units of frequency.

Units and symbols	Values
Kilohertz (KHz)	10^3 Hz
Megahertz (MHz)	10^6 Hz
Gigahertz (GHz)	10^9 Hz
Terahertz (THz)	10^{12} Hz
Petahertz (PHz)	10^{15} Hz
Exahertz (EHz)	10^{18} Hz

2.9 Quantization of Blackbody Radiation

Section 2.8.1 demonstrates the law of distribution of energy in the normal spectrum. In this view, the energy of electromagnetic waves is discreted rather than to be continuous spectrum. In this regard, each temperature has a maximum intensity of radiation that is emitted in a blackbody object, corresponding to the peaks (Figure 2.18). Accordingly, the intensity does not obey a smooth curve as the temperature grows, as prophesied by conventional physics.

In this regard, a blackbody is an idealized object, which absorbs and radiates all photon frequencies. Conventional physics can be implemented to derive an equation that describes the intensity of blackbody radiation as a function of frequency for a fixed temperature—the result is known as the Rayleigh-Jeans law. Although the Rayleigh-Jeans law works for low frequencies, it diverges as f^2; this divergence for high frequencies is called the ultraviolet catastrophe.

Consequently, energy could be gained or vanished merely in integral multiples of the particular tiniest unit of energy, a quantum (the tiniest probable unit of energy). In other words, energy can be gained or lost only in integral multiples of a quantum. In this

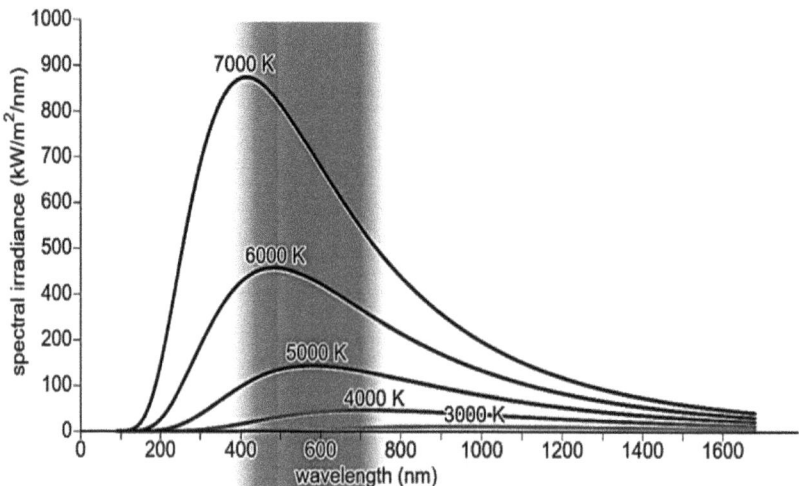

Figure 2.18. Relationship between the temperature of an object and the spectrum of blackbody radiation it emits.

understanding, Max Planck described the spectral distribution of blackbody radiation as a consequence of oscillations of electrons. Likewise, oscillations of electrons in an antenna create radio waves. Max Planck, therefore, concentrated on modelling the oscillating charges that must exist in the oven walls, radiating heat inwards and— in thermodynamic equilibrium—themselves being driven by the radiation field. He found he could account for the observed curve if he required these oscillators not to radiate energy continuously, as the classical theory would demand, but they could only lose or gain energy in chunks, called quanta, of size hf, for an oscillator of frequency f. In Figure 2.18, the quanta of radiance L is delivered by the temperature variation curve in the following dimensions: $\dfrac{Quantum\ Energy}{unit\ \text{area.wavelength.solid angle}}$; or units of watts/(m² μ ster). In this view, the quanta radiance equation is given by:

$$L_f = \frac{2hf^3}{c^2}\left[e^{\frac{hf}{KT}} - 1\right]^{-1} \qquad \left[\frac{W}{\text{sr m}^2\ \text{Hz}}\right] \tag{2.28}$$

$$L_\lambda = \frac{2hc^2}{\lambda^5}\left[e^{\frac{hc}{\lambda KT}} - 1\right]^{-1} \qquad \left[\frac{W}{\text{sr m}^2\ \text{m}}\right] \tag{2.29}$$

Planck's radiation energy density Equations 2.28 and 2.29 can be expressed in terms of frequency f and wavelength λ, respectively. In other words, the quanta *spectral radiance* of an object at a given temperature is a function of frequency L_f or wavelength L_λ. Here, $c = 3 \times 10^8$ m/s, $h = 6.626 \times 10^{-34}$ joules per second (J/s), and $k = 1.38 \times 10^{-23}$ joules per Kelvin (J/K). Real materials will differ from the idealized blackbody in their emission of radiation. The emissivity of the surface is a measure of the efficiency with which the surface absorbs (or radiates) energy and lies between 0 (for perfect reflector) and 1 (for a perfect absorber). A body that has $\varepsilon = 1$ is called a "black" body. In the infrared, many objects are nearly black bodies, in particular, vegetation. Materials that have $\varepsilon < 1$ are called grey bodies. Emissivity ε will vary with wavelength [20–22].

Let us multiply these functions by the total solid angle of a sphere (4π steradian), in which the spectral irradiance (E_f and E_λ) is obtained as:

$$E_f = \frac{8\pi hf^3}{c^2}\left[e^{\frac{hf}{KT}} - 1\right]^{-1} \qquad \left[\frac{W}{\text{m}^2\ \text{Hz}}\right] \tag{2.30}$$

$$E_\lambda = \frac{8\pi hc^2}{\lambda^5}\left[e^{\frac{hc}{\lambda KT}} - 1\right]^{-1} \qquad \left[\frac{W}{\text{m}^2\ \text{m}}\right] \tag{2.31}$$

These functions describe the power per area per frequency *and* the power per area per wavelength. When either of these functions is integrated over all possible values from zero to infinity, the result is the irradiance or the power P_O per area A is given by:

$$E = \int_0^\infty E_\lambda d\lambda = \int_0^\infty E_f d\lambda = P_O A^{-1} \tag{2.32}$$

Substitute Equations 2.30 and 2.31 into Equation 2.32, we obtain:

$$P_O A^{-1} = \left[2\pi^5 K^4\right]\left[15\hbar^3 c^2\right]^{-1}\left[T^4\right] \tag{2.33}$$

This term $\left[2\pi^5 K^4\right]\left[15\hbar^3 c^2\right]^{-1}$ is known as *Stefan's constant σ*:

$$\sigma = \left[2\pi^5 K^4\right]\left[15\hbar^3 c^2\right]^{-1} = 5.67040 \times 10^{-8} \frac{W}{m^2} K^4 \tag{2.34}$$

Therefore, the essence of the *Stefan-Boltzmann law* is obtained by multiplying the irradiance by the area:

$$P_O = \sigma A T^4 \tag{2.35}$$

Equation 2.35 reveals that emissivity is a function of a material's ability to absorb and reradiate energy. It is defined as the ratio of radiant flux from a body to that from a blackbody at the similar kinetic temperature (a "blackbody" is a perfect absorber and a perfect emitter of radiant energy). Objects that absorb and reradiate large amounts of energy have an emissivity near but always less than 1.0. Objects with low emissivity do not easily absorb and reradiate energy. Pure water has the highest natural emissivity (0.993), an offshore petroleum slick has an emissivity of 0.972, asphalt has an emissivity of 0.959, granite has an emissivity of 0.815, and a polished aluminium surface has an emissivity of 0.06.

Apply the first derivative test to the wavelength form of Planck's law to determine the peak wavelength as a function of temperature.

$$\frac{d}{d\lambda} E_\lambda(\lambda_{max}) = 0 \tag{2.36}$$

Then

$$\lambda_{max} = 2.898 \times 10^{-3} \, (m/K) T^{-1} \tag{2.37}$$

Equation 2.37 is well-known as Wien's displacement law, which delivers the maximum wavelength at the temperature peak. Besides, the Wien constant is 2.898 $\times$ 10^{-3} (m/K) for known temperature T in K, and λ_{max} is in meters. Therefore, *Wien's frequency constant* can be given by:

$$f_{max} = 58.7892 \, (GHz/K) T \tag{2.38}$$

According to the above perspective, no object radiates a mathematically perfect blackbody radiation spectrum. The temperature-spectrum curve will continuously have lumps. Therefore, the area below the intensity-wavelength curve for a real source of radiation is equivalent to the area beneath the intensity-wavelength curve for an ideal blackbody. The effective temperature of an object is the temperature of an ideal blackbody that would radiate energy at the same rate as the real body. Different parts of the Sun are at different temperatures. When combined, the Sun has an effective temperature of 5772 K. Moreover, the total energy of a blackbody radiator, not as a continuous, infinitely divisible quantity, but as a discrete quantity, is composed of an integral number of finite equal parts [22].

According to the thermal radiation, darker rocks become warmer and remain warmer than lighter rocks. In this regard, rock density is similarly a factor in that dense

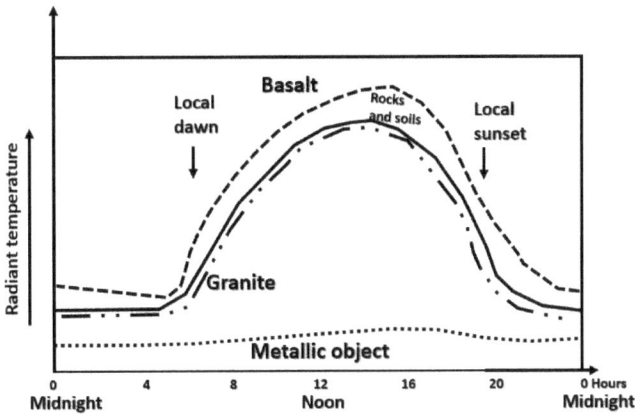

Figure 2.19. Some daily common materials' radiant temperature.

rocks cool further gradually, which have higher thermal inertia than less dense rocks. Therefore, "thermal inertia" is the rate at which the temperature of a body approaches that of its surroundings and is a function of absorptivity, the material's specific heat, its thermal conductivity, its dimensions, colour, density, and other factors. In predawn thermal imagery, for example, basalt is warmer than granite owing to its dark colour. Moreover, a dense grey sandstone appears warmer than a grey shale in predawn imagery because of the density contrast (Figure 2.19). In this way, thermal imagery can discriminate lithologic units and help map structure [22–24].

2.10 Quantization of Spectral Signature

In the classical definition of spectral signature, it is the variation of quantum energy reflectance or emittance of a material with respect to wavelength. In this regard, every substance would have its own unique pattern of spectral lines to be discriminated from each other. In other words, the ratio of discrete quantized reflected photon energy to incident photon energy is a function of discrete wavelength. This definition can raise a significant question: can the Feynman diagram be used to explain the spectral signature curve of minerals?

Richard Feynman understood that light is formed in the configuration of particles called photons. In other words, light behaves as particles as it behaves as waves. In this scenario, Feynman hypothesized that both photon and electron travel from one position and time to another position and time. Following the photoelectric concept, the reflectance signature from the mineral zones is considered to be the amount of the electronic energy that oscillates on its specific mineral atoms owing to the amount of photonic energy absorbed and caused such ejected electron from the surface of substances toward the sensor. In this understanding, the quantum energy of photoelectron would differ from one mineral to another owing to different atom structures that belong to each mineral that absorbs a photon. In this regard, the novel definition of the spectral signature can be introduced as the quantum energy that is generated due to the interaction of different photon frequency or wavelengths with different atom structures in which the induced signature of photoelectron energy

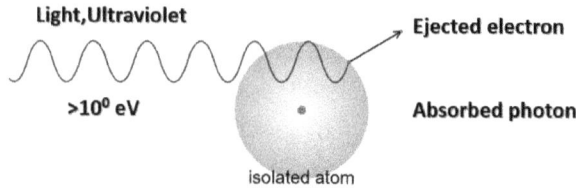

Figure 2.20. Feynman diagram for simplification of the spectral signature concept.

spectral differs from object to object due to different atomic structure of each object [10–13]. The simplification of this novel definition can be demonstrated by the Feynman diagram as shown in Figure 2.20.

Consequently, the spectral signature is a function of the "quantum efficiency". In this sense, the quantum efficiency perhaps delivers either as a function of wavelength or as energy. If all photons of a certain wavelength are absorbed and the resulting minority carriers are collected, then the quantum efficiency at that particular wavelength is unity. The quantum efficiency for photons with energy below the bandgap is zero. In this understanding, the spectral signature is defined as the ratio of the number of carriers collected by the minerals or objects to the number of photons of a given energy incident on the minerals or objects.

In some materials such as iron oxide Fe-O, the highest photon energy of Ultra-Violet–visible spectra is absorbed in which causing maximum energy level of electrons. In this circumstance, electrons spectra transfer between adjacent ions but do not turn out to be fully mobile. In this view, Fe-O absorbs photons of shorter wavelength energy. In this case, Fe-O has a red colour in the visible range owing to the reflectance photon wavelength, which has an abrupt fall-off towards the photon in the blue wavelength. This is well-known as a charge-transfer effect. In this context, the uranyl ion UO_2^{2+} in carnotite, absorbs all energy less than 0.5 μm, resulting in the yellow colour of the mineral.

In sulphides, approximately all the photons of energy larger than a specific threshold value are absorbed, in which sulphides act as semi-conductors, causing a sharp edge effect of spectral quantum signature energy owing to increment of the transmission band level of electron's energy. Therefore, the transition metals such as Ferrous ion, Feme ion, manganese, copper, Nickle, and chromium allow for electron process transition with different absorption of photon wavelengths (Table 2.5).

Table 2.5. *Transition* metals with their *transition* wavelength bands.

Metals	Transition wavelength (μm)
Ferrous ion	11, 1.8–2.0 and 0.55–0.57
Feme ion	0.87 and 0.35; sub-ordinate bands around 0.5
Manganese	0.34, 0.37, 0.41, 0.45 and 0.55
Copper	0.8
Nickel	0.4, 0.74 and 1.25
Chromium	0.35, 0.45 and 0.55

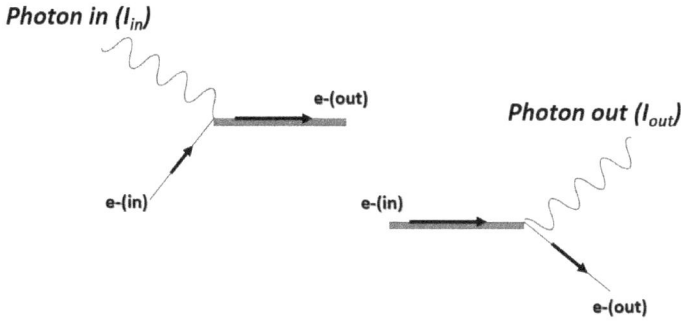

Figure 2.21. Feynman diagram for photon absorption and emission.

In other words, photon and electron are fluctuating in a dynamic system, which is a function of time. Moreover, an electron absorbs or emits a photon at a certain position and time. These dynamic behaviours are presented in a Feynman diagram (Figure 2.21). Conversely, the Feynman diagram provides a wavy line for photon propagation, and a straight line for electron ejection. However, a junction of two straight lines and a wavy one for a vertex indicating emission or absorption of a photon by an electron, which becomes a quark-antiquark pair.

The quantum spectra signature is the quantum of reflected particles due to the interaction between a photon and substance's electons. Therefore, a fermion is simply described as a process of photonic absorption or emission as a function of electron numbers exist in every substance. In this regard, let us assume that the reflectance quantity occupies each remote sensing image pixel owing to photon interaction with the Earth's minerals, which can be mathematically expressed as:

$$I_{in,out} = D_M \left(e\bar{\psi}(\bar{x}, \bar{y}) \right) (\phi, -\vec{A}) \psi(x, y) \tag{2.39}$$

where ψ presents the Dirac spinor, which is defined as a bi-spinor of the first rank, realizing the irreducible linear representation of the general Lorentz group on $\mathbb{R}^4$ equipped with the pseudo-Euclidean metric $(| \,|, | \,|) \, \mathbb{R}^4 \times \mathbb{R}^4 \to \mathbb{R}$ given by:

$$\forall \underbrace{\psi(x^0, x^1, x^2, x^3)}_{x}, \underbrace{\psi(y^0, y^1, y^2, y^3)}_{y} \in \mathbb{R}^4 : \quad \psi(x, y) \overset{df}{=} \sum_{\alpha, \beta=0}^{3} \left[\eta_{\alpha\beta} \right] x^\alpha y^\beta \tag{2.39.1}$$

$$\text{where } \left[\eta_{\alpha\beta} \right] \overset{df}{=} \begin{bmatrix} 1 & 0 & 0 & 0 \\ 0 & -1 & 0 & 0 \\ 0 & 0 & -1 & 0 \\ 0 & 0 & 0 & -1 \end{bmatrix} \tag{2.39.1.1}$$

D_M is Dirac matrices, which form part of the Dirac equation, and are defined up to an arbitrary unitary transformation so that the Dirac spinor is also defined up to such a unitary transformation. This property makes it possible to select the most physically convenient representation of the Dirac matrices and, consequently, of the

Dirac spinor. In this regard, the general formula for Dirac matrices is mathematically expressed as:

$$\left(\Box-m^2\right)E\psi(x,y) = \left(\sum_{k=0}^{3}\gamma^k\frac{\partial}{\partial x^k} - mE\right)\left(\sum_{l=0}^{3}\gamma^l\frac{\partial}{\partial x^l} + mE\right)\psi(x,y) = 0 \qquad (2.40)$$

$\Box$ is d'Alembert operator, which is considered the simplest form when solving the one-dimensional wave equation for photon propagation. In the words, it is the second-order differential operator in Cartesian coordinate, which is given by:

$$\Box \overset{df}{=} \Delta() - C^{-2}\frac{\partial^2}{\partial t^2}, \qquad (2.40.1)$$

where Δ and C are the Laplace operator and a constant, respectively. α_k, β and γ^k are matrices where $k \in \{0,1,2,3\}$. Therefore, E and m are energy and mass of the electron, respectively; refer to Section 2.8.5. Moreover, e presents the electric charge of the fermion and the term $(\phi, -\vec{A})$ is known as the gauge freedom or gauge invariance of the electromagnetic fields. In this regard, the vector and scalar potentials satisfy:

$$\nabla.\vec{A} = 0 \qquad (2.41)$$

$$\phi = 0 \qquad (2.42)$$

2.11 How can We Establish a New Definition of Remote Sensing for Mineral Identification?

Consistent with the above perspective, author can establish a new definition of remote sensing. It can be said that remote sensing is the quantum information collected from any object on the ground or space owing to the reflection or backscattering of photons, which is a function of the physical properties of the element atoms. In this understanding, different element atoms would interact with photon, i.e., absorption and emission as a function of its different wavelengths and frequencies.

For a specific definition of minerals remote sensing, it can be said that it is searching for the quantum information of physical properties of every element atoms. This quantum information is presented every place in the universe owing to the interaction between photon and any element atoms of universe through absorption or emission. In this understanding, the spectral signature is just the photon interaction signature with element atom characteristics.

In other words, this chapter delivers a novel explanation of the mineral physical atom characteristics with remote sensing photons from the point of view of quantum mechanics. This concept is named as quantization of mineral atom electron energy-remote sensing photon interaction. In this understanding, such a novel definition of mineral remote sensing imagine techniques is addressed. This novel definition states that remote sensing of mineralogy is the mobilizing of the spectra signature of fluctuation of the valence electron energy levels that are different from one element to another, which is based on the interaction of different photon spectra with different level of electron energy in each element. This novel definition can be explained using the quantum mechanics theory, which can be named as quantization of spectral signature of different elements or objects as based on the quantization of the black body radiation.

The next chapter will deal with the conventional methods of quantum image processing and their possible uses for mineral identifications in remote sensing data. In this chapter, it can be said that it is impossible to escape talking about quantum mechanics for dealing with remote sensing and minerals. Without dealing perfectly with quantum mechanics, it is impossible to understand the mechanism of photon remote sensing interaction with minerals. This chapter has demonstrated the novelty of quantum mechanics in remote sensing imagine mechanisms for minerals.

References

[1] Pauling L. The nature of the chemical bond. Application of results obtained from the quantum mechanics and from a theory of paramagnetic susceptibility to the structure of molecules. Journal of the American Chemical Society. 1931 Apr; 53(4): 1367–400.

[2] Kaji M. DI Mendeleev's concept of chemical elements and the principles of chemistry. Bulletin for the History of Chemistry. 2002; 27(1): 4–16.

[3] Fujihala M, Sugimoto T, Tohyama T, Mitsuda S, Mole RA, Yu DH, Yano S, Inagaki Y, Morodomi H, Kawae T, Sagayama H. Cluster-based haldane state in an edge-shared tetrahedral spin-cluster chain: Fedotovite $K_2Cu_3O(SO_4)_3$. Physical Review Letters. 2018 Feb 12; 120(7): 077201.

[4] Romanov D, Kaufmann G, Dreybrodt W. Modeling stalagmite growth by first principles of chemistry and physics of calcite precipitation. Geochimica et Cosmochimica Acta. 2008 Jan 15; 72(2): 423–37.

[5] Basdevant JL, Dalibard J. The Quantum Mechanics Solver: How to Apply Quantum Theory to Modern Physics. Springer Science & Business Media; 2005 Dec 14.

[6] Moore JW, Stanitski CL, Jurs PC. Principles of Chemistry: The Molecular Science. Hampshire: Brooks/Cole Cengage Learning; 2010.

[7] Brockett RW, Liberzon D. Quantized feedback stabilization of linear systems. IEEE Transactions on Automatic Control. 2000 Jul; 45(7): 1279–89.

[8] Bogoli'Ubov NN, Shirkov DV. Introduction to the theory of quantized fields. Introduction to the theory of quantized fields, by Bogoli'ubov, NN; Shirkov, DV New York, Interscience Publishers, 1959. Interscience Monographs in Physics and Astronomy; v. 3. 1959.

[9] Van Wees BJ, Van Houten H, Beenakker CW, Williamson JG, Kouwenhoven LP, Van der Marel D, Foxon CT. Quantized conductance of point contacts in a two-dimensional electron gas. Physical Review Letters. 1988 Feb 29; 60(9): 848.

[10] Williams MO. Quantum mechanics of hydrocarbon chains, from users.physics.harvard. edu/~mwilliams/.../Quantum-Mechanics-of-Hydrocarbons.pdf; 2011 Apr. 3.

[11] Peres A. Quantum Theory: Concepts and Methods. Springer Science & Business Media; 2006 Jun 1.

[12] Shao Y, Molnar LF, Jung Y, Kussmann J, Ochsenfeld C, Brown ST, Gilbert AT, Slipchenko LV, Levchenko SV, O'Neill DP, DiStasio Jr RA. Advances in methods and algorithms in a modern quantum chemistry program package. Physical Chemistry Chemical Physics. 2006; 8(27): 3172–91.

[13] Lowe JP, Peterson K. Quantum Chemistry. Elsevier; 2011 Aug 30.

[14] Kuznetsov V. Geophysical field disturbances and quantum mechanics. In E3S Web of Conferences 2017 (Vol. 20, p. 02005). EDP Sciences.

[15] Hummel RE. Electronic Properties of Materials. Springer Science & Business Media; 2011 Jun 15.

[16] Baraban L. Capped Colloids as Model Systems for Condensed Matter. PhD diss., University of Konstanz, German, 2008.

[17] Spaldin NA. Magnetic Materials: Fundamentals and Applications. Cambridge university press; 2010 Aug 19.

[18] Kasap SO. Principles of Electronic Materials and Devices. McGraw-Hill; 2006.

[19] Hummel RE. Electronic Properties of Materials. Springer Science & Business Media; 2011 Jun 15.

[20] Reddy MA, Reddy A. Textbook of Remote Sensing and Geographical Information Systems. Hyderabad: BS Publications; 2008.
[21] Elachi C, Van Zyl JJ. Introduction to the Physics and Techniques of Remote Sensing. John Wiley & Sons; 2006 May 11.
[22] Measures RM. Laser Remote Sensing: Fundamentals and Applications (Book). New York, Wiley-Interscience, 1984, 521: 1984.
[23] Vincent RK. Fundamentals of Geological and Environmental Remote Sensing. Upper Saddle River, NJ: Prentice Hall; 1997 Jan.
[24] Aggarwal S. Principles of remote sensing. Satellite remote sensing and GIS applications in agricultural meteorology. 2004 Jul 7: 23–38.

CHAPTER 3
Quantum Computing of Image Processing

The keystone question is: what is meant by quantum image processing. Prior to answering this question, the basis of quantum computing must be understood. Many academic institutions deal with mineral image processing as software that facilitates obtaining results without a perfect understanding of how the image processing algorithms are operating. In this view, if the same algorithm is applied in a certain area, it will deliver different results in different other areas. This is considered as uncertainty in applying the same algorithm. Many academic institutions are practising this sort of uncertainty research and education, especially in the nations that are trying to occupy advanced world educational ranks. The commercial software of ERDAS, ENVI, etc., has never helped to develop the field of image processing for geological applications.

In this understanding, a new wave of algorithm computing is required to distinguish between real world-ranked academic institutions and false academic institutions that flooded the high impact journals by publication bazaars of unuseful and repeated scientific papers. The new wave of algorithms would be established as based on quantum mechanics to deliver a quick solution for the uncertainty of mineral detections in remote sensing data.

3.1 What is Meant by Quantum Computing?

Quantum Computing was initially envisioned by Richard Feynman in 1982. In this regard, Feynman extended the quantum mechanical phenomena to the world of conventional computing. In this view, Feynman proposed to implement the classical algorithm solutions with quantum circuits. In this scenario, only the implementation of quantum probability calculations is an excellent approach to navigate the locations of multiple electrons and to comprehend configurations of electrons. In this sense, Feynman believed that quantum computers could supremely replicate the quantum performance as it would have naturally appeared. Nevertheless, it is impossible to be achieved even using immensely parallel conventional computers [1].

For instance, let us assume the probability detection of multiple mineral elements in remote sensing data. If we have two mineral elements (Cu and Fe) constrained to being at two groups of pixels (p and q), then there are 4 possible probabilities of their locations (both at p, one p—one cluster q, one q—one cluster p, both clusters at q, etc.). For three elements (Cu, Fe, and So_4), there are 8 probabilities, for 10 elements, there are 1,024 probabilities, and for 20 elements, there are 1,048,576 probabilities. Consequently, it is easy to find that measurements proceed of hand for classical physical systems with numerous different minerals. Accordingly, quantum computer studies began and continuing aims for the field of quantum computing have developed [2–4].

3.2 What is Meant by Quantization?

Generally, quantization is the practice of changeover from a conventional thought of physical phenomena to a novel interpretation acknowledged as quantum mechanics. In this view, it is a technique for creating a quantum field theory beginning from a conventional field theory. In other words, quantization is the scientific philosophy that a physical quantity can have merely specific discrete values. As demonstrated in Chapter 2, for instance, minerals are quantized because they are comprised of discrete particles that cannot be split; it is impossible to acquire half an electron of any mineral. Similarly, in the atoms, the energy levels of electrons are quantized [2, 4].

In a quantum system, all quantities as part of both quantization and objects function are constrained to their discrete values and wave-particle duality. In a quantum environment, therefore, the value of a physical quantity is possible to be foreseen before its measurement using Heisenberg's Uncertainty Principle in the circumstances of delivering the set of initial conditions.

The "state-vectors" are used to denote the states in a quantum mechanical system. In this view, separable Hilbert space is used to describe every quantum system, which acts as the state space of the system and is specified to be a complex number of norm 1. In this regard, possible states are points in the projective space of a Hilbert space (Figure 3.1). To this end, quantum particles can acquire the discrete eigenvalues of the Hilbert space since the eigenstate of the evidence matches the eigenvector of the operator.

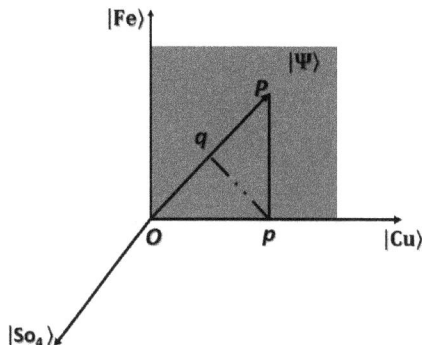

Figure 3.1. Hilbert space for three different minerals.

Let us assume $|\Psi\rangle$ is the wave function of the three mineral states owing to wave-particle duality, In quantum mechanics, a probability amplitude is a complex number used in describing the behaviour of systems. The modulus squared of this quantity represents a probability density. For example, we could measure the probability of the spectral signature of different minerals being encoded in optical electromagnetic spectral regions. Wave collapse, therefore, is frequently a matter when it approaches quantum measurement; subsequently, the probability information of the different mineral states can collapse from the initial state to a particular eigenstate. Nowadays, scientists avoid this matter through the practice of quantum tomography, a procedure depleted to recreate a particular quantum state through a combination of present mineral quantum states. In this regard, Heisenberg's uncertainty principle enhances the mineral quantum imaging mechanisms through the mannar of probabilistic functions. Indeed, the more precisely the pixels of some minerals is determined, the less precisely its quantity can be predicted from initial conditions, and vice versa [3–5].

3.3 What are Quantum Computers and How do they Work?

It is well-recognized that quantum computing is the discipline that assimilates quantum mechanical phenomena into a computing device. Quantum computers, therefore, deplete characteristics like superposition and entanglement to achieve computation.

3.3.1 Qubits and Superposition

Quantum state is grounded on the quantum information, which is exploited for the investigation of computational operation. In this regard, quantum information is expressed in the formula of "qubits", which can be believed as the computational similarity for bits. Besides, qubits are the basic units of information for quantum computing. The qubit can be designated as a vector-state in the 2-D Hilbert space. In this view, a conventional bit has 2 states, 0 and 1. In this respect, the state for a qubit correspondingly has 2 ground states – $|0\rangle$ and $|1\rangle$. Consequently, quantum states can be acquired as probability functions and as state-vectors. The practice of $|0\rangle$ and $|1\rangle$ strengthens this notion. In this understanding, the most important states for a qubit are modeled as:

$$|0\rangle = \begin{bmatrix} 1 \\ 0 \end{bmatrix}, \quad |1\rangle = \begin{bmatrix} 0 \\ 1 \end{bmatrix} \tag{3.1}$$

A qubit also can perform quantum superposition: it perhaps holds numerous states at once. In this view, superposition is contributing to the processes of a quantum computer. Subsequently, it is what delivers exponential accelerations in both memory and processing rapidity for a quantum computer. The superposition of a qubit, therefore, can be embodied as a probability function reliant on the amplitudes of the qubit in its Hilbert space (p and q), which is given by:

$$|\phi\rangle = p|0\rangle + q|1\rangle \tag{3.2}$$

Here, p and q represent the amplitude probabilities and are typically complex numbers. Therefore, $|\phi\rangle$ is a superposition of a qubit. Consequently, Equation 3.3

signifies every qubit as a linear sequence of $|0\rangle$ and $|1\rangle$ [1,3,6]. In this regard, Born Rule is used to describe the qubit measurement as:

$$|\alpha|^2 + |\beta|^2 = 1 \tag{3.3}$$

Bloch sphere is used to understand the visualization of a qubit. Indeed, the probabilities p and q can be demonstrated by the Bloch sphere. This sphere reveals the absolute factors for the qubit and delivers 3 degrees of freedom, which are initially 4 degrees of freedom but 1 degree is eliminated by the normalization constraint through the Born Rule. In this regard, 2-D Bloch Sphere visualization is shown in Figure 3.2.

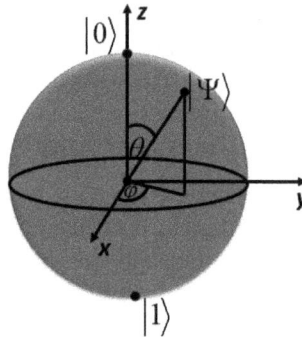

Figure 3.2. Bloch sphere visualization.

3.3.2 *Quantum Registers*

The quantum register has a significant role in quantum computing. In a quantum computer, quantum registers are merely deliberated to be systems of numerous qubits. Consequently, the quantum analogue of a conventional computing register is all considered. In this view, all quantum computations are achieved by operating qubits within the register. In this perspective, the quantum register has a size reliant on the number of qubits in its system [6–9]. Consequently, the Hilbert space of a quantum register develops with an enlarged quantity of qubits. In this regard, the 2-qubit quantum register can be mathematically represented as:

$$|\phi\rangle = |\phi 1\rangle \otimes |\phi 2\rangle = |\phi 1\rangle |\phi 2\rangle = |\phi 1 \phi 2\rangle \tag{3.4}$$

Equation 3.4 demonstrates the concept of qubits by tensor products $\otimes$, which is termed also as Kronecker product. In this sense, $\otimes$ combines the dual quantum states of a qubit. Therefore, the disagreement between a conventional register and a quantum register is realized in the content of every respective environment. In this understanding, a conventional register would accumulate N "flip-flops", while a quantum register would amass N qubits. Moreover, quantum register has the ability to accumulate the data in a state of superposition. According to this perspective, the states that a quantum register may retrieve are termed as Hilbert spaces with 2^n states. In this view, the Hilbert space is what supplies the power to a quantum computer by applying superposition on various qubits.

3.3.3 Quantum Gates

It is well-known that through logic gates, classical computers deploy bits. In other words, logic gates basically convert the value that a bit stores. In this understanding, they are the construction blocks of the somewhat digital system. Logic gates, therefore, are manipulated with certainty tables to form alterations of the state into conventional circuits. Here the cornerstone question is: what is meant by quantum gates? The quantum gates or the quantum logic gates are the elementary quantum circuits functioning on a slight number of qubits. Consequently, they are the correspondents for quantum computers to conventional logic gates for traditional digital computers. In this regard, quantum logic gates are adjustable, dissimilar to numerous conventional logic gates. Therefore, a unitary operator that operates on a tiny number of qubits is frequently known as a 'gate', analogical to traditional logic gates identical to AND, OR, and NOT [1, 4, 6].

3.3.3.1 NOT Gate

A simple NOT gate is known as bitflip gate, or X. If the input of this gate is $I_0|0\rangle + I_1|1\rangle$, the output would be $I_1|0\rangle + I_0|1\rangle$, which can be expressed in matrix form as:

$$X = \begin{pmatrix} 0 & 1 \\ 1 & 0 \end{pmatrix} \tag{3.5}$$

3.3.3.2 Controlled-NOT Gate

Let us identify two-qubit gates as the controlled-not gate or CNOT. In this view, CNOT does not operate if the initial bit is 0 and reverses the subsequent bit of its input if the initial bit is $|1\rangle$. Consequently, CNOT is a quantum gate, which is dissimilar to NOT and the output is subject to the initial input. In this regard, the initial qubit is termed as the control qubit, while the second qubit is termed as the target qubit, which can be expressed in the matrix form as:

$$CNOT = \frac{1}{\sqrt{2}} \begin{pmatrix} 1 & 0 & 0 & 0 \\ 0 & 1 & 0 & 0 \\ 0 & 0 & 0 & 1 \\ 0 & 0 & 1 & 0 \end{pmatrix} \tag{3.6}$$

3.3.3.3 Hadamard Gate

Hadamard gate is another significant quantum gate, which is manipulated to generate the quantum superposition ($\pi/2 - pulse$), quantified by $H|0\rangle = \frac{1}{\sqrt{2}}|0\rangle + \frac{1}{\sqrt{2}}|1\rangle$ when beginning from zero states. Hadamard gate can be expressed in the form of a unitary matrix as:

$$H = \frac{1}{\sqrt{2}} \begin{pmatrix} 1 & 1 \\ 1 & -1 \end{pmatrix} \tag{3.7}$$

In general, a quantum state, which may be operated by a quantum gate, is archetypally demonstrated as "kets" or a bra-ket ($|0\rangle$ and $|1\rangle$). In this view, the transformation can be modelled and directed by a quantum gate (U) on a state (ψ) as the following:

$$U|\psi_1\rangle = |\psi_2\rangle \tag{3.8}$$

Consequently, Equation 3.8 multiplies the initial quantum state vector ψ_1 by the unitary quantum gate operator (U) to accomplish a new quantum state ψ_0 [2, 4, 9].

3.3.4 Entanglement

In a quantum computing system, the correlation between two qubits is known as entangled. In this circumstance, when one qubit is in one specific state, the other one has to be in another specific state. In this regard, if dual electrons develop entangled, their spin states are interrelated. In this view, if one of the electrons has a spin-up, then the other one has a spin-down after measurement. Consequently, the concept of the Bell states allows us to model the quantum entanglement. In this regard, the Bell states indicate four certain extremely entangled quantum states of two qubits [7, 9, 10]. Consequently, they are in a superposition of 0 and 1, that is, a linear combination of the two states. In this view, the two Bell States signify two maximally entangled 2-qubits states that can be given by:

$$\left|\Phi^{\pm}\right\rangle = \frac{1}{\sqrt{2}}\left(|0\rangle_A \otimes |0\rangle_B \pm |1\rangle_A \otimes |1\rangle_B\right) \tag{3.9}$$

$$\left|\Psi^{\pm}\right\rangle = \frac{1}{\sqrt{2}}\left(|0\rangle_A \otimes |1\rangle_B \pm |1\rangle_A \otimes |0\rangle_B\right) \tag{3.10}$$

Both Equations 3.9 and 3.10 demonstrate that the qubits held by subscripts "A and B" can be 0 as well as 1. In this circumstance, the consequence, would be all random, either 0 or 1 having probability 1/2. Accordingly, the qubits held by subscript "B" have random outcomes. However, two qubits held by "A and B" can be perfectly correlated although they have random outcome [10–12].

Finally, entanglement is valuable to quantum computing in that it can be operated as a fragment of quantum circuits to employ real-world data since correlations occur between real-world minerals in conjunction with those of a quantum system.

3.4 Quantum Image Processing

The critical question is what is meant by quantum image processing? Perhaps the simple answer is quantum image processing (QIMP) is mainly offered to exploit quantum computing and quantum information processing to generate and work with quantum images. This directs us to another important question: what is a quantum image? Quantum image is programming the image information into quantum-mechanical systems as a replacement for conventional ones and substituting traditional with quantum information processing, perhaps improve some of these challenges such as accuracy of automatic detection of object physical characteristics and constructions. In this view, the laws of quantum mechanics allow us to diminish the mandatory resources for some tasks by countless orders of magnitude if the image data are encoded in the quantum state of a suitable physical system. For instance, achieving an appropriate technique for encoding image data, and exploiting a new quantum algorithm that can detect boundaries among portions of an image with a single logical quantum computing procedures to preserve edges between different clusters [3, 6, 10, 13].

Venegas-Andraca and Bose's Qubit Lattice are the pioneering research that delivered the birth of quantum image processing in 2003. Subsequently, the pioneer scientist Latorre proposed Real Ket, which encoded quantum images as an element for further applications in quantum image processing. Precisely, the subsequent studies grounded on these pioneering work can be categorized into three core clusters: (i) quantum-assisted digital image processing, (ii) Optics-based quantum imaging, and (iii) classically-inspired quantum image processing (QIP) [6, 9, 10, 14]. In this regard, classically-inspired QIP originates its stimulation from the prospect that quantum computing hardware will momentarily be physically understood and, later, such achievement aims at encompassing conventional image processing responsibilities and applications to quantum computing context [15–19]. The second framework of optics-based quantum imaging focuses on formulating novel techniques for optical remote sensing imaging and corresponding information processing at the quantum level by manipulating the quantum nature of electromagnetic spectra and the intrinsic correspondence of optical signals. Consequently, quantum-assisted digital image processing applications aim at operating some of the characteristics accountable for the power of quantum computing algorithms to develop some well-known digital or classical image processing tasks and applications as we will demonstrate in the next chapters of this book.

3.5 Flexible Representation for Quantum Images

The flexible representation for quantum images (FRQI) [20], which is analogous to the pixel sign for images on conventional computers, is presented. FRQI, therefore, depicts the essential information; for instance, the colors and the corresponding geographical positions of every pixel in an image and assimilates them into a quantum state, which can be mathematically defined as:

$$\left|I(n)\right\rangle = \frac{1}{2^n}\sum_{i=0}^{2^{2n}-1}\left(\cos\vartheta_i\left|0\right\rangle + \sin\vartheta_i\left|1\right\rangle\right)\otimes\left(\left|y\right\rangle\left|x\right\rangle\right),\tag{3.11}$$

where

$$\vartheta_i\in\left[0,\frac{\pi}{2}\right],\ i=1,2,3,\ldots\ldots,2^{2n}-1,\tag{3.12}$$

Equations 3.11 and 3.12 demonstrate that the 2-D quantum images involve row and column similar to classical remote sensing. Therefore, in 2n-qubit systems for preparing quantum images, the vector $(\left|y\right\rangle\left|x\right\rangle)$ is given by:

$$\left(\left|y\right\rangle\left|x\right\rangle\right) = \left|y_{n-1}y_{n-2}\ldots\ldots y_0\right\rangle\left|x_{n-1}x_{n-2}\ldots\ldots x_0\right\rangle,\tag{3.13}$$

where

$$x_i, y_j \in \{0, 1\}, \tag{3.14}$$

In Equation 3.14, for every individual pixel of j, i.e., $j = 0,1,....,n$ encodes the image column utilizing the first n-qubit $y_{n-1}y_{n-2}.....y_0$. Similarly, the pixels in row images are encoded by using the second n-qubit $x_{n-1}x_{n-2}.....x_0$. Consequently, the mathematical description of the normalized FRQI is identified by:

$$\left\| I(n) \right\| = \frac{1}{2^n} \sqrt{\sum_{i=0}^{2^{2n-1}} \left(\cos^2 \vartheta_i + \sin^2 \vartheta_i \right)} = 1. \tag{3.15}$$

where ϑ is the vector of angles encoding colors. Figure 3.3 demonstrates 2×2 FRQI quantum image with its quantum state, which involves the variation of image color vector intensities ϑ across pixels of the column and row, respectively.

The significant question is : can FRQI state represent the unitary transform $\mathcal{P}$? Hadamard transforms and controlled-rotation transforms are used to achieve the unitary transform $\mathcal{P}$, from which the initialized state $|0\rangle^{\otimes 2n+1}$ would be replaced by $|H\rangle$ and then change to $|I(n)\rangle$ (Figure 3.4).

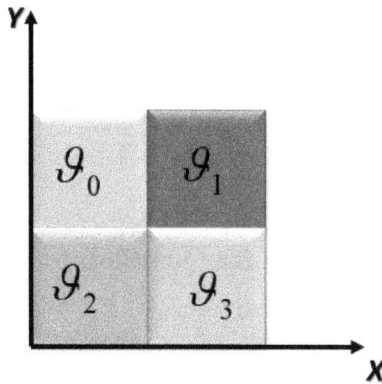

Figure 3.3. A 2×2 FRQI quantum image.

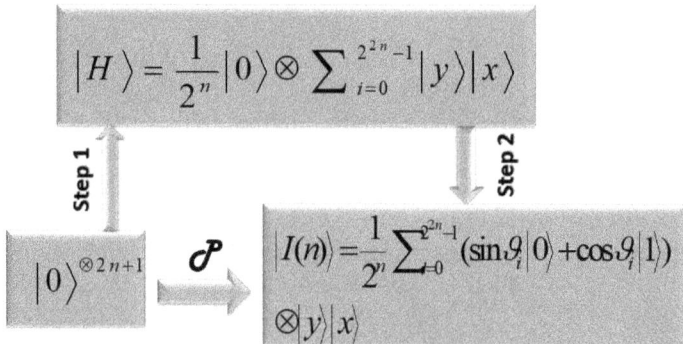

Figure 3.4. Hadamard transforms for FRQI state through the unitary transform $\mathcal{P}$.

Let us assume that the 2D identity remote sensing image can be presented through Hadamard transforms $|H\rangle$ by:

$$\mathcal{H}\left(|0\rangle^{\otimes 2n+1}\right) = \frac{1}{2^n} \otimes \sum_{i=0}^{2^{2n}-1} |y\rangle|x\rangle = |H\rangle. \tag{3.16}$$

Let us consider that the rotation quantum remote sensing image is $R_y(2\vartheta_i) = \begin{pmatrix} \cos\vartheta_i & -\sin\vartheta_i \\ \sin\vartheta_i & \cos\vartheta_i \end{pmatrix}$, and controlled rotation quantum remote sensing image $R_i(i = 0, 1, ..., 2^{2n} - 1)$ which is mathematically described by:

$$R_i = \left(I \otimes \sum_{j=0, j \neq i}^{2^{2n}-1} |j\rangle\langle j| \right) + \begin{pmatrix} \cos\vartheta_i & -\sin\vartheta_i \\ \sin\vartheta_i & \cos\vartheta_i \end{pmatrix} \otimes |i\rangle\langle i|. \tag{3.17}$$

Then, R_k, and $R_l R_k$ are applied through Hadamard transforms $|H\rangle$ as given by:

$$R_k = \frac{1}{2^2}\left[|0\rangle \otimes \left(\sum_{i=0, i \neq k}^{2^{2n}-1} |i\rangle\langle i| \right) + \left(\cos\vartheta_k |0\rangle + \sin\vartheta_k |1\rangle\right) \otimes |k\rangle \right], \tag{3.18}$$

$$R_l R_k |H\rangle = \frac{1}{2^n}\left[|0\rangle \otimes \left(\sum_{i=0, i \neq k, l}^{2^{2n}-1} |i\rangle\langle i| \right) + \left(\cos\vartheta_k |0\rangle + \sin\vartheta_k |1\rangle\right) \otimes |k\rangle + \left(\cos\vartheta_l |0\rangle + \sin\vartheta_l |1\rangle\right) \otimes |l\rangle \right]. \tag{3.19}$$

It is significantly noticeable from Equation 3.19 that

$$R|H\rangle = \left(\prod_{i=0}^{2^{2n}-1} R_i \right)|H\rangle = |I(n)\rangle. \tag{3.20}$$

Equation 3.20 reveals that the unitary transform turns the quantum computers [22] from the initialized state to the FRQI state as demonstrated through the mathematical Equations 3.11 to 3.20.

3.6 Fast Geometric Transformations on FRQI Quantum Images

In quantum images, the operations that focus on deploying the accurate geographical position information of the FRQI quantum images are known as geometric transformations on quantum images (GTQI) [20–24]. Exploitation of the GTQI processes, traditional-like transformations, for example, two-point exchange, flip, co-ordinate exchange and orthogonal rotations, can be achieved on the quantum images encoded in the FRQI image exploitation of the elementary quantum gates. In this understanding, the geometric transformations G_I, on FRQI quantum images can be mathematically described as:

$$G_I\left(|I(n)\rangle\right) = 2^{-n} \sum_{i=0}^{2^{2n}-1} |c_i\rangle \otimes G\left(|y\rangle|x\rangle\right), \tag{3.21}$$

In Equation 3.21, $G(|y\rangle|x\rangle)$ is the unitary transformation performing geometric exchanges based on the position information $|y\rangle|x\rangle$. In the circumstance of an N-sized image, the comprehensive analysis of quantum circuits demonstrates that the

$$\left| I\left(n\right) \right\rangle \qquad\qquad G_I\left(\left| I(n)\right\rangle\right)$$

colour

Y-axis
y_{n-1}
y_{n-2}
$\vdots$
y_0

x_{n-1}
x_{n-2}
X-axis
$\vdots$
x_0

G

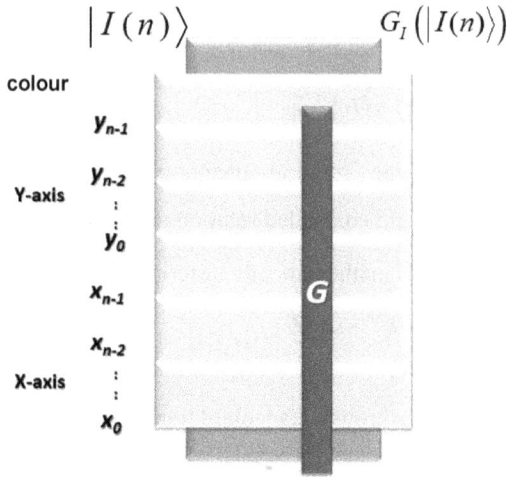

Figure 3.5. Generalized circuit design for geometric transformations on quantum images.

complexity is $O(\log^2 N)$ for two pixels swapping and $O(\log N)$ for other procedures of GTQI (Figure 3.5).

3.7 Efficient Colour Transformations on FRQI Quantum Image

According to the above perspective, the GTQI transformations are focused merely on the spatial content encoded in the FRQI quantum image. Therefore, the FRQI representation in Equation (3.10) is considered for the single-qubit that encodes the chromatic (color) content of the image unchanged. In colour transformation circumstances, pointing the single-qubit chromatic content of FRQI quantum images (CTQI) is considered. In this view, this transformation is subsequently focused on control repair procedures on the colour chain of the FRQI image, as shown in Figure 3.6. In this approach, the colours of each controlled pixel in the image are modified mathematically as:

$$C_I\left(\left| I(n)\right\rangle\right) = 2^{-n} \sum_{i=0}^{2^{2n}-1} C(\cos \vartheta_k \left|0\right\rangle + \sin \vartheta_k \left|1\right\rangle) \otimes \left(\left| y\right\rangle \left| x\right\rangle\right), \tag{3.22}$$

where C_I is used for transferring the colour to the quantum image I and the term $(\cos \vartheta_k \left|0\right\rangle + \sin \vartheta_k \left|1\right\rangle)$ for encoding the colour information. Consequently, a single qubit gate can be described in the form of the mathematical matrix S_I as:

$$S_I = \begin{pmatrix} 0 & 1 \\ 1 & 0 \end{pmatrix}, \tag{3.23}$$

The main characteristic of S_I is speciously to shift the value $\left|0\right\rangle$ and $\left|1\right\rangle$. In other words, $S_I\left|0\right\rangle = \left|1\right\rangle$ and $S_I\left|1\right\rangle = \left|0\right\rangle$. For instance, Figure 3.7 demonstrates that the LANDSAT8 satellite data green band was acquired along tens of kilometres in southwestern Spain, the Rio Tinto River on February 2nd, 2020. It is noticeable that the green band just provides grey intensity variation as only one colour band.

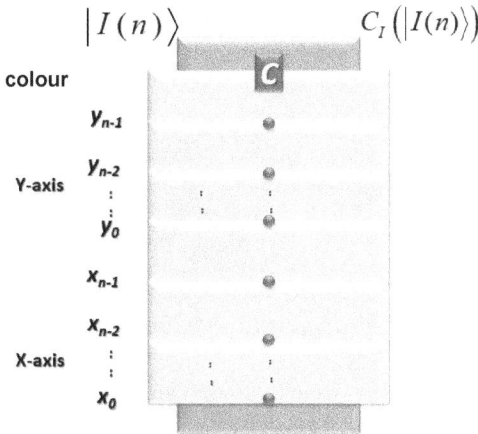

Figure 3.6. Single qubit gates applied on the colour chain.

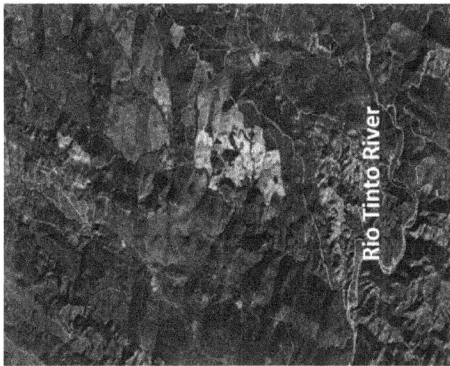

Figure 3.7. LANDSAT8 green band image along the Rio Tinto River, Spain.

Therefore, the mathematical algorithm of S_I gate is applied on LANDSAT8 green band image as given by:

$$\begin{pmatrix} 0 & 1 \\ 1 & 0 \end{pmatrix}\left(C\left(\cos\vartheta_k |0\rangle + \sin\vartheta_k |1\rangle\right)\right) = \left|c\left(0.5\pi - \vartheta_k\right)\right\rangle, \ \forall k \in 0,1,...,2^{2n} - 1, \qquad (3.24)$$

Figure 3.8 reveals the performance of the mathematical algorithm of S_I gate. There is extreme improvement in transferring the colour across the LANDSAT8 than original image (Figure 3.7). In a quantum computer, the algorithm of S_I gate extremely encodes the transfer color into original remote sensing, which is proven by clear feature distribution along with the image. Needless to say, algorithm of S_I gate, when encoding the colour in the image, also perceives the boundary between different clusters, which assist to deliver sharp features (Figure 3.8).

However, LANDSAT8 image involves more than three bands which can cause difficulties later to implement the algorithm of S_I gate. In this regard, the next section raises a critical question: how quantum image processing can deal with multichannel satellite data?

Figure 3.8. Performance of mathematical algorithm of S_I gate on LANDSAT8 green band image.

3.8 Multi-Channel Representation for Quantum Images

The multi-channel representation for quantum images, MCQI, is aimed at capturing the RGB channel information, which is achieved by conveying three qubits to encode colour information about satellite images. In other words, by encompassing the grayscale information encoded in an FRQI image to a colour representation, MCQI exploits R, G, and B channels to signify the different colour information about the remote sensing images. In this view, the scientific explanation of MCQI can be mathematically written as:

$$\left| I(n)_{mc} \right\rangle = \frac{1}{2^{n+1}} \sum_{i=0}^{2^{2n}-1} \left| C^i_{RGB\alpha} \right\rangle \otimes |y\rangle |x\rangle, \tag{3.25}$$

$\left| C^i_{RGB\alpha} \right\rangle$ is for the colour information that is encoded in FRQI remote sensing data and can be given by:

$$\begin{aligned} \left| C^i_{RGB\alpha} \right\rangle &= \cos \vartheta^i_R |000\rangle + \cos \vartheta^i_G |001\rangle + \cos \vartheta^i_B |010\rangle + \cos \vartheta_\alpha |011\rangle \\ &+ \sin \vartheta^i_R |000\rangle + \sin \vartheta^i_G |001\rangle + \sin \vartheta^i_B |010\rangle + \sin \vartheta_\alpha |011\rangle, \end{aligned} \tag{3.26}$$

where ϑ^i_R, ϑ^i_G, and ϑ^i_B are three angles encoding the colours of the R, G and B channels of image pixel i respectively. Moreover, ϑ_α is set as zero to form the two coefficients constant ($\cos \vartheta_\alpha = 1$ and $\sin \vartheta_\alpha = 0$) to bring no information. For ease of demonstration, let us assume that c_1, c_2, and c_3 are colour qubits that encode RGB colour information for an image (Figure 3.9), and the remaining 2n qubits (y_{n-1}, y_{n-2},....., y_0) and (x_{n-1}, x_{n-2},...., x_0) are manipulated to encode the pixel location information along column Y-axis and row X-axis, respectively, about pixels of a $2^n \times 2^n$ image.

Therefore, a normalized state of MCQI ($\left\| I(n)_{mc} \right\rangle \right\|$) is used for retrieving quantum colour quantum information images, which is casted as:

$$\left\| I(n)_{mc} \right\rangle \right\| = \frac{1}{2^{n+1}} \sqrt{ \sum_{i=0}^{2^{2n}-1} \begin{array}{l} \cos^2 \vartheta^i_R + \sin^2 \vartheta^i_R + \cos^2 \vartheta^i_G + \sin^2 \vartheta^i_G + \cos^2 \vartheta^i_B, \\ + \sin^2 \vartheta^i_B + \cos^2 0 + \sin^2 0 \end{array} } \tag{3.27}$$

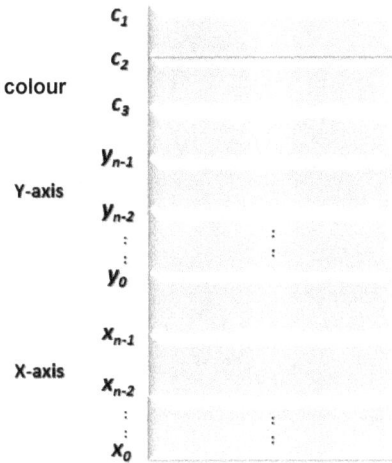

Figure 3.9. MCQI quantum images.

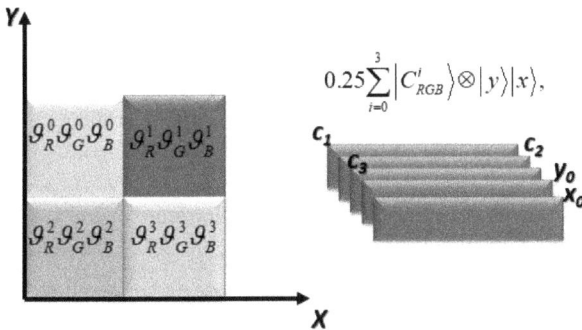

$$0.25\sum_{i=0}^{3}\left|C_{RGB}^{i}\right\rangle\otimes|y\rangle|x\rangle,$$

Figure 3.10. All RGB information about an MCQI image.

Equation 3.27 demonstrates that all RGB information about an MCQI image is stored simultaneously (Figure 3.10) and design colour image operators with much lower complexity. Figure 3.11 proves that MCQI encoded three qubits of RGB perfectly than traditional RGB LANDSAT8 image, which MCQI reveals to be the sharpest RGB colour representation than the conventional display of LANDSAT8 data.

In this view, MCQI representation provides a solution using many or fewer qubits to encode R, G and B channel information in normalized quantum states. Moreover, an MCQI image is stored in the preparation process exploiting the multi-channel PPT (MC-PPT) that encompasses the vector in (FRQI) [20–23] PPT to three vectors of angles. In this circumstance, the computational complexity of the entire preparation for MCQI image is similar to that expected for an FRQI image, i.e., $O(2^{4n})$.

RGB LANDSAT 8   MCQI quantum image

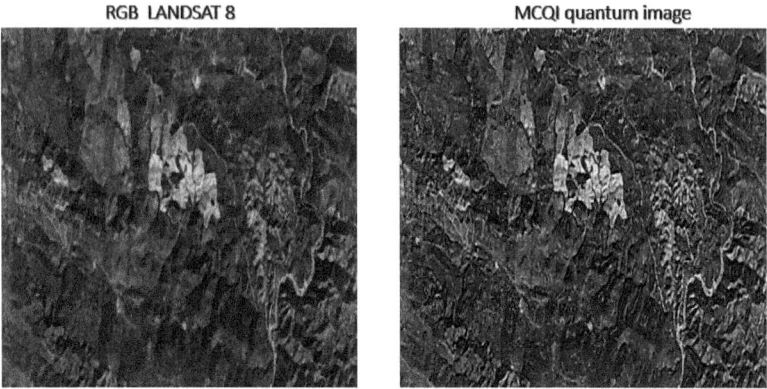

Figure 3.11. Output results of the normalized state of MCQI.

3.9 Novel Enhanced Quantum Image Representation (NEQR)

This procedure aims to provide an extremely enhanced remote sensing image. This procedure is based on encompassing the storage of the grayscale value of every pixel rather than an angle encoded in a qubit in FRQI to the original state of a qubit sequence. Consequently, to store the digital image in NEQR representation exploitation quantum mechanics, dual entangled qubit sequences are operated to store the entire image, which embodies the grayscale and positional information of entirely the pixels.

In this view, let us consider that the grayscale range of the LANDSAT8 image is 2^q with a binary sequence $C_{yx}^{q-1} C_{yx}^{q-2} ..... C_{yx}^0$ (Figure 3.12). The encoding grayscale value of LANDSAT8 is given by:

$$\left| C_{yx} \right\rangle = \left| C_{yx}^{q-1} C_{yx}^{q-2} ..... C_{yx}^0 \right\rangle, \tag{3.28}$$

Equation 3.28 reveals that $C_{yx}^k \in \{0,1\}$ and $C_{yx} \in [2^q - 1]$. To formulate an NEQR image, the first phase is similar to the grounding of an FRQI image. The second step, therefore, is partitioned into 2^{2n} sub-operations to store the grayscale information for every LANDSAT8 image pixel. Figure 3.13 demonstrates an example of implementation of Equation 3.28 on the normalized state of MCQI, which

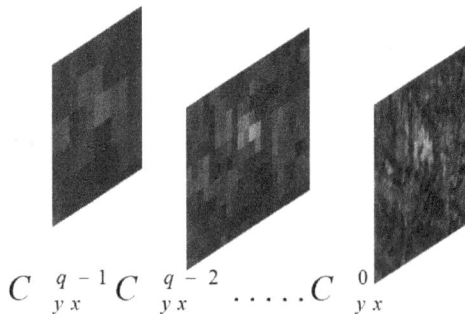

$$C_{yx}^{q-1} C_{yx}^{q-2} ..... C_{yx}^0$$

Figure 3.12. Example of the quantum binary store of LANDSAT8 image.

Figure 3.13. NEQR of LANDSAT8 image.

reveals extremely sharper boundary features than MCQI image. Indeed, the NEQR representation exploits the foundation state of a qubit categorization to signify the grayscale values of pixels instead of the probability amplitude of a single qubit used in the FRQI representation [20, 23].

This chapter demonstrates the principle of quantum image processing. The discrimination between bit and qubit is addressed. Indeed, the keystone of quantum image processing is based on using qubit, as discussed clearly in this chapter. Some basic quantum image processing such as quantum image storage, quantum image colour transfer and novel enhanced quantum image representation (NEQR) are introduced. These principles of quantum image processing are the cornerstone in dealing with quantum image processing tool for mineral exploration in remote sensing data.

This chapter yet does not deliver any quantum image processing procedures to deal with multispectral wavelength. The next chapter will answer how to establish a quantum algorithm to deal with multispectral and hyperspectral remote sensing data for minerals mapping.

References

[1] Vlaso AY. Quantum computations and images recognition. Federal Center for Radiology, arXiv:quant-ph/9703010 May. 1997.

[2] Venegas-Andraca SE, Bose S. Storing, processing and retrieving an image using quantum mechanics. Proceedings of SPIE in Quantum Information and Computing, 2003.

[3] Latorre JI. Image Compression and Entanglement. arXiv:quant-ph/0510031, 2005.

[4] Le PQ, Dong F, Hirota K. A flexible representation of quantum images for polynomial preparation, image compression, and processing operations. Quantum Information Process. 2011; 10(1): 63–84.

[5] Wineland DJ. Superposition, entanglement, and raising Schrödingers cat. Annalen der Physik. Jun. 2013; 525(10–11): 739–752.

[6] Nour Abura'ed, Faisal Shah Khan, Harish Bhaskar. Advances in the quantum theoretical approach to image processing applications. ACM Computer Surveys. 2017; 49(4), Feb. 2017. DOI: https://doi.org/10.1145/3009965.

[7] Deutsch D, Jozsa R. Rapid solutions of problems by quantum computation. Proceedings of the Royal Society of London. 1992; 439: 553–558.

[8] Grover LK. A fast quantum mechanical algorithm for database search. pp. 212–219. *In*: Proceedings of the 28th Annual ACM Symposium on the Theory of Computing, 1996.

[9] Venegas-Andraca SE, Ball JL. Processing images in entangled quantum systems. Quantum Information Processing. Feb. 2010; 9(1): 1–11.

[10] Reilly DJ. Engineering the quantum-classical interface of solid-state qubits. Quantum Information. Aug. 2015. http://dx.doi.org/10.1038/npjqi.2015.11.

[11] Latorre JI. Image compression and entanglement. arXiv:0510031, 2005.

[12] Quang PL, Dong F, Arai Y, Hirota K. Flexible representation of quantum images and its computational complexity analysis. pp. 185–185. *In*: Proceedings of the Fuzzy System Symposium, 2009.

[13] Le PQ, Iliyasu AM, Dong F, Hirota K. A flexible representation and invertible transformations for images on quantum computers. Studies in Computational Intelligence. 2011b; 372: 179–202.

[14] Liu Y. An exact quantum search algorithm with arbitrary database. International Journal of Theoretical Physics. Aug. 2014; 53(8): 2571–2578.

[15] Miyajima H, Fujisaki M, Shigei N. Quantum search algorithms in analog and digital models. IAENG International Journal of Computer Science. 2012; 39(2): 182–189.

[16] Bulger Dd. Combining a local search and grover's algorithm in black-box global optimization. Journal of Optimization Theory & Applications. June 2007; 133(3): 289–301.

[17] Zhang Y, Lu K, Xu K, Gao Y, Wilson R. Local feature point extraction for quantum images. Quantum Information Processing. 2015; 14(5): 1573–1588.

[18] Ruan Y, Chen H, Tan J, Li X. Quantum computation for large-scale image classification. Quantum Information Processing. July 2016; 15: 4049–4069.

[19] Le PQ, Dong F, Hirota K. A flexible representation of quantum images for polynomial preparation, image compression, and processing operations. Quantum Information Processing. 2011 Feb; 10(1): 63–84.

[20] Zhang Y, Lu K, Gao Y, Wang M. NEQR: a novel enhanced quantum representation of digital images. Quantum Information Processing. 2013 Aug; 12(8): 2833–60.

[21] Khan RA. An improved flexible representation of quantum images. Quantum Information Processing. 2019 Jul; 18(7): 1–9.

[22] Le PQ, Iliyasu AM, Dong F, Hirota K. A flexible representation and invertible transformations for images on quantum computers. pp. 179–202. *In*: New Advances in Intelligent Signal Processing 2011. Springer, Berlin, Heidelberg.

[23] Yan F, Iliyasu AM, Guo Y, Yang H. Flexible representation and manipulation of audio signals on quantum computers. Theoretical Computer Science. 2018 Dec 15; 752: 71–85.

[24] Lu Z, Wang X, Shang J, Luo Z, Sun C, Wu G. A multimedia image edge extraction algorithm based on flexible representation of quantum. Multimedia Tools and Applications. 2019 Sep; 78(17): 24067–82.

CHAPTER 4

Quantum Spectral Libraries of Minerals in Optical Remote Sensing Data

Uncertainties regarding spectral reflectance of mineral identifications implemented in commercial and open sources software are dominant problems in the field of remote sensing mineral applications. Numerous spectral libraries exist and are developed by different well-known organizations such as the USGS (United States Geological Survey) spectral libraries. Yet, there is no great promise for extremely high accuracy applications over different world regions. In other words, a similar application of spectral library delivers a different level of accuracy from one region to another. Indeed, spectral reflectance of minerals such as USGS spectral library examined in the tropical region would have to be different than one examined in the arid region. Numerous factors regulate the spectra reflectance of minerals over different geographical regions such as geographical variations of atmospheric temperature, amount of rain, amount of solar radiation, the intensity of cloud covers, etc.

Avoiding specific uncertainties in mineral identifications and detections, each study would establish its spectral libraries by gathering numerous reflectance samples of different minerals over different spaces and time using, for instance, spectrometer. This chapter devotes to developing a novel approach for mineral identifications and detections exploiting quantum computing algorithm.

4.1 How do Spectral Libraries Build Up?

Speculation of building up such spectral libraries for thousands of minerals is based on the real or near real-time fieldwork during satellite or airborne overpasses. The promise accuracy of spectral reflectance has validated by spectroscopy field data collection and laboratory facilities. In this view, an inclusive range of materials is examined: minerals, soils (including rocks, and mineral mixtures), coatings on rock surfaces, vegetation and other biologic materials, etc., throughout X-ray diffraction, and electron probe micro-analysis. For countless samples, wavelength exposure bridges the ultraviolet, visible, near-infrared, mid-infrared, and far-infrared regions (0.2 to 200 microns (μm)) [1–3].

A significant question now arises: how does a spectrometer work? Triggering the reflected light from minerals or other materials throughout a spectrometer and then breaking into its spectral components is the keystone of the functionality of a spectrometer. Therefore, the reflectance electromagnetic wavelengths from different minerals encode into the binary signal of spectral signatures of the diversity of minerals.

The spectral resolution, therefore, plays an extremely vital role in remote sensing and spectrometer. In this sense, it determines the ability of the system in acquiring the maximum number of spectral peaks that the spectrometer can resolve. For instance, if a spectrometer with a wavelength range of 200 nm had a spectral resolution of 1nm, the system would be capable of resolving a maximum of 200 individual wavelengths (peaks) across a spectrum. Consequently, the determined resolution $S_0(\lambda)$ is the convolution of the two sources [2, 4, 5], the spectral resolution of the spectrometer (R) and the linewidth of the signal (S_r), which is casted of:

$$S_0(\lambda) = S_r(\lambda) * R(\lambda). \tag{4.1}$$

Equation 4.1 reveals that if the signal linewidth is pointedly larger than the spectral resolution, the impact can be discounted and one can undertake that the measured resolution is the equivalent of the signal resolution. On the contrary, when the signal linewidth is meaningfully finer than the spectrometer resolution, the detected spectrum will be restricted merely by the spectrometer resolution.

In disseminative array spectrometers, the spectral resolution of a spectrometer is determined by the slit, the diffraction grating, and the detector (Figure 4.1). The slit regulates the minimum image size that the optical bench can create in the detector plane. A lamp delivers the cause of photon beam, i.e., light. The beam of photon strikes the diffraction grating, which operates identically to a prism and splits the light into its component wavelengths, i.e., red, green, blue, etc. The grating, consequently, is gyrated accordingly that every specific individual wavelength of light reaches the departure slit. At that point, the light interacts with the sample. In this concern, the detector gauges the transmittance and absorbance of the sample. Accordingly, transmittance denotes the number of photons that passes entirely through the sample and assaults the detector. On the other hand, absorbance is an amount of photon that

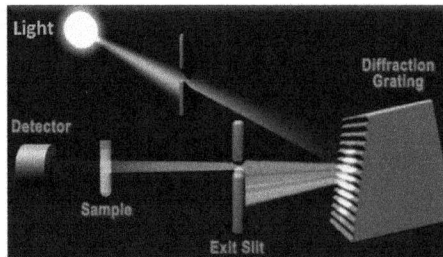

Figure 4.1. Array structure of spectrometers.

is absorbed by the sample. The detector, subsequently, senses the incident photon through the sample and commutes this information into a digital display [3].

For most appliances, it is safe to presume that one of these restraints photon signal conditions are in use, but for specific applications; for instance, high-resolution Raman spectroscopy: this convolution is impossible to be ignored. When a spectrometer, for instance, has a spectral resolution of ~ 3 cm^{-1} and exploits a laser with a linewidth of ~ 4 cm^{-1}, the detected signal would have a linewidth of ~ 5 cm^{-1} as long spectral resolutions are accordingly approaching each other. In this circumstance, it is imperative to guarantee that the detected signal is extensively narrow when endeavouring to compute the spectral resolution of a spectrometer to declare that the quantity is resolution constrained [3, 5].

4.2 Jablonski Energy Diagram

The third chapter has demonstrated the most of principles of quantum mechanics, in which atoms and molecules have a discrete energy spectrum and energy can merely be absorbed or radiated in quantities that are equivalent to the variance between two energy levels. Since the energy spectrum for every chemical species is inimitable, the tolerable transitions between these energy levels deliver a "fingerprint" for that species, which is well-known as the spectral signature. Exploitation of such a fingerprint to detect and quantify the species is an initial task of all chemical spectroscopies. For a molecule of identified configuration, the vibrational spectrum can correspondingly be depleted to identify the symmetry of the molecule and the force quantities allied with the property vibrations.

Absorption of energy by fluorochromes arises between the strictly spaced vibrational and rotational energy levels of the excited states in different molecular orbitals. The various energy levels involved in the absorption and emission of light by a fluorophore are classically presented by a Jablonski energy diagram, named in honour of the Polish physicist Professor Alexander Jablonski. Absorption of light extremely occurs rapidly (almost a femtosecond, the time needed for the photon to travel a single wavelength) in discrete amounts labelled quanta and matches the electronic energy transition of the fluorophore from the ground state to an excited state. Likewise, an emanation of a photon through fluorescence or phosphorescence is correspondingly quantified in terminologies of quanta.

In absorption spectroscopies, when the incident photon strikes the mineral samples, their chemical species experience transitions that are permitted by the suitable range regulations among rotational, vibrational, or electronic circumstances (Figure 4.2). Along with this perspective, let us assume that an electronic transition is compatible with a blue wavelength of 440 nm. The equivalent electronic energy shift is $\Delta E_e = hc(\lambda)^{-1} = 2.82$ eV (see Chapter 2). At this moment, if the vibrational modes permit energy levels that are equivalent to the electronic transition $\pm$ a vibrational level, the energy shift can be $\Delta E_e \pm \Delta E_v$. The corresponding wavelengths are then $\lambda = hc(\Delta E_e \pm \Delta E_v)^{-1}$, in which $\lambda = 0.434$ and 0.446 μm. In other words, a wavelength shift of 6 nm occurs to either side of the 400 nm or 440 nm electronic transition line.

For easy description, Figure 4.3 demonstrates that the thin horizontal lines in the left panel signify the further molecular energy levels as soon as both electronic and

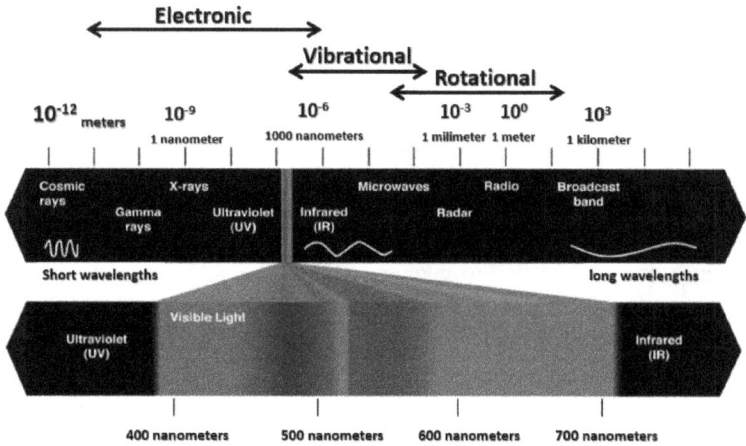

Figure 4.2. Representation of rotational, vibrational, or electronic transition line in electromagnetic spectral.

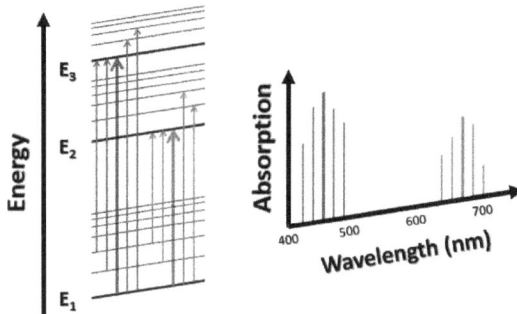

Figure 4.3. Electronic and vibrational energy levels in a mineral's molecule.

vibrational levels are involved in the energy diagram. The thin blue and red arrows in the left panel, therefore, exemplify how the existence of vibrational modes permits further probable electronic transitions to arise in the wavelength neighbourhoods of the electronic transitions between subshells. These transitions, subsequently, increase extra spectral lines to the absorption spectrum, as shown by the thin lines in the right-hand panel (Figure 4.3).

As well as the vibrational modes, molecules correspondingly have rotational modes. Explicitly, the entire molecule can spin about an axis with various frequency, which is once more quantized. Electrifying a molecule from one rotational mode to another stereotypically requires matching slight energy (in the 10^{-3} to 10^{-6} eV range). The intention of microwave radiation (λ in the millimetre to meter range) is acceptable. Figure 4.4 reveals that once the further quantized energy statuses allied with rotational modes are comprised in the energy diagram, the permitted energy levels acquire extremely precisely spaces as shown by the dashed lines in the left panel (Figure 4.4). In this view, the equivalent fluctuations in absorption line wavelengths are in the sub-nanometer range when equated to electronic

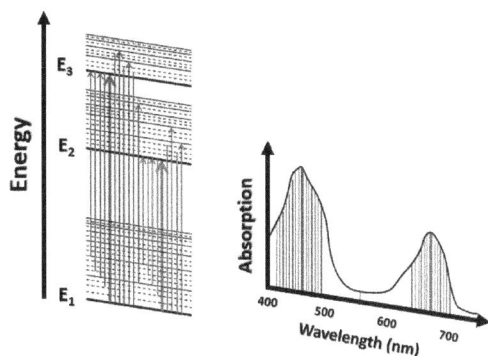

Figure 4.4. The quantized energy level of the molecular absorption spectrum.

or electronic-vibrational transitions. Consequently, the net outcome of having electronic, vibrational, and rotational modes is that the consequential molecular absorption spectrum is such a dense assembly of absorption lines that it seems like an incessant function of wavelength once gauged by instruments with spectral reactions larger than a nanometer.

In general, electronic transitions (those that move electrons into other orbitals) are typically the most energetic and UV (and a few in the visible) wavelengths are required. These transitions are critical for breaking bonds and thereby driving atmospheric photochemistry. Rotational transitions on the other hand require weak photons in the far IR (20 μm and longer). Needless to say that electronic transitions involve absorption spectroscopy, and emission spectroscopy. In this regard, in absorption spectroscopy, a photon is absorbed ("lost") as the molecule is raised to a higher energy level. However, in emission spectroscopy, a photon is emitted ("created") as the molecule falls back to a lower energy level (Figure 4.5).

Therefore, if dissimilar enormous molecules, have infinite numbers of rotational states, their rotational states are quantized and only certain transitions are allowed. Rotational transitions ensure wavelengths longer than ~ 20 μm. At temperatures characteristic of Earth's atmosphere, most molecules are rotationally excited. As a consequence, for molecules that have a handle with which to 'grab' a photon (a permanent dipole moment, p), numerous spectrally-overlapping transitions occur

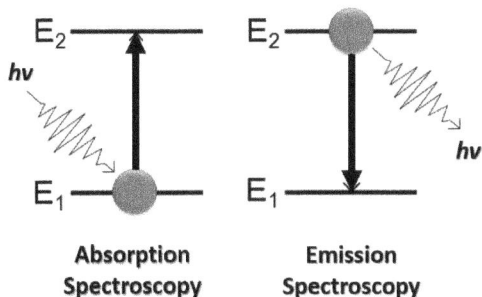

Figure 4.5. Electronic transitions.

and the far IR is nearly opaque at all wavelengths. H_2O is by far the furthermost vital molecule for blackening the spectral region longward of 20 μm (500 cm^{-1}). The larger the dipole moment, the more intense the transition will be achieved.

The bond in a molecule creates an equilibrium separation between the atoms. Regarding this equilibrium separation, a molecule can vibrate. Like rotational states, vibrational states are similarly quantized. Excitation of vibrational modes in a molecule necessitates a photon with a wavelength stereotypically shorter than 20 μm. Besides, quantum mechanical algorithms cause transitions that similarly excite (or de-excite) a single quantum of rotation is more intense than those have rotational energy. Consequently, the exchange in the atom energy (potential energy ↔ kinetic energy exchange) form energy quantization.

4.3 Infrared Absorption Spectroscopy

Let us assume that the incident photon of intensity $I_0(\lambda)$ is attenuated in passing a sample mineral thickness dl. Determination of absorption quantity is a function of the concentration of the absorber C and intensity of the transmitted photon departing the cell $I_t(\lambda)$. This is well-known as the Beer-Lambert Law. In other words. the Beer-Lambert Law expresses the correlation between the absorbance of a sample, its concentration, and its thickness. The law can be written as:

$$dI(\lambda) = -\varepsilon(\lambda) \times C \times I(\lambda)dl \qquad (4.2)$$

Equation 4.2 says ε is the molar absorptivity (this is a constant, which relies on the nature of the absorbing system and the wavelength passing through), dl is the path length (the width of the sample cell or cuvette, frequently 1 cm), and C is the concentration of the sample and $dl(\lambda)$ is the absorbance of the sample. In this regard, the information on the discrete energy spectrum of the chemical species in the cell is contained in the wavelength dependence of the molar absorption coefficient $\varepsilon(\lambda)$.

Consequently, a quantity of the absorbance of a substance can yield its concentration when the molar absorptivity and path length are known. Moreover, the integration of Equation 4.2 delivers:

$$I(\lambda) = I_0(\lambda)e^{-\varepsilon(\lambda)Cl} \qquad (4.3)$$

Needless to say that Equation 4.3 reveals that the strength of the absorption is proportional to $\dfrac{I(\lambda)}{I_0(\lambda)}$, which increases with the concentration of the sample (C); molar absorption coefficient ($\varepsilon(\lambda)$); and the path (l); respectively. In this view, $\varepsilon(\lambda)$ is a measure of how strongly a chemical species or substance absorbs light at a particular wavelength. It is an intrinsic property of chemical species that are dependent upon their chemical composition and structure.

In forensic investigations, the most popular practice of Ultra Violet/Visible spectral (UV/VIS) spectroscopy is the acquiring of precise colours. In precise explanation, let us assume a chip of paint from an accident scene that is needed to get to a suspect's car; UV/VIS spectroscopy can be implemented to find out the wavelength of maximum absorbance and that will deliver the precise colour of the paint chip for comparison. UV/VIS spectroscopy, therefore, is similarly implemented for finding out the absorption of illegal substances in mixtures. In this regard, UV/VIS spectroscopy can identify not only the hidden heroin in sugar but correspondingly quantify it; subsequently, the sugar does not absorb light at the same wavelength as the heroin. Identification of the illegal substance, consequently, can be confirmed by other spectroscopic techniques; for instance, Mass Spectrometry or IR spectroscopy.

4.4 Spectral Regions Relevant to Mineralogy

Now the question is: what are spectral regions that are relevant for mineral identifications? Visible and near-infrared (VNIR) is one of the vital parts of the electromagnetic spectrum consisting of both the visible (350–780 nm) and near-infrared (780–2500 nm) (Figure 4.6). Therefore, VNIR intersects with the optical radiation range (100–1000 nm) (Figure 4.6). Occasionally, the 350–1000 nm wavelength range is denoted as VNIR (visible-near-infrared), and the 1000–2500 nm range is referred to as the SWIR (short-wave infrared). In this understanding, each part of these spectra is only absorbed by specific minerals in which the spectral signature is the main keystone for accurate identification of those minerals [1–3].

The third chapter demonstrated in detail the quantum mechanics in electromagnetic wave interactions with different metals and minerals. Briefly, when the photons strike the minerals, they will either be absorbed or reflected from their surfaces. Here photons convey discrete quantities of energy termed quanta, which can be moved into atoms and molecules when photons are absorbed. Reliant on the frequency of the electromagnetic energy, photons, consequently, in the UV or visible ranges of the electromagnetic spectrum can carry out adequate energy to excite electrons. As soon as those electrons relax back to their ground states, photons will be released, and the atom or molecule will emit visible light of specific frequencies. This interchange of an electron from a lower energy level to a higher energy level, or from a higher energy yield to a lower energy level, is acknowledged as a transition. In this view, the energy of the photon absorbed must be greater than or equal to the difference in energy between the 2222 energy levels to achieve a transition. Once the electron is in the excited, but higher energy level, it is in a more unsteady situation than it was when it was in its relaxed, ground state. Intrinsically, the electron can rapidly retreat down to the lower energy level and, in doing so, it emanates a photon with an energy equivalent to the variance in energy levels (Figure 4.7).

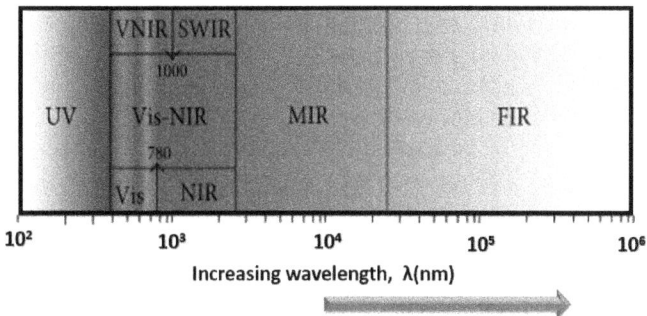

Figure 4.6. Optical spectrum.

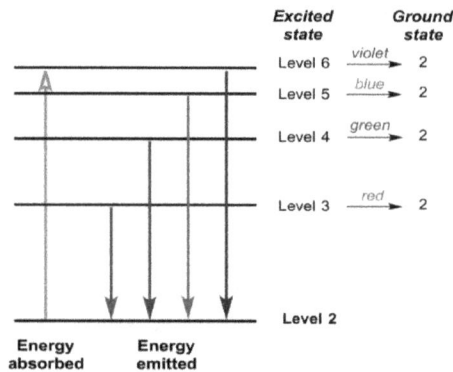

Figure 4.7. Different levels of photon transitions.

In other words, the higher the transition between energy levels, the further energy the absorbed/emanated. Greater frequency photons, consequently, are allied with higher energy transitions. If an electron, for instance, cascades from the third energy level to the second energy level, it emanates a photon of red light (wavelength of about 700 nm). On the contrary, when an electron falls from the sixth energy level to the second energy level (larger transition), it emanates a photon of purple light (wavelength of about 400 nm), which has a higher frequency and thus greater energy than red light [2–4].

These atomic emission spectra can be used (often informally by using a flame test) to gain insight into the electronic structure and identity an element. For instance, iron oxides (Haematite, Goethite, Jarosite) are sensed in the VNIR spectrum ranges of 400–1000 nm.

Owing to unfilled d-orbitals of Fe-oxide minerals, a common electronic process is discovered in the visible region. Fe-oxide is an extremely regular transition element in minerals. In this understanding, electron energy levels are shaped by numerous factors, involving the valence state of the atom (e.g., Fe^{2+} and Fe^{3+}), the sort of ligands, the asymmetry of the position it engages, the space between the metal ion and the ligand, and the distortion level of the position. This procedure is well-known for the crystal-field effects [4].

The significant question is: does the charge transfer occurs during photon-mineral interactions?

Charge transfer is ruled by mineralogy as it is considered a hundred times more powerful than the crystal-field impacts [3–5]. It is the foremost cause of the red colour of hydroxides and Fe oxides. Furthermore, the conduction bands and colour centres can correspondingly be origins of the electronic transitions in some minerals [6].

In a crystal lattice or molecule, the bonds vibrate identical to springs. In this regard, the strength of each molecular bond and mass govern their vibration frequency. For instance, in the VNIR range, the absorption bands are observed as a consequence of molecular vibrations [5]. In particular, soil minerals (e.g., phyllosilicate and carbonate minerals) have unique absorption features in the VNIR region as a result of implications and vibrational amalgamations affiliated with the stretching and bending of the molecular bonds; for instance, O-H, C-H, C-C, and N-H [5, 7].

On the other hand, SWIR spectroscopy within the 1000–2500 nm wavelength range includes the recognition of the reflected energy modified by molecular vibrations, rotational transitions, bending and stretching of bonds. In this array, minerals' clusters, for example, phyllosilicates, sorosilicates, inosilicates, carbonates, and zeolites demonstrate distinguishing spectra with analytical absorption features of precise molecular clusters (e.g., OH, H_2O and NH^4) and cation hydroxyl bonds (e.g., Al-OH, Fe-OH and Mg-OH) within the mineral crystal lattice. In SWIR spectroscopy, the energy of absorption relates to discrete energy levels and take place in well-classified wavelength sites. Each molecular cluster has physical absorption features arising at a precise wavelength range. For instance, the occurrence of OH and H_2O features in 1400 and 1900 nm, respectively. Equally, the occurrence of diagnostic features related to Al-OH, Fe-OH and Mg-OH are revealed at 2200 nm, 2250 nm and 2330 nm wavelength range, respectively [8–10].

In an almost—perfect matching set of mineral classification, Long-Wave Infrared (7000–1300 nm), LWIR, allocates recognition of mineral clusters; for instance, tectosilicates (e.g., feldspars, quartz, albite), inosilicates (e.g., diopside, amphiboles), phosphates (e.g., apatite), and sulphates [8, 9, 11, 12].

The Mid or Thermal Infrared of the EM spectrum of 6000–14000 nm is characterized by spectral features exhibited by numerous rock-forming mineral groups, e.g., silicates, carbonates, oxides, phosphates, sulphates, nitrates, nitrites, and hydroxyls.

Calcite is the most common identified carbonate occurring between 2340–2345 nm wavelength range. Therefore, its spectral feature in some investigations overlaps with those of chlorite, epidote and amphibole. Yet, the practice of inimitable features in other minerals, e.g., 1540 nm feature of epidote, can be implemented to distinguish them. For instance, calcite occurs from a 700 m downhole in association with chlorite, epidote and amphibole.

Owing to the existence of SiO_4-tetrahedron, the silicate reveals vibrational spectral features in the TIR region. Given the spectra of regular silicates, for example quartz, feldspars, muscovite, augite, hornblende, and olivine, the succeeding common looks can be delineated. In the spectrum region of 7–9 μm, what is termed the Christiansen peak occurs, in which felsic minerals occur at 7 μm, while ultramafic minerals are found at 9 μm. Consequently, in the spectrum region of 8.5 to 12 μm, an intense silicate absorption band occurs. In this sense, 10 μm is usually nominated as the Si-O vibration absorption (Reststrahlen) region. Its precise site, however, is delicate to the silicate structure and swings from nearly 9 μm (structure silicates or felsic minerals) to 11.5 μm (sequence silicates or mafic minerals). Therefore, the sensitive spectrum region for silicate and aluminium-silicate structure of tectosilicate form is 12 to 15 μm. On the contrary, other silicates having constructions of sheet, chain or ortho sorts do not exhibit absorption features in this spectrum region. The absorption patterns, therefore, in the form of quantities and scene of peaks are dissimilar for diverse feldspars, hence consenting to probable distinguishing.

4.5 Entanglement by Absorption

Consistent with the above perspective, the significant question is: what is the correlation between emitted and absorbed photon? Entanglement can modify photon-matter interaction impacts and, conversely, these interactions can modify the non-classical correlations existing in the spectroscopy system. In this view, the complete absorption of a single-photon pulse occurs in an optically dense medium owing to photon-atom interaction. Since the pulse is absorbed in the medium, an entangled state of the atoms and the field is created; as soon as the pulse is absorbed, the atoms depart in an entangled state engaging a single excitation. In other words, the energy absorbed by the field at any time is formed an entangled state of atoms and field.

In the case of the large separation of the atoms, with no spatial overlap between them, they intermingle with the photon. Let us consider conventional light in the linear (deprived of various absorptions) system. The modification of atomic absorption rate does not require further entanglement with field or interchanging into nonlinear regime.

Consequently, the beams must encompass the absorption frequencies of the dual atoms; it can exploit light beams with dissimilar frequencies or a single wideband beam. By way of low intensity of the beams being assumed (linear regime), after the interaction the atomic states develop as:

$$|\phi_j > A \ |g > A \rightarrow \alpha|\bar{\phi}_j > A|e > A + \beta|\phi_j > A|g > A \qquad (4.4)$$

In Equation 4.4, $|g >_i$, $i = A, B$ represents the electronic ground state of the atom, and $j = L, R$ denote opposite travelling directions either from left or right, respectively. Consequently, the mathematical expression of Equation 4.4 after absorption is given by:

$$|\varphi_j > B \ |g > B \rightarrow \gamma|\bar{\varphi}_j > B|e > B + \delta|\varphi_j > B|g > B \qquad (4.5)$$

Here $|e >$ denotes the exciting internal state and $\bar{\phi}$ and $\bar{\varphi}$ are the wave functions for emission and absorption states, respectively. In this view, $|\alpha|^2 + |\beta|^2 = 1$ and $|\gamma|^2 + |\delta|^2 = 1$. Both Equations 4.4 and 4.5 are only valid in the low-intensity beam approximation, which is quantitatively expressed by the relations $|\alpha|^2 \ll 1$ and $|\gamma|^2 \ll 1$, signifying that merely single absorptions are appropriate in the problem. Therefore, the probability of multi-absorption processes is extremely low and can be ignored. In the presence of multi-photon absorptions, a non-linear regime would be considered and new terms would be combined with Equations 4.4 and 4.5. Subsequently, the mathematical expression of the final state after multi-photon interaction with mineral atoms is given by:

$$|\psi_f > = \frac{1}{\sqrt{2}} \Big(\alpha\gamma|\bar{\phi}_L > A|\bar{\varphi}_R > B + \gamma\alpha|\bar{\phi}_R > A|\bar{\varphi}_L > B \Big)|e > A|e > B + |....> \qquad (4.6)$$

In Equation 4.6, $|....>$ comprises the relaxation terms, which do not allow double absorptions. Therefore, the probabilities of double absorptions can be computed from the expression of $|\psi_f >$. In this regard, two alternatives participate in the probability of dual absorption: (i) absorption by an atom of kind A in L and by one of kind B in R and, (ii) absorption by an atom of sort B in L and by one of sort A

in *R*. However, these alternatives are not distinct (both atoms can absorb at both sides of the array); consequently, consistent with the regulations of quantum theory, probability amplitudes must be added instead of probabilities. Lastly, the probability of double absorption is casted as:

$$P_d = \left| \frac{1}{\sqrt{2}} \alpha\gamma + \frac{1}{\sqrt{2}} \gamma\alpha \right|^2 = 2|\alpha\gamma|^2 \qquad (4.7)$$

Subsequently, the probability of the double mixture absorption can be given by:

$$P_d^{mix} = |\alpha\gamma|^2 \qquad (4.8)$$

The probability of dual absorption in the entangled state doubles that in product ones. Needless to say that the absorption probabilities in multi-atom systems rely on entanglement. In ultraviolet frequency-entangled photons, molecules causing photochemical reactions of interest frequently have electronic transition energies in the ultraviolet region. For instance, one powerful candidate for creating such ultraviolet frequency-entangled photons is a technique utilizing localized surface plasmon (LSP) (Figure 4.8). Consistent with this perspective, LSP is quantized plasma oscillation shaped near the surface of a nanometal and is created by directly exposing light with wavelengths in the ultraviolet region, specifically, about 350 nm for silver, and around 200 nm for aluminium [16]. Consequently, LSP can be treated as an emitter with rapid radiative decay and high directivity, which are useful for a broadband frequency-entangled photon source.

Under this condition, LSP can coherently and efficiently interact with incident photons through cavity modes and can be saturated by one-photon absorption. Therefore, either one of two input photons is absorbed by the LSP and the other remains unabsorbed owing to the absorption saturation effect of the LSP. Photon entanglement is formed as a consequence of the interference between the unabsorbed photon and the photon re-emitted from the LSP [16].

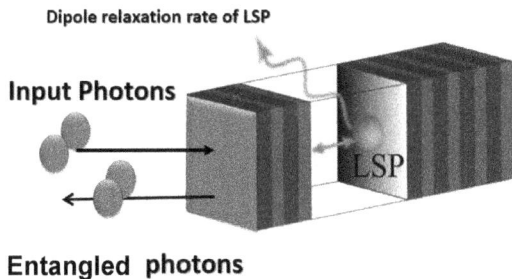

Figure 4.8. Entanglement of photon absorption and reflection.

4.6 How Does Entanglement Form Spectral Libraries?

The practice of dual-photon absorption (TPA), the light-caused transition between dual-energy phases of a medium arbitrated by the absorption of dual photons, is a construction block of particular technologies designed at searching the construction of atoms and molecules; for instance, dual-photon microscopy and dual-photon spectroscopy. In particular, nonlinear dual-photon spectroscopy has turned out to be a precious tool, where the ability of TPA is demoralized to attain information about an experiment that would not be reachable else [16, 23].

The absorption of dual photons by an atom or a molecule causes a transition between dual of photons' energy phases fit the general energy of the incident photons. In this regard, the quantum mechanical computation of the TPA transition probability reveals that its rate can be appreciated as a weighted total of numerous energy non-conserving atomic transitions, which is also known as virtual-state transitions between energy phases. At that point, the virtual-state transitions, a signature of the medium, can be revealed empirically by presenting a delay between the dual absorbed photons (Figure 4.9). Then, averaging is accounted for completed different experimental recognitions with diverse sequential links between the photons [24].

The significant question now is: can the sought-after information (energy level structure) be retrieved with somewhat sort of frequency relationships between the photons? In multidimensional pump-probe spectroscopy, it can be recommended that spectroscopic information resident in the TPA signal is fundamentally similar, irrespective of the presence or not of relationships between the photons absorbed [16, 25].

However, it ruins the unanswered question: to what limit these impacts create genuine entanglement impressions and whether they are possibly imitated, for example, by formed or stochastic conventional pulses? To reveal the exact function of entanglement in virtual-state spectroscopy, two ingredients are depleted. Primarily, a complete quantum formalism is applied to the dual-photon status. Accordingly, the amount of entanglement existing between the photons can be recognized evidently. Secondly, a general form of the dual-photon state is considered, which permits dissimilar sorts of relationships and spectral shapes of the photons to be determined.

In the case of the combination of entanglement and a specific form of the frequency correlations between photons, the realization of two-photon virtual-state spectroscopy is possible. However, this consequence is immeasurable since it requires the sort of dual-photon source that has to be implemented to empirically achieve the dual-photon virtual-state spectroscopy technique.

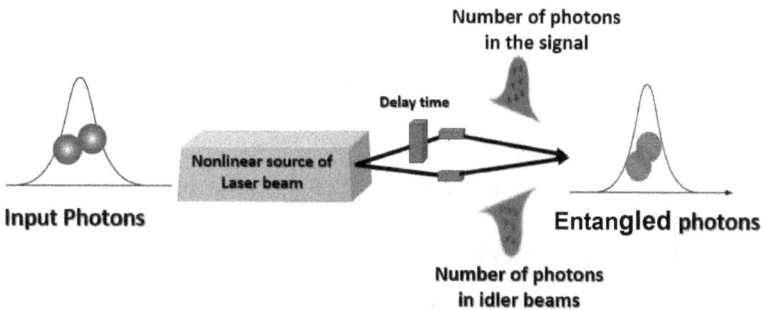

Figure 4.9. Concept of dual-photon absorption.

The existence of a high degree of frequency entanglement between the photons is the significant component that allows retrieving accurate average energy levels using dual-photon virtual-state spectroscopy.

Nevertheless, the practice of highly entangled photons does not promise the operative reclamation of spectroscopic information of the minerals. Comparatively, the application of a precise spectral form of the frequency correlations is what makes up the recognition of dual-photon virtual-state spectroscopy possible. However, when quasi-uncorrelated pairing photons are counted, a low degree of entanglement occurs. On the other hand, illuminating the function of entanglement, under the circumstance of matching photons with a low degree of entanglement, but with the appropriate sine series of the spectral shape, guarantees the successful identification of virtual-state spectroscopy (Figure 4.10). This implies that entanglement by itself is not the key ingredient to experimentally perform virtual state spectroscopy with a delay time of approximately 15 ns [23, 25].

According to the above perspective, entanglement is a superior sort of correlation, which occurs between spectroscopy beams and the mineral atoms. Therefore, in the sum-frequency generation, the flux of generated photons expands with the bandwidth of the incident energetic photons (Figure 4.11). The bandwidth of the absorbed photons can be assembled extremely with the properly well-organized source of photon fluxes, which at the identical time generates entangled photons with an extremely great level of entanglement [16, 24, 25].

Yet, the requirement of the instability rate on the bandwidth spreads correspondingly to conventional pulses. Now, the question raises: is the entangled photons function of frequency and energy?

Answering this requires the implementation of the concept of Spontaneous parametric down-conversion (also known as SPDC, parametric fluorescence or parametric scattering). In this sense, SPDS is a nonlinear instant optical process that converts one photon of higher energy (namely, a pump photon) into a pair of photons

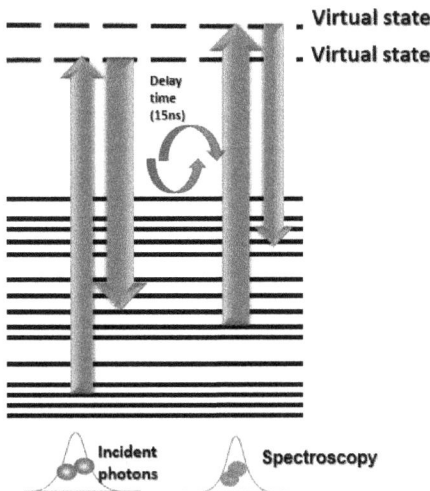

Figure 4.10. Virtual state spectroscopy.

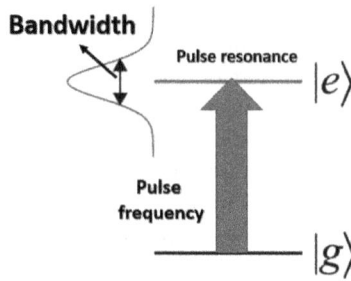

Figure 4.11. The bandwidth of incident photon.

(namely, a signal photon, and an idler photon) of lower energy, following the law of conservation of energy and law of conservation of momentum. It is an important process in quantum optics, for the generation of entangled photon pairs, and single photons [23, 25].

In this view, paired photons formed in SPDC can correspondingly be depleted to standardize detectors. In this circumstance, the significant enabling factor is the existence of dual photons, since SPDC generates dual photons, their frequencies do not require to be entangled. When one photon is detected from minerals and the other is not, we can infer that this is owing to the disorganization of the detectors. Consequently, by acquiring the quantity of photons detected in each detector, and the coincidence counts of paired photons detected in both detectors, we can measure the efficiency of each detector. Finally, several protocols proposed for spectroscopy similarly produce utilization of frequency correlations between photons rather than entanglement. This is closely related to the demonstration of the possibility to exploit thermal (or pseudothermal), and thus non-entangled, radiation for two-photon imaging experiments. Entangled and non-entangled sources can illustrate extraordinarily comparable performances when traversing the similar optical system, categorized by a particular transfer function, provided that certain characteristics of the frequency associations between photons are similar for both sources [17, 23, 25].

Owing to photons-mineral atom interaction, the correlation between specific photons frequency and specific mineral electron energy is entangled, which is demonstrated as well-known spectral signatures or spectral libraries.

4.7 How Does Quantum Teleportation Establish the Spectral Libraries?

In quantum teleportation, the properties of quantum entanglement are used to send a spin state (qubit) between sensors without physically moving the involved particle. The particles themselves are not teleported, but the state of one particle is destroyed on one side and extracted on the other side, so the information that the state encodes is communicated (Figure 4.12). The process is not instantaneous, because information must be communicated classically between sensors as part of the process. The usefulness of quantum teleportation lies in its ability to send quantum information arbitrarily to far distances without exposing quantum states to thermal decoherence from the environment or other adverse effects.

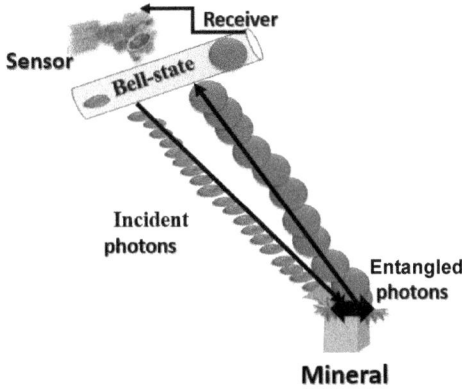

Figure 4.12. Remote sensing quantum teleportation.

Although quantum teleportation can in principle be used to teleport macroscopic objects (in the sense that two objects in the same quantum state are identical), the number of entangled states necessary to accomplish this is well outside anything physically achievable, since maintaining such a massive number of entangled states without decohering is a difficult problem. Quantum teleportation is, however, vital to the operation of quantum computers, in which manipulation of quantum information is of paramount importance. Quantum teleportation may eventually assist in the development of a "quantum remote sensing" that would function by transporting information between satellite or sensors and ground stations using quantum teleportation.

Developing a novel theory to understand the quantization of the spectral library, let us assume an entangled pair of electrons with spin states of sensors (spectroscopy and remote sensing) and minerals, in a particular Bell-state:

$$|\Phi_0\rangle = \frac{1}{\sqrt{2}}\left(|\uparrow\rangle_s \otimes |\uparrow\rangle_m + |\downarrow\rangle_s \otimes |\downarrow\rangle_m\right). \tag{4.9}$$

Equation 4.9 says that the entangled electrons occur between sensors and minerals. In this regard, the sensor measures the "Bell-state" of the sensor and its quantum state, entangling sensor and its quantum state. Therefore, the sensor sends the result of its reflectance measurement of minerals through the specific conventional technique of photon-electron interaction. Consequently, the Pauli matrices are cast-off to hypothetical Bell states, an orthonormal foundation of entangled states for the tensor product space of spin--$\frac{1}{2}$ particles:

$$|\Phi_0\rangle = I \otimes \sigma_0 |\Phi_0\rangle = \frac{1}{\sqrt{2}}\left(|\uparrow\rangle_s \otimes |\uparrow\rangle_m + |\downarrow\rangle_s \otimes |\downarrow\rangle_m\right). \tag{4.10}$$

$$|\Phi_1\rangle = I \otimes \sigma_1 |\Phi_0\rangle = \frac{1}{\sqrt{2}}\left(|\uparrow\rangle_s \otimes |\uparrow\rangle_m + |\downarrow\rangle_s \otimes |\uparrow\rangle_m\right). \tag{4.11}$$

$$|\Phi_2\rangle = I \otimes \sigma_2 |\Phi_0\rangle = \frac{i}{\sqrt{2}}\left(|\uparrow\rangle_s \otimes |\downarrow\rangle_m - |\downarrow\rangle_s \otimes |\uparrow\rangle_m\right). \tag{4.12}$$

$$|\Phi_3\rangle = I \otimes \sigma_3 |\Phi_0\rangle = \frac{1}{\sqrt{2}} \left(|\uparrow\rangle_s \otimes |\uparrow\rangle_m - |\downarrow\rangle_s \otimes |\downarrow\rangle_m \right). \tag{4.13}$$

From Equations 4.10 to 4.13, the Pauli matrices can be mathematically expressed as:

$$\sigma_0 = I = \begin{pmatrix} 1 & 0 \\ 0 & 1 \end{pmatrix}, \ \sigma_1 = \begin{pmatrix} 0 & 1 \\ 1 & 0 \end{pmatrix}, \ \sigma_2 = \begin{pmatrix} 0 & -i \\ i & 0 \end{pmatrix}, \ \sigma_3 = \begin{pmatrix} 1 & 0 \\ 0 & -1 \end{pmatrix}. \tag{4.14}$$

Consequently, the spin operators along each axis are demarcated as $\frac{\hbar}{2}\sigma_0$, σ_1, σ_2, and σ_3 for the x,y,z, axes, respectively.

Either remote sensing satellite or spectroscopy measures the quantity of the entanglement between photons and mineral atoms interaction. The rate of entanglement is different from one spectrum band to another with mineral atoms interactions. In this understanding, the speculation of quantum teleportation identifies the remote sensing sensor spectral measurement of minerals in separate specific bands. In other words, spectral measurement disentangles sensors and minerals and entangles reflectance in different separated bands. Relying on what particular entangled state sensor determines, every separated spectral band will know exactly how the mineral is disentangled (m_r) and can manipulate the mineral to take the state that reflectance spectra originally have. Thus, the state of the different spectral channels or bands is "teleported" from sensors to minerals, which now has a state that looks identical to how the stored spectral channels or bands originally looked. It is important to note that state varieties of the binary spectral signatures are not preserved in the processes: the no-cloning and no-deletion theorems of quantum mechanics prevent quantum information from being perfectly replicated or destroyed. The state of spectral signatures $|\Psi\rangle_{sm_r}$, which is determined in the separated individual binary bands or channels, can be given by:

$$|\Psi\rangle_{sm_r} = \sum_{i=0}^{n} 0.5 |\Phi\rangle_{sm_r} \otimes \begin{pmatrix} 0 & 1 \\ 1 & 0 \end{pmatrix} |\Psi\rangle_m \tag{4.15}$$

Consequently, mineral atoms have changed the spin state of particle owing to photon-atom interaction, which is expressed as:

$$|\Psi\rangle_m = c_1 |\uparrow\rangle_m + c_2 |\downarrow\rangle_m, \tag{4.16}$$

Using Equation 4.16 into 4.15, we obtain the full scenario of how the remote sensing operates to identify mineral atoms as:

$$|\Psi\rangle_{sm_r} = \sum_{i=0}^{n} 0.5 |\Phi\rangle_{sm_r} \otimes \begin{pmatrix} 0 & 1 \\ 1 & 0 \end{pmatrix} c_1 |\uparrow\rangle_m + c_2 |\downarrow\rangle_m. \tag{4.17}$$

Equations 4.17 reveals the state to be teleported and the state of one of the entangled particles. The state to be teleported is the quantum state of mineral spectral reflections and the state to be entangled is photon-molecule interactions (Figure 4.13). In this understanding, the final spin state of photon-atom interaction looks similar to the mineral spectral reflectance state.

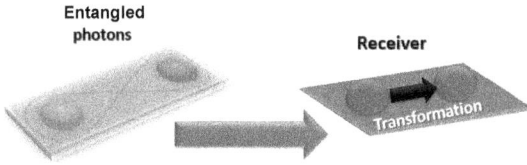

Figure 4.13. Teleported state of quantum mineral spectral entanglement.

4.8 Modeling of Quantum Mineral Spectral Libraries

Estimation of mineral spectral reflectance from optical remote sensing is assumed to be a form of the quantum linear system. If any optical remote sensing is assumed to be presented as a Hermitian matrix I^H and $DN \in I^H$, the quantum spectral radiance can be estimated as:

$$|\Psi\rangle_{sm_r} = \left(\frac{|L(\lambda)\rangle_{max}}{254} - \frac{|L(\lambda)\rangle_{min}}{255} \right) \otimes |DN\rangle\langle DN| + |L(\lambda)\rangle_{min} \qquad (4.18)$$

Equation 4.18 reveals the quantum state of the spectral reflectance in optical remote sensing image, in which both quantum states of $|DN\rangle\langle DN| \in I^H$ and spectral reflectance $|\Psi\rangle_{sm_r}$ are entangled. This concept is demonstrated previously in Section 4.7.

Moreover, the quantum gain state is presented in $\left(\dfrac{|L(\lambda)\rangle_{max}}{254} - \dfrac{|L(\lambda)\rangle_{min}}{255} \right)$ and the quantum state of minimum spectral radiance is presented by $|L(\lambda)\rangle_{min}$. Consequently, the spectral reflectance of a single band of the optical DNs image can be formulated as:

$$|\Psi\rangle_{sm_r} = \left(\frac{|L(\lambda)\rangle_{max} - |L(\lambda)\rangle_{min}}{|DN\rangle_{max} - |DN\rangle_{min}} \right) \otimes \left(|DN\rangle - |DN\rangle_{min} \right) + |L(\lambda)\rangle_{min} \pm |\varepsilon\rangle \qquad (4.19)$$

4.9 Image Storage

In Equation 4.19, the gate complexity is the number of 2-qubit gates, which is exploited. In this understanding, an algorithm is gate-efficient if its gate complexity is larger than its query complexity only by a logarithmic factor. Formally, an algorithm with query complexity Q is gate-efficient if its gate complexity is $O(Q.\text{poly}(\log(qN)))$. Subsequently, the runtime of the quantum algorithm for obtaining precise spectral reflectance can be demonstrated implementing $O(\kappa^2 \log N\varepsilon^{-1})$, where $\varepsilon > 0$ is the error and κ is condition number. Therefore, $\kappa = \dfrac{max_i |\psi\rangle_{smr}}{min_i |\psi\rangle_{smr}}$, in which, if the condition number is not much larger than one, the estimated spectral reflectance of the mineral is well-conditioned, and it means that its inverse can be computed with good accuracy.

In quantum computing, each pixel's grey level is encoded as qubit, which represents the quantum spectral signature of every pixel in optical remote sensing data.

In the quantum form, this could be, therefore, understood by exploitation of polynomials to designate the code in a conventional approach. Nevertheless, the quantum spectral signature can be identified as the $|0\rangle$ and $|1\rangle$, but the 0s are suppressed.

Then, the equivalence of polynomial zeros is interpreted as the coefficient of the base state $|0\rangle$, and the equivalence of polynomial ones as the coefficient of the base state $|1\rangle$ as follows. For instance, for an 8 Bit Binary Code, 8−BBC = 01111111, if 0s and 1s are distinguished in the 8-BBC as:

Zero	one
0	1111111

(4.20)

For an 8-bit encoding, one has $2^8 = 256$ levels and represents a scale from 0 to 255. The polynomials would be as follows:

Quantum superposition state positive

$$(0)\cdot128+(1)\cdot64+(1)\cdot32+(1)\cdot16+(1)\cdot8+(1)\cdot4+(1)\cdot2+(1)\cdot1 \ = \ 255. \qquad (4.21)$$

Then, based on Equation 4.20, the normalization condition can be given by:

$$(0).\frac{128}{255}+(1).\frac{64}{255}+(1).\frac{32}{255}+(1).\frac{16}{255}$$
$$+(1).\frac{8}{255}+(1).\frac{4}{255}+(1).\frac{2}{255}+(1).\frac{1}{255}, \qquad (4.22)$$

Both Equations 4.20 and 4.21 can be further merged as the succeeding form:

$$(0).\frac{128}{255}+(1).\frac{127}{255}=0.501\,960\,78\times(0)+\ 0.498\,039\,21\times(1) \qquad (4.23)$$

Satisfying the quantum spectral reflectance of the minerals $|\Psi_{sm_r}|_0^2 + |\Psi_{sm_r}|_1^2 = 1$, let us consider, for instance, $|\Psi_{sm_r}|_0^2 = 0.501\,960\,78$ and $|\Psi_{sm_r}|_1^2 = 0.498\,039\,2$. Then, the wave function of the spectral reflectance is $|\Psi_{sm_r}\rangle = 0.501\,960\,78|0\rangle + 0.498\,039\,2|1\rangle$. Simultaneously, it has dual states of the observable reflectance spectra quantity, the positive and negative of the pixel taken.

Therefore, remote sensing image can reconstruct the spectral reflectance or original image with the sequences of quantum superposition states. In this view, it is used to form dual images [18]. The first image is allied with the state 0, which is formed from every coefficient of the state $|0\rangle$ of all pixels. The second image is associated with state 1, which is created from every coefficient of the state $|1\rangle$ of all pixels (Figure 4.14). In this regard, the original image and the equivalent image are created with sequences of quantum superposition states of the positive and negative.

4.10 Tested Remote Sensing Data

Examined remote sensing satellite data is from WorldView-3 (WV-3), which began on August 13, 2014, and is modern in a constellation of commercial high-spatial-resolution Earth-imaging satellites that is industrialized by DigitalGlobe

Original-positive

Original-negative

Quantum superposition
states-positive

Quantum superposition
states-negative

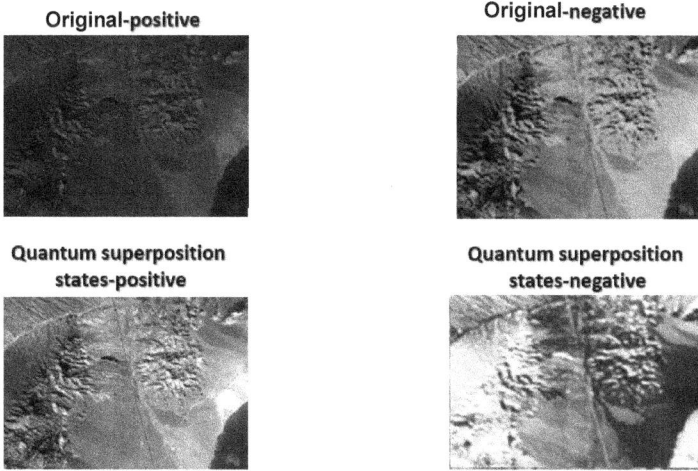

Figure 4.14. Example of quantum image generations.

Inc. (Longmont, Colorado, USA) [20]. In 2007, WorldView-1 (WV-1) data involve a panchromatic (PAN) imaging system with the spatial resolution of 0.5-m. Consequently, in 2009, WorldView-2 (WV-2) departed, which delivers high-spatial-resolution PAN data at a 0.46-m pixel size, besides eight spectral bands at a 1.85-m that involves visible and near-infrared (VNIR) imagery (0.4 to 1.04 μm). WV-3, therefore, is an incremental development to the earlier sensors, carrying basically the similar PAN and VNIR multispectral proficiencies (notwithstanding at 0.31- and 1.24-m spatial-resolution, respectively). In this regard, further eight shortwave infrared (SWIR) bands, which mentioned as bands S1 to S8 are used to enhance the imaging mechanism of WV-3 sensor. Both S1 and S8 range from approximately 1.2 to 2.33 μm at 3.7-m spatial resolution. Nonetheless, they are presently launched merely at 7.5-m spatial resolution. The system correspondingly comprises a further 12 bands; for instance, "Clouds, Aerosols, Water Vapour, Ice, and Snow" (CAVIS) at the 30-m resolution for atmospheric compensation. Needless to say that WV-3 sensor contains different 29 spectral bands covering the spectral ranges of VNIR-SWIR. Presently, SWIR is a high-resolution multispectral Earth-imaging satellite in orbit [19–21].

Examined image is acquired along with the site—Cuprite mining district, Nevada, which is situated approximately 200 km northwest of Las Vegas, Nevada, USA, along U.S. Highway 95. The WV3 g bands S3, S6, S8 (1.661, 2.202, 2.329 μm) are acquired between the latitude of 37° 30' N and 37° 34' N and the longitude of 117° 10' W and 117° 14' W (Figure 4.15).

4.11 Example of Reflectance Spectra

SWIR quantum reflectance spectra of the muscovite, calcite, kaolinite, alunite, hydrothermal silica, and buddingtonite is shown in Figure 4.16. Therefore, these spectra illustrate absorption, asymmetrical, highly varied and distinct features; the highly varied and distinct nature of the SWIR features for these three minerals

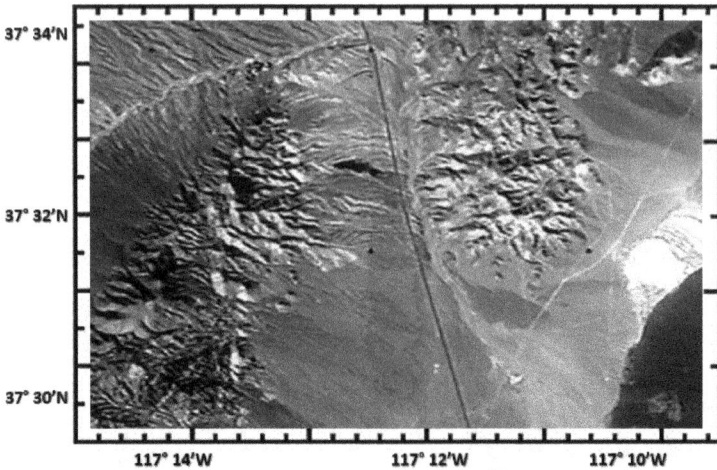

Figure 4.15. WV3 acquired data.

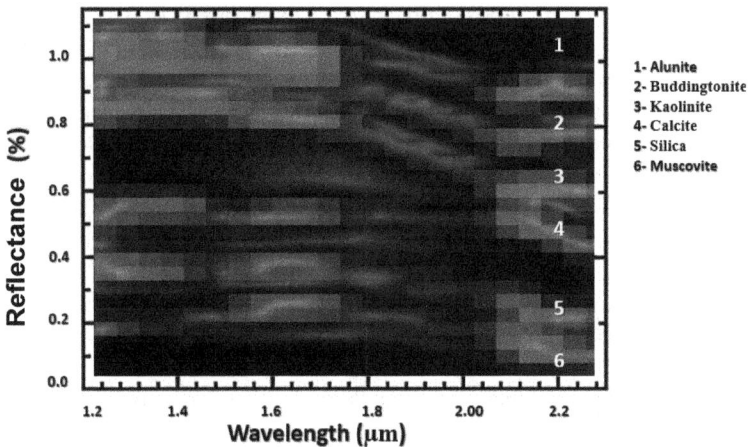

Figure 4.16. SWIR quantum reflectance map of WV3 Satellite data.

(Figure 4.17). Strong vibrational absorptions are included at the 2.2 and 2.3 μm in muscovite.

Buddingtonite is still distinct, although it usually comprises a broad weak band near 2.1 μm, but is shifted to 2.2 μm since WV-3 does not contain a precise band at 2.1 μm. Consequently, alunite and kaolinite (Figure 4.17) are illustrated in an irregular reflectance pattern of 2.16 μm and 2.2, respectively. However, the calcite has the strongest vibrational absorptions involved with spectral reflectance of 2.33 μm. These results agree with the study of Kruse et al. [22].

Direct identifications of materials and chemical measurements are possible to be achieved by WorldView-3's spectral bands. Based on the mineral content, dissimilar surfaces absorb precise wavelengths of electromagnetic radiation and reflect others. In other words, photon energy is either reflected away from mineral surfaces or

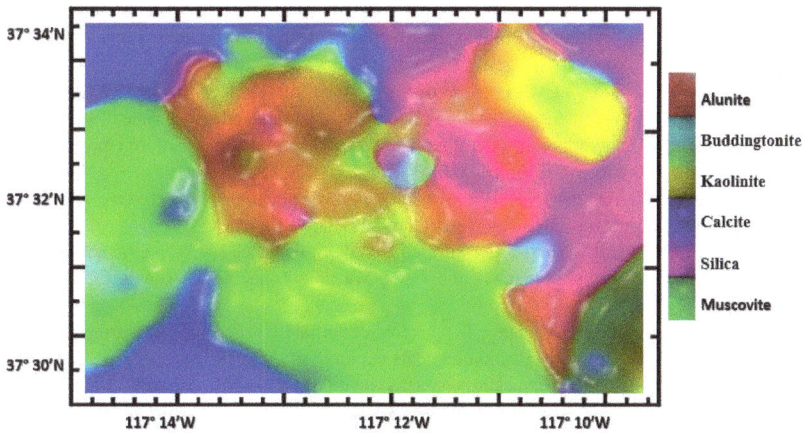

Figure 4.17. SWIR quantum reflectance map of WV3 Satellite data.

absorbed into its molecules, for example as heat. These patterns of reflection and absorption, consequently, form distinctive spectral signatures, which are known as entanglement spectral signatures. In other words, different minerals are entangled with different photons either to reflect or absorb as a function of photon-molecules' interaction mechanisms. In this understanding, absorptions in the SWIR wavelengths can be entangled with atoms of iron as explained earlier in this chapter.

In this chapter, demonstration of quantum spectral modelling is addressed. However, the limitation of this chapter is accuracy determination. The quantum computing of spectral signature is used to identify minerals in optical remote sensing data such as WV-3. In this regard,two practical theories of quantum entanglement and quantum teleportation are discussed. The next chapters will discuss in detail the accuracy of quantum image processing algorithms for mineral identifications and automatic detection; for instance, iron, gold, and copper in multispectral and hyperspectral remote sensing data.

References

[1] Roush TL. Estimation of visible, near-, and mid-infrared complex refractive indices of calcite, dolomite, and magnesite. Icarus. 2021; 354: 114056.

[2] Clark RN. Spectral properties of mixtures of montmorillonite and dark carbon grains: Implications for remote sensing minerals containing chemically and physically adsorbed water. Journal of Geophysical Research: Solid Earth. 1983 Dec 10; 88(B12): 10635–44.

[3] Adams JB. Imaging spectroscopy: Interpretation based on spectral mixture analysis. Remote Geochemical Analysis: Elemental and Mineralogical Composition. 1993: 145–66.

[4] Adams JB. Visible and near-infrared diffuse reflectance spectra of pyroxenes as applied to remote sensing of solid objects in the solar system. Journal of Geophysical Research. 1974 Nov 10; 79(32): 4829–36.

[5] Fang Q, Hong H, Zhao L, Kukolich S, Yin K, Wang C. Visible and near-infrared reflectance spectroscopy for investigating soil mineralogy: A review. Journal of Spectroscopy. 2018 Jan 1; 2018.

[6] Clark RN. Spectroscopy of rocks and minerals, and principles of spectroscopy. Manual of Remote Sensing. 1999 Jun 25; 3: 3–58.

[7] Clark RN, King TV, Klejwa M, Swayze GA, Vergo N. High spectral resolution reflectance spectroscopy of minerals. Journal of Geophysical Research: Solid Earth. 1990 Aug 10; 95(B8): 12653–80.

[8] Kamau M, Hecker C, Lievens C. Use of Short-Wave Infrared Reflectance (SWIR) spectroscopy to characterize hydrothermal alteration minerals in Olkaria geothermal system, Kenya. pp. 1–15. Proceedings, 45th Workshop on Geothermal Reservoir Engineering Stanford University, Stanford, California, February 10–12, 2020 SGP-TR-216.

[9] Thompson AJ. Alteration mapping in exploration: Application of short wave infrared (SWIR) spectroscopy. Economy Geology Newsletter. 1999; 30: 13.

[10] Simpson MP, Rae AJ. Short-wave infrared (SWIR) reflectance spectrometric characterisation of clays from geothermal systems of the Taupō Volcanic Zone, New Zealand. Geothermics. 2018 May 1; 73: 74–90.

[11] Brown A, Walten M, Cudahy T. Short-wave infrared reflectance investigation of site of paleobiological interest application for Mars exploration. Astrobiology. 2004; 4(3): 359–379.

[12] Kraal KO, Ayling B. Hyperspectral characterization of Fallon FORGE Well 21–31: new data and technology applications. In Proceedings, 44th Workshop on Geothermal Reservoir Engineering Stanford University, Stanford, California 2019.

[13] Sancho P. Entanglement in absorption processes. European Journal of Physics. 2018 Nov 30; 40(1): 015404.

[14] Lü XY, Si LG, Hao XY, Yang X. Achieving multipartite entanglement of distant atoms through selective photon emission and absorption processes. Physical Review A. 2009 May 21; 79(5): 052330.

[15] Kurpiers P, Magnard P, Walter T, Royer B, Pechal M, Heinsoo J, Salathé Y, Akin A, Storz S, Besse JC, Gasparinetti S. Deterministic quantum state transfer and remote entanglement using microwave photons. Nature. 2018 Jun; 558(7709): 264–7.

[16] Oka H. Generation of broadband ultraviolet frequency-entangled photons using cavity quantum plasmonics. Scientific Reports. 2017 Aug 14; 7(1): 1–0.

[17] de J, León-Montiel R, Svozilik J, Salazar-Serrano LJ, Torres JP. Role of the spectral shape of quantum correlations in two-photon virtual-state spectroscopy. New Journal of Physics. 2013 May 15; 15(5): 053023.

[18] Laurel CO, Dong SH, Cruz-Irisson M. Equivalence of a bit pixel image to a quantum pixel image. Communications in Theoretical Physics. 2015 Nov 1; 64(5): 501.

[19] DigitalGlobe Inc. Worldview-1 specifications. https://www.digitalglobe.com/sites/default/ files/ WorldView1-DS-WV1.pdf (19 November 2020).

[20] DigitalGlobe Inc. Worldview-2 specifications. https://www.digitalglobe.com/sites/default/ files/ DG_WorldView2_DS_PROD.pdf (19 November 2020).

[21] DigitalGlobe Inc. Worldview-3 specifications. https://www.digitalglobe.com/sites/default/ files/ DG_WorldView3_DS_forWeb_0.pdf (19 November 2020).

[22] Kruse FA, Baugh WM, Perry SL. Validation of DigitalGlobe WorldView-3 Earth imaging satellite shortwave infrared bands for mineral mapping. Journal of Applied Remote Sensing. 2015 May; 9(1): 096044.

[23] Saleh BE, Jost BM, Fei HB, Teich MC. Entangled-photon virtual-state spectroscopy. Physical Review Letters. 1998 Apr 20; 80(16): 3483.

[24] Peřina J, Saleh BE, Teich MC. Multiphoton absorption cross-section and virtual-state spectroscopy for the entangled n-photon state. Physical Review A. 1998 May 1; 57(5): 3972.

[25] Lee DI, Goodson T. Entangled photon absorption in an organic porphyrin dendrimer. The Journal of Physical Chemistry B. 2006 Dec 28; 110(51): 25582–5.

CHAPTER 5

Quantum Multispectral and Hyperspectral Remote Sensing Imaging of Alteration Minerals

In literature, mineral explorations by remote sensing are restricted to conventional image processing techniques, which are embedded to available commercial software; for instance, ERDAS, PCI Geomatica, and The Environment for Visualizing Images (ENVI). Numerous remote sensing geology works in a developed country are ill-defined. Since the beginning of remote sensing mineralogy applications, there is no precise algorithm developing to state the quantity of detected mineralogies. Up to date, there is no study able to model the percentage amount of mineral such as iron in the optical satellite images. This chapter, therefore, is devoted to deliver a novel algorithm in computing the percentage of any specific minerals in multispectral and hyperspectral satellite data. The author also derived a new algorithm based on quantum mechanics to determine some of the mineral alterations.

5.1 What is an Alteration?

This question arises always for non-geologists. A complex process involving mineralogical, chemical and textural changes, resulting from the interaction of hot aqueous fluids with the rocks through which they circulate, under evolving physicochemical conditions is well-known as a hydrothermal alteration (Figure 5.1). In this sense, hydrothermal fluids chemically pounce on the mineral elements of the wall rocks, which cultivate to re-equilibrate by creating innovative mineral accumulations that are in steadiness with the innovative circumstances.

Two keystone sectors of wall rock alteration involve the mode of development of hypogene alteration and supergene alterations. Consequently, hypogene alteration is instigated by ascending relatively high-temperature hydrothermal solutions. Supergene alteration by relatively low temperature descending meteoric water reacts with the previously mineralized ground. For instance, biotite is instigated with a high temperature higher than 400°C, while chlorite is created by an intermediate temperature between 200 and 400°C. However, smectite is formed under the lowest

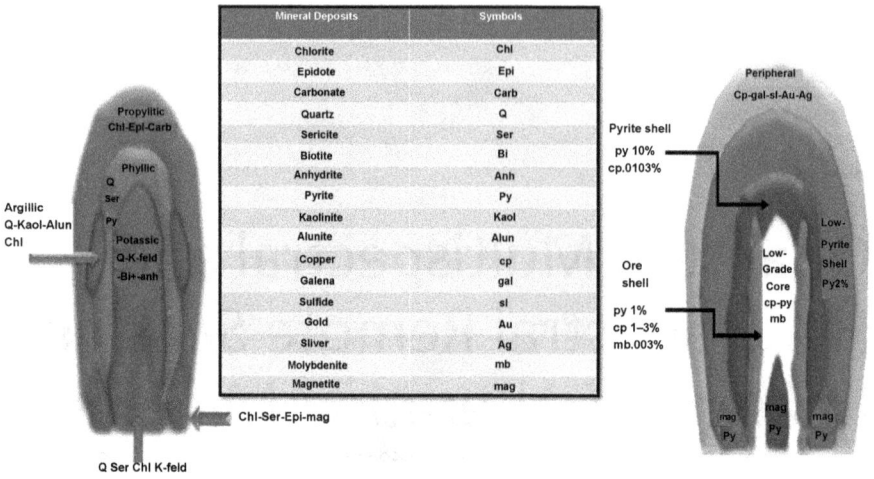

Figure 5.1. Hydrothermal alteration zones associated with porphyry copper deposit.

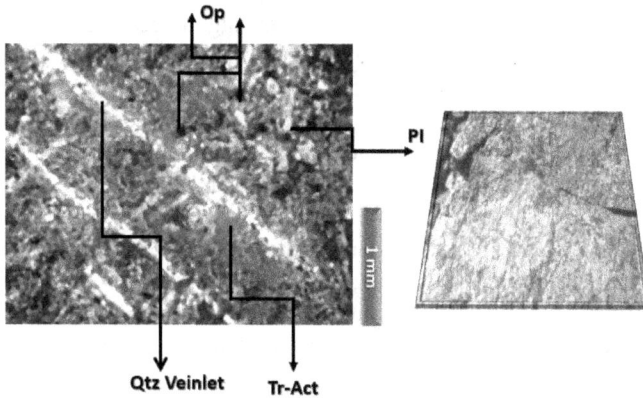

Figure 5.2. Thin section of potassic alteration and silicified vein in diorite with opaque-minerals.

temperature less than 150°C. In this understanding, temperature variations can form dissimilar sort of alterations [1–3].

At temperatures in the region of 600–450°C, potassium silicate alteration is formed as a proxy of plagioclase and mafic silicate minerals. In this regard, potassic (or K-silicate) alteration is categorized by the creation of new K-feldspar and/or biotite (green coloured and Fe-rich), regularly associated with minor sericite, chlorite, and quartz (Figure 5.2). Therefore, potassic alteration is particularly common and vital in porphyry and epithermal systems, where it arises in the high-temperature core zones [2–4].

5.1.1 Potassic Alteration

Perhaps the occurrences of accessory amounts of magnetite/hematite and anhydrite may be allied with the potassic alteration accumulation. In this regard, pyrite, molybdenite and chalcopyrite are common sulfides. Subsequently, a common mineral in potassic alteration in porphyry systems is anhydrite. Generally, a distinction of potassic alteration includes extensive accumulation of Na and Ca, which are termed as sodic and calcic alteration and embodied by abundant albite, epidote and actinolite. Owing to minute hematite inclusions, the K-feldspars of the potassic zones are naturally reddish in colour [1, 3, 5].

5.1.2 Propylitic Alteration

Propylitic alteration encompasses primarily composite of chlorite and epidote with fewer amounts of clinozoisite, calcite, zoisite, and albite. Therefore, propylitic alteration is distinguished by the accumulation of H_2O and CO_2, and locally S, with no considerable H^+ metasomatism. Moreover, it is a mild form of alteration epitomizing low to middle range of temperatures of 200 to 350°C and low fluid/rock ratios. In this circumstance, it embodies the margins of porphyry Cu deposits, besides epithermal precious metal ores. Lastly, in zones of sericite, Fe-oxides, montomorillonite and zeolite are commonly existed [1, 3, 5].

5.1.3 Phyllic (Sericitic) Alteration

The destabilisation of feldspars mainly forms sericitic by hydrolysis (H ion metasomatism) in the presence of OH, K and S. This process can also form quartz, sericite (fine-grained white mica), pyrite, chlorite, and some chalcopyrite. In this regard, sulphide content can be up to 20% by volume. Consequently, this process can leach out Na, Mg, Ti, Fe and also K. Generally, phyllic alteration is allied with porphyry Cu deposits, but similarly with mesothermal precious metal ores and volcanogenic massive sulfide deposits in felsic rocks.

Consequently, these altered minerals are rating into the potassic sorts by growing of the quantities of (K, aK) K-feldspar and/or biotite. Moreover, they also sorted into the argillic alteration by cumulative quantities of (H, aH) clay minerals [1–6].

5.1.4 Argillic Alteration

Argillic alteration is regularly segmented into (i) intermediate argillic and (ii) advanced argillic categories, which is reliant on the passion of constituents of clay minerals or host mineral breakdown.

Intermediate argillic alteration, therefore, distresses primarily plagioclase feldspars and is experienced by the generation of clay minerals kaolinite and the smectite category, which is known as montmorillonite. Less than the temperature of 250°C, it is mostly formed by H^+ metasomatism and exists on the fringes of porphyry systems.

Advanced argillic alteration, therefore, suggests an extreme structure of base discharge somewhere because rocks are dominated by alkali components, which is formed by identical acidic fluids energetic through the rock structures.

It is experienced by kaolinite, pyrophyllite, or dickite and alunite together with lesser quartz, topaz, and tourmaline as a function of temperature fluctuations. In this regard, it is regularly allied with epithermal precious metal deposits anywhere where alteration is combined with boiling fluids and condensation of volatile-rich vapours to develop enormously acidic solutions [3, 5, 7].

5.1.5 Silicification

In the course of alteration, silicification denotes precisely the creation of new quartz or amorphous silica minerals in a rock. Therefore, it is regularly formed as a consequence of isochemical hydrolysis reactions, where Si is domestically grown. In this circumstance, the Si is regularly developed by leaching of the local rocks in which the fluids are mingling. Nevertheless, intense silicification is created as a consequence of cation metasomatism. Generally, the mainstream of fractures through which hydrothermal fluids have passed are at least partially saturated with quartz to construct veins [1, 4, 7, 9].

5.1.6 Carbonatization and Greisenization

In alteration of a rock, carbonatization usually denotes the generation of carbonate minerals, which involves calcite, dolomite, magnesite, siderite, etc. Therefore, it is disseminated by fluids illustrated by high partial pressures of carbon dioxide (PCO_2) and neutral to alkaline pH [5, 8].

However, greisens reveal an alteration accumulation, which includes primarily quartz, muscovite, and topaz. Besides, it is allied with lesser tourmaline and fluorite, and typically creates an alteration zone, which is adjacent to quartz–cassiterite–wolframite veins [7–10].

5.2 Multispectral and Hyperspectral Remote Sensing Sensors

The question now is: what is multispectral and hyperspectral remote sensing? Multispectral imagery is produced by sensors that measure reflected energy within several specific sections (also called bands) of the electromagnetic spectrum (Figure 5.3). Hyperspectral sensors measure energy in narrower and more numerous bands than multispectral sensors. This perspective leads to another question: what are the main differences between multispectral and hyperspectral sensors?

The main difference between multispectral and hyperspectral is the number of bands and how narrow the bands are. Multispectral imagery (MSI) generally refers to 3 to 10 bands. Hyperspectral imagery (HSI) consists of much narrower bands (10–20 nm). A hyperspectral image could have hundreds or thousands of bands (Figure 5.4). In these concerns, multispectral and hyperspectral remote sensing sensors are used for geological applications, ranging from a few spectral bands to

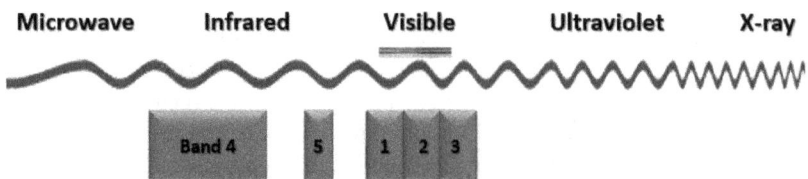

Figure 5.3. Multispectral spectral bands.

Microwave   **Infrared**   **Visible**   **Ultraviolet**   **X-ray**

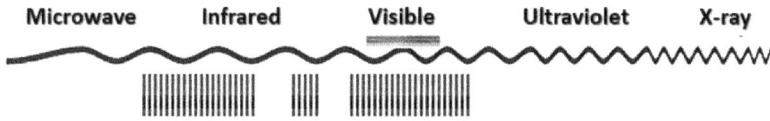

Figure 5.4. Hyperspectral spectral bands.

more than 100 contiguous bands, covering the visible to the shortwave infrared regions of the electromagnetic spectrum [11].

Having a higher level of spectral detail in hyperspectral images gives a better capability to see the unseen. For example, hyperspectral remote sensing distinguished between 3 minerals because of its high spectral resolution. But the multispectral Landsat Thematic Mapper could not distinguish between the three minerals. But one of the downfalls is that it adds a level of complexity. If you have 200 narrow bands to work with, how can you reduce redundancy between channels? Hyperspectral and multispectral images have many real-world applications. For example, we use hyperspectral imagery to map invasive species and help in mineral exploration. There are hundreds of more applications where multispectral and hyperspectral enable us to understand the world. For example, we use it in the fields of agriculture, ecology, oil and gas, atmospheric studies, and more. Consequently, the output from a hyperspectral sensor is called a 'data cube'. This can be thought of as a stack of hundreds of pictures with each image representing a specific wavelength. It is the data cube, which represents the unique fingerprint or 'spectral signature' of an object, considerably improving its identification and classification [11, 12].

According to the above perspective, an MSI array system is analogous in numerous techniques to an HSI array system, with key discrepancies. On the contrary, HIS array uses to continuous wavelengths, MSI focuses on numerous wavebands that are selected based on the request. Common RGB sensors are commonly used to illustrate the discrepancies between HSI and MSI systems. In an RGB sensor, a Bayer pattern—comprising of red, green, and blue filters—is coated over the pixels. In this view, the filters permit wavelengths from precise colour bands to be absorbed by the pixels while the rest of the light is attenuated. Therefore, these bandpass filters have transmission bands in the range of 100 to 150 nm and have a slight spectral overlap (Figure 5.5). In other words, the RGB camera quantum efficiency curve views the overlap between red, green, and blue [14].

The images captured, consequently, are then purified with false colour to approach what the human eye realizes. In most MSI requests, the wavelength bands are pointedly narrower and more numerous. Commonly, the wavebands are on the order of tens of nanometers and are not entirely a part of the visible spectrum. Reliant on the application, UV, NIR, and thermal wavelengths (midwave IR) can have isolated channels as well.

Particular trust, MSI is merely a lower system of HSI with abridged spectral resolution. In truth, HSI is best suited for applications that are sensitive to subtle variances in signal along an incessant spectrum. These small signals, therefore, could be ignored by a system that samples larger wavebands. Yet, certain systems necessitate that substantial quantities of the electromagnetic spectrum be congested and that electromagnetic beam is just selectively acquired (Figure 5.6). The other

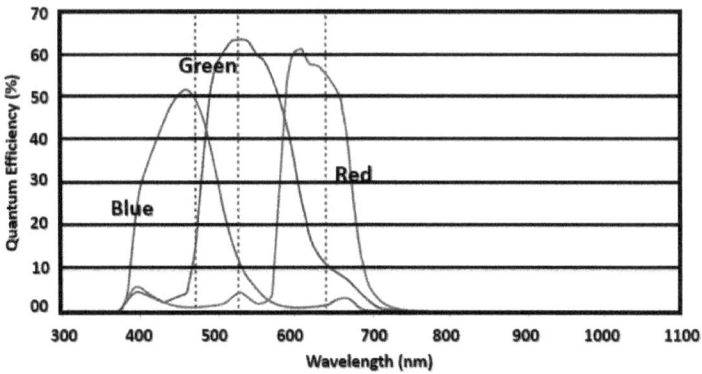

Figure 5.5. An RGB camera quantum efficiency curve.

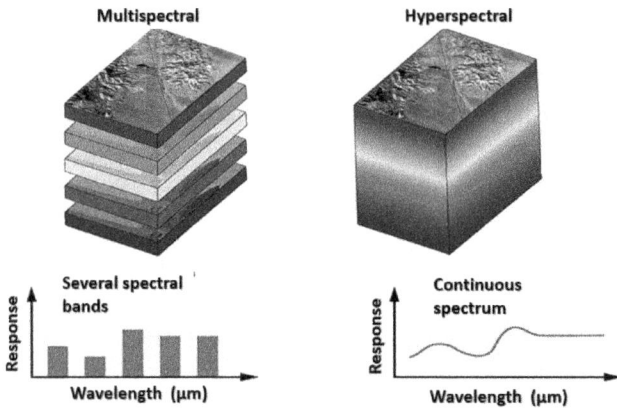

Figure 5.6. Multi- and hyperspectral stacks of images.

wavelengths could induce remarkable noise that could potentially trash measurements and observations. Correspondingly, if less spectral information is comprised in the data cube, the image formation processing, and analysis can perform rapidly [11–14]. In this view, a comparison of the image stacks in MSI, in which several images are taken across various discrete wavelength regions, and HSI, in which images are taken over a larger continuous range of wavelength bands (Figure 5.6).

5.3 Mineral Exploration from Space

Applications of optical remote sensing in geology date back to some early studies using the Earth Resources Technology Satellite-1, the predecessor of the Landsat satellite program. In the 1980s, the seventh channel in the short-wave infrared (SWIR) of the Landsat thematic mapper program was added, as a result of spectroscopic mineral studies by geologists. A subsequent satellite-borne instrument, the Advanced Spaceborne Thermal Emission and Reflection Radiometer (ASTER), launched in 1999, had specific bands in the SWIR and thermal infrared dedicated to mapping mineral groups.

Much of modern geologic remote sensing is built on [16] and [17], the authors of which meticulously analyzed the spectra of minerals and rocks and designed the first spectral libraries linking absorption features to mineralogy. In that same era, NASA's Jet Propulsion Laboratory started to develop and fly airborne imaging spectrometers from the early Airborne Imaging Spectrometer to the present Airborne Visible Infrared Imaging Spectrometer (AVIRIS) and AVIRIS Next Generation (AVIRIS-NG) [18, 7]. In 2000, NASA launched the first imaging spectrometer.

In orbit, Hyperion EO-1 was decommissioned in 2017 [19]. There are plans for several more imaging spectrometers in space, including the German EnMAP mission [20], the NASA-led HyspIRI mission [21], and the CHIME mission (the European Space Agency's hyperspectral imaging satellite planned as *Sentinel-10* for launch in 2024) [15].

5.3.1 Multispectral Satellite Sensors

One of the most beneficial and operative remote sensing multispectral sensors in mineral exploration is Landsat family. This involves Landsat Multi-Spectral Scanner (MSS), Landsat Thematic Mapper (TM), and Satellite Pour l'Observation de la Terre (SPOT) with four to seven spectral bands, which have been implemented in geological mapping [22].

The TM system records three wavelengths of visible energy-blue, green, and red- and three bands of reflected IR energy. These visible and reflected IR bands have a spatial resolution of 30 m. Band 6 records thermal IR energy 10.5 to 12.5 mm with a spatial resolution of 120 m. Each TM scene records 170 by 185 km of terrain [23].

The main key question is: why was Band 7 so important to geologists? In fact, Band 7 records the presence of clay, carbonate, chlorite, gypsum—but not individually. Therefore, in combination with Band 1, which detected iron oxides, there is potential to identify alteration, which is an indicator of mineralisation.

In other words, Landsat Thematic Mapper/Enhanced Thematic Mapper+ (TM/ETM+) images are implemented for sensing alteration mineral accumulations allied with epithermal gold and porphyry copper mineralization and lithological signifying. In this regard, shortwave infrared bands (bands 5 and 7) of TM/ETM+ are mostly implemented as a tool to distinguish hydroxyl-bearing minerals in the exploration phases of copper/gold exploration. Moreover, a band ratio of 5/7 is sensitive to hydroxyl (OH) minerals, which occur in the alteration zones [21, 23].

Landsat 8 was launched in 2013 (California, USA), which returns 400 scenes daily with spatial resolution coverage of 15–100 m. It possesses two sensors (Operational Land Imager (OLI) and Thermal Infrared (TIR)), eleven bands, spatial range of 0.435–12.5 μm, and image size (185 km × 180 km) by the swath of 185 km. Consequently, OLI bands have higher spatial resolution than TIR bands. In this view, TIR bands have a lower spatial resolution of 100 m, while the OLI bands have a 30 m. Subsequently, the Landsat-8 data have a high signal to noise (SNR) radiometer functioning. In this regard, TIR bands have 12-bit quantization, which allows the quantity of indirect inconsistency in surface environments. In this understanding, lithological identification can be performed in TIR bands owing to their high radiometric sensitivity [15, 20, 23].

The spatial resolution of the Landsat 8 data is comparable to that of the Landsat 7 data. Concerning spectral resolution, six of the Landsat 8 bands have spectral sensitivities comparable to Landsat 7, but they have been refined somewhat. For example, the NIR band has been fine-tuned to decrease the effects of atmospheric absorption. Landsat 8 orbits at the same altitude as Landsat 7. Both satellites complete an orbit in 99 minutes, and complete close to 14 orbits per day. This results in every point on Earth being crossed every 16 days. But, because the orbits of the two satellites are offset, it results in repeat coverage every 8 days. Approximately 1000 images per day are collected by Landsat 7 and Landsat 8 combined. That is almost double the images collected when Landsat 5 and Landsat 7 were operating concurrently.

5.3.2 Hyperspectral Satellite Sensors

The primary hyperspectral satellite sensors that are launched efficaciously are the Hyperion and AC systems, which are designated as EO-1 spacecraft and the CHRIS sensor. Therefore, the EO-1 Hyperion sensor technically delivers 242 spectral bands of data included the wide range of wavelengths that vary from the 0.36- to 2.6-μm. Subsequently, each of them has a width of 0.010 to 0.011 μm. Certain bands, chiefly those at the lower and upper ends of the range, reveal a deprived signal-to-noise ratio. As a consequence, through Level 1 processing, merely 198 of the 242 bands are standardized; for furthermost data products, radiometric values in the remaining bands are aligned to 0. Consequently, Hyperion comprises of dual dissimilar linear array sensors, one with 70 bands in the UV, visible, and near IR, and the other with 172 bands in the near IR and mid-IR (with some overlap in the spectral sensitivity of the two arrays). Thus, the spatial resolution of this sensor is 30 m, and the swath width is 7.5 km [24]. Quantum simulated Hyperion image false colour and true colour (refer to Chapter 3) are illustrated in Figure 5.7.

The primary sensor aboard the Proba satellite is CHRIS with narrow spectral bands at a resolution of 34 m. Therefore, CHRIS can additionally function at a spatial resolution of 17 m. On the other hand, its typical nadir image is 13 km × 13 km in size and comprises of data from 18 spectral bands, respectively, in the higher resolution mode. Besides, the CHRIS operates in the visible/near-infrared range (400 to 1050 nm) with a minimum spectral sampling interval ranging between 1.25 (at

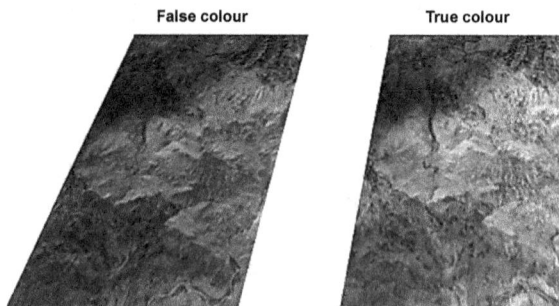

False colour   True colour

Figure 5.7. Quantum simulated Hyperion image.

400 nm) and 11 nm (at 1050 nm). Moreover, in both the along- and across-track directions, the CHRIS sensor can correspondingly be directed. Consequently, its great advantage is within a time of 2.5 min, number of five dissimilar images of an area can be acquired at viewing angles of −55°, −36°, 0°, 36°, and 55° in the along-track direction, which permits the bidirectional reflectance characteristics of features to be measured.

The hyperspectral PRISMA sensor assembles data exploitation dual spectrometers: a visible/near IR (VNIR) spectrometer assembling 66 bands of data over a spectral range of 0.4–1.01 μm, and a SWIR spectrometer assembling 171 bands of data over a spectral range of 0.92–2.50 μm. In this manner, the bandwidth for every band is 10 nm. The panchromatic band, therefore, functions over the spectral range of 0.40–0.70 μm. The satellite employs a sun-synchronous orbit at an altitude of approximately 770 km, an inclination of 98.19°, with a 10:30 A.M. equatorial crossing time. However, the data collections focus on an area of interest covering all of Europe and the Mediterranean region [24].

WorldView-3 satellite, for example, has such high spectral and radiometric quality that we can measure ethane and methane gas leakage in the atmosphere after the proper atmospheric corrections have been applied to the data. WorldView-3, launched in 2014, is a satellite constellation that was developed by DigitalGlobe and built by Ball Aerospace & Technologies. The WorldView-3 remote sensing platform was designed, in part, for geological exploration. Its single panchromatic (pan) spectral band is used to rapidly collect high-resolution imagery, which is particularly useful for capturing sharp image detail (30 cm/12 in. pixel resolution). The visible and near-infrared (VNIR) system collects eight high-resolution (1.2 m/4 ft., 1 in. pixel resolution) multispectral bands used primarily for iron minerals, rare earth elements, vegetation health, and coastal and land-use applications. The pan and VNIR systems are complemented by eight shortwave infrared (SWIR) bands (3.7 m/12 ft., 2 in. pixel resolution) for the measurement and mapping of clay minerals and an atmospheric sensor is known as CAVIS (Cloud, Aerosol, Vapour, Ice, and Snow) with an additional 12 spectral bands. CAVIS bands provide very accurate atmospheric corrections of the imagery for the effects of clouds, aerosols, vapour, ice, and snow [22–24].

Metallic ore deposits and their constituent minerals have characteristic properties that are visible using different wavelengths of light beyond the visible range. Those unique properties can be evaluated to map the distribution of specific minerals [25].

For large regional areas, Landsat imagery would be an excellent choice. For instance, to map all of southern Per, it would doubtless be approximately 30 Landsat scenes. In this view, this would deliver an extensive view for mineral exploration as well as the geological structural, rock and soil are visible during the satellite overpasses accompanied by any visible alteration in the geology that offers more detailed mapping, geochemical sampling, or drilling.

Other hyperspectral sensors, for example, HyMap and the Airborne Visible/IR Image Spectrometer (AVIRIS) with more than 100 continuous bands in the shortwave infrared region are also exploited to acquire precise information regarding hydrothermal alteration mineral collections. This sensor collects data in

224 contiguous spectral bands with a bandwidth of 0.10 μm. Each 20-m square cell in the scene has a continuous spectrum over the range from 0.4 to 2.5 μm. The scene covers the Cuprite mining district in western Nevada, USA. Rocks in several areas of the scene have been altered and mineralized by hot water solutions (hydrothermal alteration), creating concentrations of alteration minerals that provide good targets for spectral analysis. However, the main problems of airborne-based hyperspectral data are costly mobilization and having a small coverage area; besides, they are not eagerly obtainable data [14, 24].

5.4 Why Does The Spectral Analyst Tool Work Properly in Some Cases and Not At All in Others?

In nature, there is relatively an alteration between the spectral libraries restrained on powders of extremely pure minerals and natural geological objects influenced by numerous effects associated with wetness, micro shades, surface conditions, and granularity. For this reason, there is a great curiosity about gauging *in situ* reflectances in correlation with a remote sensing campaign and constructing precise site interrelated spectral libraries. On the contrary, the spectral signature samples collected in the laboratory would be different than in situ measurements. Moreover, the geographical mineral and rock distributions are also controlled by age of formations. On several occasions, the physical characteristics of a similar mineral would be different from one place to another owing to the different period of formation, deposition, and age of rocks. For instance, the granite age of 600 million years must be different than one of age 1000,0000,000 years, in addition to the degree of crystallization that impacts the energy of similar mineral spectral.

In other words, all granites are not shaped at a similar time as other rocks with which they may be adjacent and that some granite bodies are younger than other granites. The circumstance that granite forms interferes with other rocks by filling in cracks, for instance, to create dikes. Therefore, it designates that the other rocks are older than the granite. In some places, if one sort of granite accumulations diagonally discontinue other granite forms, which also illustrates that approximately granites are younger than others. This plays a tremendous role in delivering dissimilar library mineral/rock spectral of same mineral/rock over different places.

Consequently, the impact of tectonic deformation in the mineral formation and deposition varies from one zone to another, which leads to spatial fluctuation of similar mineral spectral signature. In this understanding, every nation or country must have its own mineral/rock spectral signatures.

The spectral mixture of minerals composing a rock, consequently, cannot be obtained by linear unmixing. In this regard, across the grains, the optical path of photon energy is complicated and results in a non-linear mixed signature, not linearly correlated to the profusion of the constituents. In this circumstance, a mixture of 80% of feldspars and 20% of pyroxenes, for instance, will illustrate a spectral signature closer to one of the pyroxenes than to the feldspar. Naturally, the abundance distributions of pyroxenes are extremely varied geographically than the feldspar. Nevertheless, the existence of a mineral in a rock can be evaluated by the existence of its exact spectral features; for instance, an absorption peak at 2.2 μm for clay minerals, and at 2.34 μm for carbonates.

5.5 Quantization of Multispectral and Hyperspectral Data

In the 1980s, Feynman proposed the concept of quantum computer, which can be implemented as speculation for quantum image processing. Yet, no research has been conducted in quantum image processing for mineral mapping in multispectral and hyperspectral data. In previous chapters, remote sensing mineral quantization theories and the basis of quantum image processing were established. Prior to quantum image processing for mineral exploration in remote sensing data, image transformation into qubit format is required.

In this view, the differences between Bit and Qubit must be perfectly understood. Therefore, the bit is a minimum unit, utilized to amass spectral information (either 0 or 1) in conventional computers or classical devices and to embody a pixel in an image too. Nevertheless, the minimum unit to store spectral information in a quantum computer is the qubit.

In this understanding, let us consider binary possible states $|0\rangle$ or $|1\rangle$ matching to the states 0 and 1 for a conventional bit [26]. Therefore, the superposition of a qubit quantum state can be simply expressed as:

$$|\Psi\rangle = \alpha|0\rangle + \beta|0\rangle \tag{5.1}$$

with

$$|\alpha|^2 + |\beta|^2 = 1 \tag{5.2}$$

Equation 5.2 says that the wave function $|\psi\rangle$ allied with a pixel is correlated to the probability of retrieving the pixel in the circumstance of the gray level 0 (gl0), gray level 1 (gl1), etc., which typically denotes the equivalent quantum pixel image along with the value of the pixels. This process imitates an image that is allied with the acquired visual observation, which is primarily stored in quantum memory in the precise 'quantum digital' format (Figure 5.8):

$$\left|\Psi_\Sigma^D\right\rangle = (N)^{-0.5} \sum_{DN}\left|\sum(DN)\right\rangle|DN\rangle, \tag{5.3}$$

In this regard, the index register $|DN\rangle$ is given by:

$$|DN\rangle = \otimes_{l=1}^{\log_2(N)}|DN_l\rangle \tag{5.4}$$

Equation 5.4 reveals that the index register $|DN\rangle$ is represented by $\log_2(N)$ qubits. The state of two qubits, for instance, requires four complex amplitudes, which can be stored in a 4D-vector as given by::

$$|DN\rangle = DN_{00}|01\rangle + DN_{01}|10\rangle + DN_{11}|11\rangle \tag{5.5}$$

For n qubits, it is required to sustain the track of 2^n complex amplitudes. Consequently, these vectors grow exponentially with the number of qubits. This is why quantum computers with large numbers of qubits are extremely complicated to simulate. Therefore, a modern laptop can easily simulate a general quantum state of

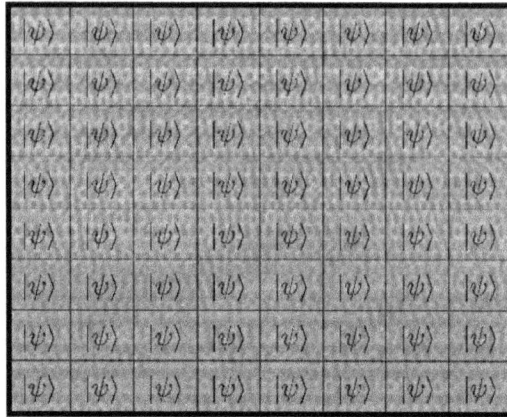

Figure 5.8. Quantum image digital format.

approximately 20 qubits, but simulating 100 qubits is so complicated for the largest supercomputers [27–29].

Figures 5.9 and 5.10 demonstrate two sources of raw multispectral and hyperspectral data that are represented in quantum image digital format. Both Landsat-8 (Table 5.1) and ASTER (Table 5.2) data are acquired along Inglefield Land, which is a regional part of northwestern of Greenland. Both raw data illustrate low-quality pixel variations. ASTER and the Landsat-8 Operational Land Imager provide a higher spatial resolution (15 m, 30 m) (Table 5.1) but with a less frequent collection. ASTER and Landsat-8 satellite data are obtained free from EROS, USGS Global Visualization Viewer (GloVis) (https:// glovis.usgs.gov/) on July 3 2003, and the U.S. Geological and Science Center (EROS) (https:// earthexplorer.usgs.gov/) on August 21 2018, respectively. In this regard, true colour Landsat-8 imagery is similar to the interpretation of aerial photography, and a natural colour image from ASTER. Natural colour imagery attempts to replicate the overall quality of a true colour image with green, blue, and greyish pixels, but ASTER lacks the specific R, G, B visible wavelength bands required for a "true" true colour image.

Figure 5.9. Raw true colour Landsat-8 image along with Inglefield Land.

Figure 5.10. Raw true ASTER image along with Inglefield Land.

Table 5.1. Landsat-8 band spectral characteristics.

Subsystem	Band number	Spectral range (µm)	Ground resolution (m)
VNIR	1	0.433–0.453	
	2	0.450–0.515	
	3	0.525–0.600	30
	4	0.630–0.680	
	5	0.845–0.885	
	Pan	0.500–0.680	15
SWIR	6	1.560–1.660	15
	7	2.100–2.300	
TIR	9	1.360–1.390	
	10	10.30–11.30	100
	11	11.50–12.50	

Table 5.2. ASTER band spectral characteristics.

Subsystem	Band number	Spectral range (µm)	Ground resolution (m)
VNIR	1	0.520–0.600	
	2	0.630–0.690	
	3	0.780–0.860	15
SWIR	4	1.600–1.700	
	5	2.145–2.185	
	6	2.185–2.225	
	7	2.235–2.285	30
	8	2.295–2.365	
	9	2.360–2430	
TIR	10	8.125–8.475	
	11	8.475–8.825	90
	12	8.925–9.275	
	13	10.25–10.95	
	14	10.95–11.65	

It is clear that when the raw image pixels transform into quantum digital format, the information of the images is stored in qubits. Storing the images into quantum digital format improves the information quality of the images. The quantum pixel image is based on the super-position code in quantum mechanics (Figures 5.11 and 5.12). The quantum states of a qubit are implemented to embody the complete gray level of each pixel $|DN\rangle$. It is probable to practice a qubit/pixel to code every pixel of both images, epitomizing the measure of the pixel over all the gray scales $|DN_{00}\rangle$, $|DN_{01}\rangle$, ..., $|DN_{NN}\rangle$. This technique is effective in transforming multispectral and hyperspectral raw images into images created by a chain of quantum superposition states and reduces the effect of the change in a grey level $|DN\rangle$ in one-bit plane to another [26–28].

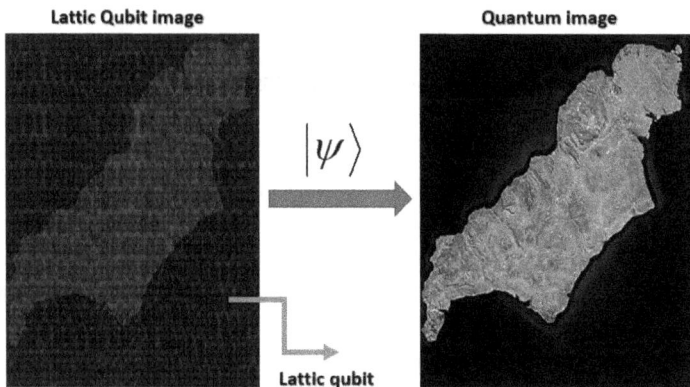

Figure 5.11. Encoded Landsat-8 image in the qubit.

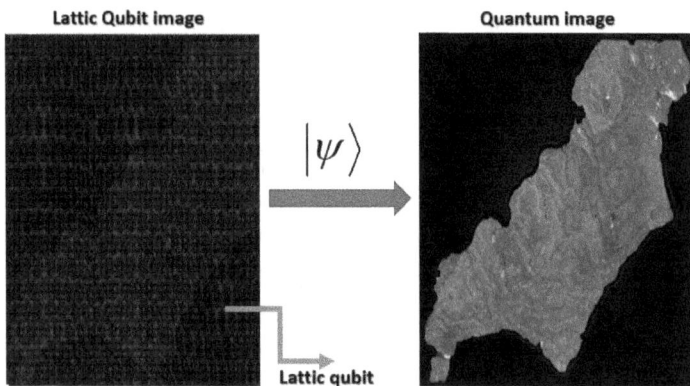

Figure 5.12. Encoded ASTER image in the qubit.

5.6 Spectral Reflectance Quantum Image Formation (SRQIF)

The mineral spectral reflectances are simulated from USGS spectral library version 7.0. To this end, flexible representation for quantum images is implemented for encoding the matched real spectral reflectance in remote sensing data [29]. In this regard, the similarity between the spectral library $|\psi\rangle_r$ and spectral reflectance of images $|\psi\rangle_{smr}$ is mathematically expressed as:

$$Sim(|\psi\rangle_{smr}, |\psi\rangle_r) = f(DN_{i,j}^0, DN_{i,j}^1, ..., DN_{i,j}^{2^{2n-1}}), \qquad (5.6)$$

where $DN_{i,j}$ is the pixel spectral information along row i and column j, respectively, and Equation 4.6 reveals that $Sim(|\psi\rangle_{smr}, |\psi\rangle_r) \in [0,1]$ [26, 28]. In this circumstance, the matched spectral library $|\psi\rangle_r$ automatically overprints $|\psi\rangle_{smr}$ and determines the specifically requested mineral as a function of the following criteria:

In the circumstance of $\forall i, DN_{i,j}^n = \dfrac{\pi}{2}$, $Sim(|\psi\rangle_{smr}, |\psi\rangle_r) = 0$ and both $|\psi\rangle_r$ and $|\psi\rangle_{smr}$ are signified as being absolutely dissimilar. However, the specifically matched minerals overprint automatically in remote sensing if $\forall i, DN_{i,j}^n = 0$, then $Sim(|\psi\rangle_{smr}, |\psi\rangle_r) = 1$, where $n = 0,1,2,.....,2^{2n} - 1DN_{i,j}^n$.

Let us consider thee cluster of different mineral spectral reflectances $Sim(|\psi\rangle_{smr}, |\psi\rangle_r)$ derived post-implementation of Equation 5.6 for remote sensing data as $C|\psi\rangle_{smr} \in Sim(|\psi\rangle_{smr}, |\psi\rangle_r)$, which is signified by

$$C|\psi\rangle_N = \left[C|\psi\rangle_1, C|\psi\rangle_2, ..., C|\psi\rangle_N \right] \qquad (5.7)$$

The centroid of every mineral cluster is given by:

$$C|\psi\rangle_{c_j} = \frac{1}{M_j} \sum_{k \in C|\psi\rangle_j} C\left(Sim(|\psi\rangle_{smr}, |\psi\rangle_r) \right)_k \qquad (5.8)$$

Equation 5.8 reveals that each $C(Sim(|\psi\rangle_{smr}, |\psi\rangle_r))$ is the data point of the cluster $C|\psi\rangle$ and M_j represents the number of data points of that cluster [27, 30].

5.7 Marghany Quantum Spectral Algorithms for Mineral Identifications (MQSA)

This section delivers a novel theory for mineral identifications based on the quantum entanglement theory. The main speculation is that mineral-photon interaction causes entanglement between the quantum energy of mineral atoms and photon energies. This novel speculation can be named as quantized Marghany's mineral spectral.

Based on the similarity of the spectral reflectance wave function, Equation 5.8 can be adjusted to detect accurately specific minerals as:

$$C|\psi\rangle_{Fe_2O_3} = \left[\frac{1}{M_j} \sum_{j=1}^{M_j} |DN\rangle_4 \otimes |DN\rangle_2^{-1} \right] \otimes |g\rangle_{Fe_2O_3} + |B\rangle_{Fe_2O_3} \qquad (5.9)$$

$$C|\psi\rangle_{FeO} = \left[\frac{1}{M_j} \sum_{j=1}^{M_j} |DN\rangle_6 \otimes |DN\rangle_4^{-1} \right] \otimes |g\rangle_{FeO} + |B\rangle_{FeO} \qquad (5.10)$$

$$C|\psi\rangle_{-OH} = \left[\frac{1}{M_j} \sum_{j=1}^{M_j} |DN\rangle_6 \otimes |DN\rangle_7^{-1} \right] \otimes |g\rangle_{-OH} + |B\rangle_{-OH} \qquad (5.11)$$

Owing to hyperspectral pattern variations in ASTER data, a different spectral reflectance band than Landsat-8 is used. In this regard, the different mineral clusters can be given by:

$$C|\psi\rangle_{-OH} = \left[\frac{1}{M_j}\sum_{j=1}^{M_j}\left(\left(|DN\rangle_3 \otimes |DN\rangle_2^{-1}\right)\otimes\left(|DN\rangle_1 \otimes |DN\rangle_2^{-1}\right)\right)\right]\otimes|g\rangle_{-OH}+|B\rangle_{-OH} \quad (5.12)$$

$$C|\psi\rangle_{KLI} = \left[\frac{1}{M_j}\sum_{j=1}^{M_j}\left(\left(|DN\rangle_4 \otimes |DN\rangle_5^{-1}\right)\otimes\left(|DN\rangle_8 \otimes |DN\rangle_6^{-1}\right)\right)\right]\otimes|g\rangle_{KLI}+|B\rangle_{KLI} \quad (5.13)$$

$$C|\psi\rangle_{ALI} = \left[\frac{1}{M_j}\sum_{j=1}^{M_j}\left(\left(|DN\rangle_7 \otimes |DN\rangle_5^{-1}\right)\otimes\left(|DN\rangle_7 \otimes |DN\rangle_8^{-1}\right)\right)\right]\otimes|g\rangle_{ALI}+|B\rangle_{ALI} \quad (5.14)$$

$$C|\psi\rangle_{CLI} = \left[\frac{1}{M_j}\sum_{j=1}^{M_j}\left(\left(|DN\rangle_6 \otimes |DN\rangle_8^{-1}\right)\otimes\left(|DN\rangle_9 \otimes |DN\rangle_8^{-1}\right)\right)\right]\otimes|g\rangle_{CLI}+|B\rangle_{CLI} \quad (5.15)$$

$$C|\psi\rangle_{QI} = \left[\frac{1}{M_j}\sum_{j=1}^{M_j}\left(\left(|DN\rangle_{11} \otimes |DN\rangle_{11}\right)\otimes\left(|DN\rangle_{10} \otimes |DN\rangle_{12}\right)^{-1}\right)\right]\otimes|g\rangle_{QI}+|B\rangle_{QI} \quad (5.16)$$

$$C|\psi\rangle_{CI} = \left[\frac{1}{M_j}\sum_{j=1}^{M_j}|DN\rangle_{13} \otimes |DN\rangle_{14}^{-1}\right]\otimes|g\rangle_{CI}+|B\rangle_{CI} \quad (5.17)$$

$$C|\psi\rangle_{MI} = \left[\frac{1}{M_j}\sum_{j=1}^{M_j}|DN\rangle_{12} \otimes |DN\rangle_{13}^{-1}\right]\otimes|g\rangle_{MI}+|B\rangle_{MI} \quad (5.18)$$

To obtain accurate mineral indexes, the gain $|g\rangle$ and bias $|B\rangle_{MI}$, respectively, for every particular mineral must be estimated. Finally, the entanglement between the predicted mineral abundances and the actual spectral reflectance of each particular mineral can be mathematically expressed as:

$$\rho(|\psi\rangle,C|\psi\rangle) = \left[\left[\frac{\sum|DN\rangle_j C|\psi\rangle_j}{M}\right]-\left[\frac{\sum|DN\rangle_j}{M}\otimes\frac{\sum C|\psi\rangle_j}{M}\right]\right]\otimes$$

$$\left(\left[\frac{\sum|DN^2\rangle_j}{M}-\left(\frac{|DN\rangle_j}{M}\right)^2\right]^{0.5}\otimes\left[\frac{\sum C|\psi\rangle_j^2}{M}-\left(\frac{C|\psi\rangle_j}{M}\right)^2\right]^{0.5}\right)^{-1} \quad (5.19)$$

Equation 5.19 reveals that Pearson Correlation Coefficient is a measure for certifying and quantifying High Dimensional Entanglement. A scheme for characterizing entanglement using the statistical measure of correlation specified by the Pearson correlation coefficient (PCC) was recently recommended that has remained unexplored beyond the qubit case.

5.8 Selected Investigation Area for MQSA Application

The tested area is selected along with Inglefield Land, a regional part of the northwestern of Greenland, which is located between the latitude of 78° 15'N to 79° 15'N and longitude of 67°W to 73°W (Figure 5.13). Inglefield land comprises garnet-sillimanite paragneiss, orthogneiss, and mafic-ultramafic rocks, which accommodate copper-gold mineralization.

Up to date, scarce geological surveys and researches are conducted in Inglefield Land. In this regard, an airborne geophysical survey, fieldwork geological mapping, mineralization investigations, and a regional stream-sediment geochemical survey are conducted by the Geological Survey of Denmark and Greenland (GEUS) in 1994 and 1995, respectively. Moreover, there is no accurate remote sensing study carried for mapping hydrothermal alteration minerals and lithological types in Inglefield land. Hence, in Inglefield Land, mineralization is experienced by copper-gold ore allied with hydrothermal alteration assemblages; for instance, sericite, chlorite, epidote, quartz (silicification), hematite, jarosite, biotite, which surprint the wall-rocks and altered zones.

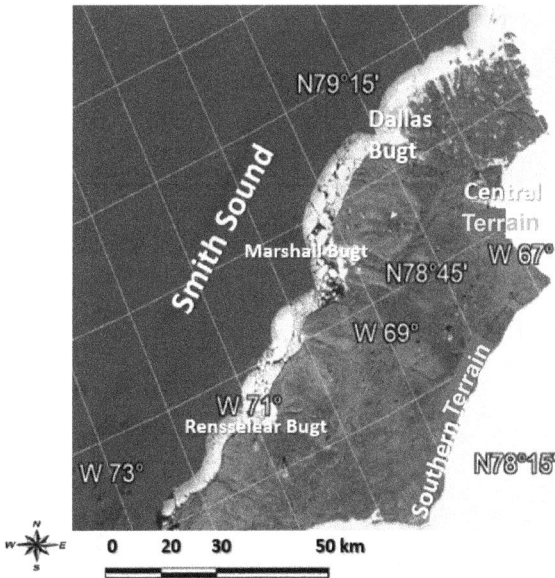

Figure 5.13. The geographical location of the investigation zone.

5.9 MQSA Application of Different Minerals in Landsat and ASTER Images

The quantum algorithm searches the different spectral signature of hydrothermal alteration mineral assemblies. Investigation of four hydrothermal alteration minerals: (i) silicification (Si-OH group (opal/chalcedony) and/or SiO_2 set), (ii) biotite and sericite (Al/Fe-OH set), (iii) hematite and jarosite (iron oxide/hydroxide set), and (iv) chlorite and epidote (Mg-Fe-OH set) is identified in Landsat-8 and ASTER data.

Equation 5.8 discriminates probably the variety of minerals in both Landsat-8 and ASTER data. The quantum spectral library for Landsat-8 reveals that hematite, jarosite, goethite, and limonite incline to have strong absorption features in 0.4 to 1.1 μm. In other words, absorption features of Fe^{3+} near 0.45 to 0.90 μm and Fe^{2+} near 0.90 to 1.2 μm corresponding to bands 2, 4, 5. Therefore, the highest quantum spectral reflectance at 1.56 μm to 1.70 μm coincided with band 6 (Figure 5.14). Consequently, the Landsat-8 retrieved quantum spectral illustrates different clusters of minerals (Figure 5.15). In this view, clusters 1 and 2 confirmed the existence of the iron oxide/hydroxide minerals in bands, 2, 3, 4, 5, and 7. In this regard, Fe^{3+} and Fe-OH absorption features have occurred at 0.45 μm to 0.70 μm, 0.80–0.90 μm, and 2.20–2.30 μm, which are coinciding with bands 2, 3, 4, 5, and 7 of Landsat-8, respectively. Consequently, quantum spectral clusters, 3, 4, and 6 demonstrate the characteristics of Al-OH/Si-OH absorption characteristics (Figure 5.15). Muscovite,

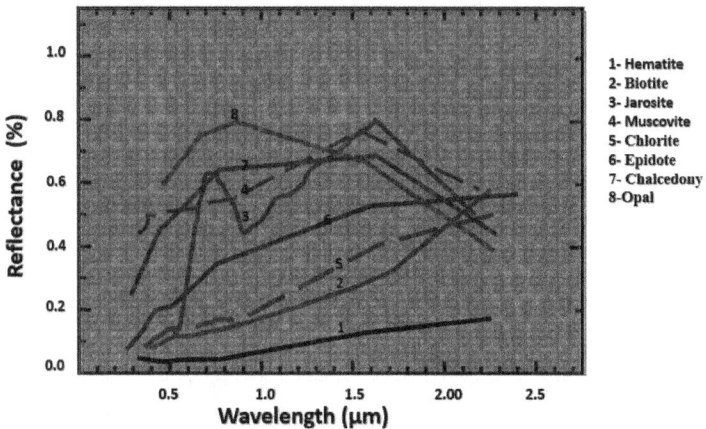

Figure 5.14. Quantum spectroscopy of mineral clusters.

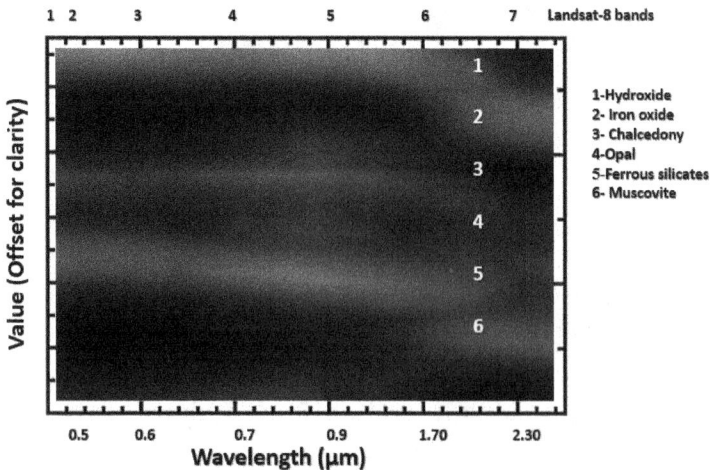

Figure 5.15. Quantum spectral of mineral clusters in Landsat-8 image.

chalcedony, and opal show high reflectance in band 6 (1.560–1.660 μm) and strong absorption in band 7 (2.100–2.300 μm) of Landsat-8. Nevertheless, quantum spectrum cluster 5 contains approximately analogous spectral signatures associated with Mg-Fe-OH alteration minerals (ferrous silicates). Consequently, iron oxide (Fe^{+2}/Fe^{+3}) absorption features in bands 2 to 3 (0.50–0.60 μm) and bands 4 to 5 (0.70–0.90 μm) and Mg, Fe-OH absorption in bands 7 of Landsat-8 are detectible for the quantum spectral 5 (Figure 5.15).

The quantum spectrum Landsat-8 image illustrates the percentage occurrences of oxide/hydroxide cluster, clay mineral groups, and ferrous silicates, respectively (Figure 5.16). In this view, iron oxide/hydroxide minerals have the highest spatial distribution of 75% than clay minerals and ferrous silicates. On the other hand, ferrous silicates have a less spatial distribution of 2% than both clay minerals and iron oxide/hydroxide minerals.

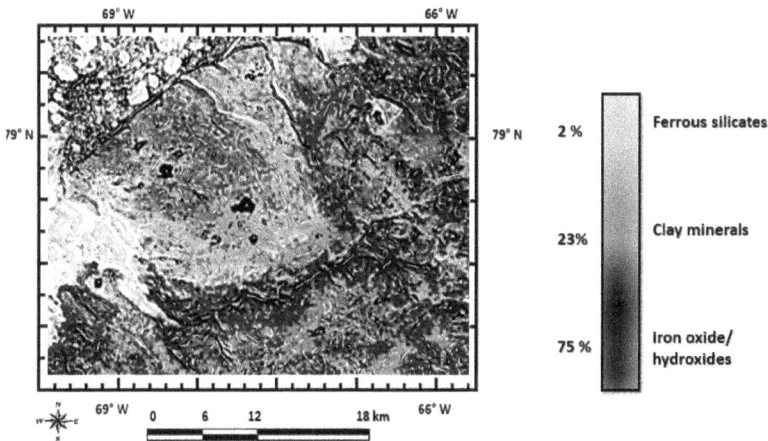

Figure 5.16. Quantum mineral spectral-spatial distributions in Landsat-8 image.

Unlike the mineral spectral signature of Landsat-8, ASTER reveals a different pattern of mineral spectral signature (Figure 5.16). In this regard, the quantum spectral mineral identification algorithm assigns cluster 1 as muscovite/kaolinite owing to the great Al-OH spectral absorption features at 2.20 μm, which is allied with band 6 of ASTER. Consequently, quantum spectral cluster 2 indicates the occurrences of the mixed hematite and jarosite. Subsequently, the occurrence of biotite is assigned as the quantum mineral spectral cluster 3. This could be attributed to slight and foremost absorption of iron and Mg, Fe-OH, respectively. Moreover, chlorite and epidote are illustrated in cluster 4 owing to a dominant absorption of Mg, Fe-OH across the spectral domain of 2.30–2.35 μm. Consequently, the mixture of jarosite and hematite is shown across quantum spectral cluster 5 as the dominant absorption features of Fe^{3+} and Fe-OH are shown through spectral absorption wavelength of 0.48 μm and 0.83–0.97 μm and 2.27 μm, respectively. In this regard, the absorption spectral features of Fe^{3+} and Fe-OH are associated with bands 1, 2, 3, and 7 of ASTER (Figure 5.17). Lastly, the quantum spectral reflectance of cluster 6 indicates the

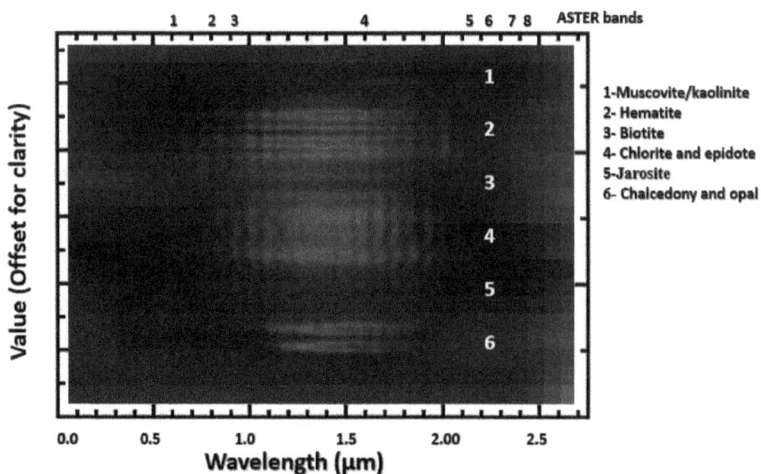

Figure 5.17. Quantum spectroscopy of mineral clusters in ASTER data.

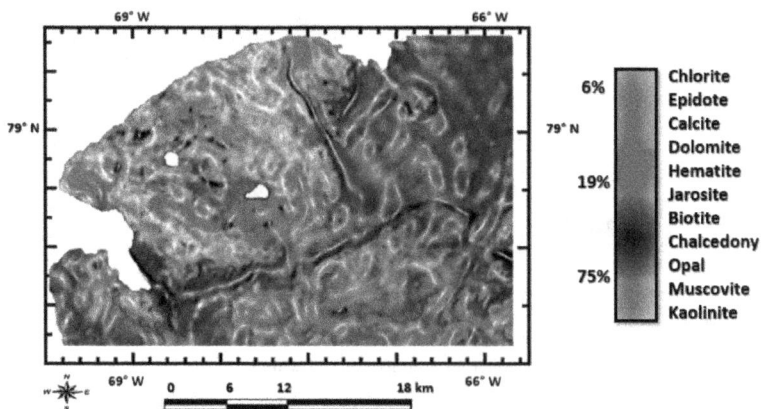

Figure 5.18. Quantum spectral of mineral clusters in ASTER data.

existence of chalcedony and opal (Figures 5.17 and 5.18). In this view, the existence of Si-OH causes a great spectral absorption through bands 7, 8, and 9 in ASTER.

The quantum mineral spectral identifications from Equations 5.12 to 5.18 deliver the percentage of mineral spatial distributions in the ASTER image. Varieties of hydrothermal alteration features are illustrated across NVIR and SWIR ASTER bands. In this view, hematite/jarosite, muscovite/kaolinite, and biotite have the highest quantum spectral reflectance than chalcedony/opal and chlorite/epidote. On the other hand, muscovite/kaolinite occupied 75% of mineral spatial distribution across the ASTER data. On the contrary, chlorite/epidote/calcite have the lowest spatial distributions of 6% than other minerals. Moreover, the moderate mineral spatial distributions of 19% are associated with dolomite/hematite/jarosite/biotite.

5.10 Why Marghany Quantum Spectral Algorithms (MQSA) Identify Accurate Quantum Mineral Images?

It is worth mentioning that both Landsat-8 and ASTER deliver accurate information in dissimilar mineral differentiations. In practice, Figure 5.19 delivers a sense of how the usefulness of quantum spectral identifications, which are approved by Pearson correlation for predicting the percentage of occurrence of spatial mineral variations with its magnitude in both Landsat-8 and ASTER data. In this regard, the quantum spectral identification algorithms indicate a high correlation with quantum spectral library computing algorithm of 90% and 95% confidence levels for Landsat-8 and ASTER images, respectively. The hyperspectral type of ASTER data with more bands than Landsat-8 improves the implementation of the quantum spectral algorithm in identifying numerous minerals than Landsat-8 data.

The novelty of the Marghany Quantum Spectral Algorithm (MQSA) mineral percentage algorithm is the first quantum computing algorithm for mineral identifications. Conventional image band ratio procedures just identify the possible occurrence of specific minerals. However, the band ratio algorithm is just a sort of image enhancement without delivering an exact quantity of minerals through optical remote sensing data. On the contrary, our new algorithms can identify the percentage of the mineral spatial variations in optical data.

The question is now why quantum mineral spectral algorithm delivers accurate information than band ratio? In practice, entangled (dependent) states are frequent exclusions, for different photon wavelength interacts with different mineral atoms, and the interaction creates correlations between them. This correlation is encoded in DN of different bands according to the ability of every band for encoding its entanglement with specific minerals. These are illustrated from Equations 5.9 to 5.18. For instance, Equation 5.9 is specified for distinguishing only the percentage of Fe_2O_3 rather than other minerals. This is caused due to the entanglement of Fe_2O_3 electrons and photon wavelength of band 2 and band 4. On the other hand, the highest level of absorption entanglement feature of Fe_2O_3 is more sensitive only in Landsat-8 wavelength ranges of 0.450–0.515 μm and 0.630–0.680 μm, respectively. On the contrary, in ASTER, bands 2 and 4 do not permit to identify Fe_2O_3 as

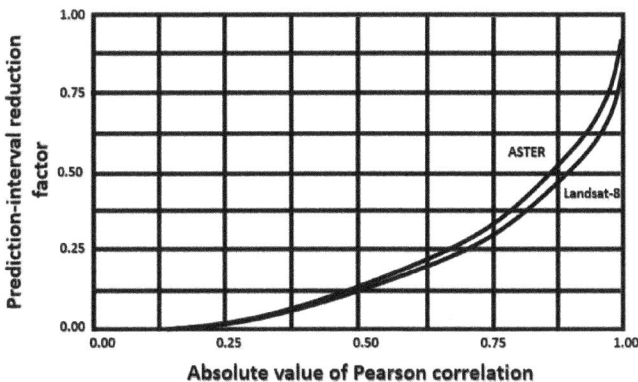

Figure 5.19. Pearson correlation for predicting the percentage of occurrence of spatial mineral variations.

unentangled of electrons creates independent states between bands' wavelengths and Fe_2O_3 electrons.

In Landsat-8, the entanglement of strong absorption is established between electrons of hematite, jarosite, goethite, and limonite and photon wavelength of 0.4 to 1.1 μm. In this circumstance, the exact identification of Fe^{3+} electrons near 0.45 to 0.90 μm and Fe^{2+} electrons close the wavelength range from 0.90 to 1.2 μm, which are entangled with the wavelengths of the photons of bands 2, 4, and 5. Consequently, strong entanglement occurs between Fe^{3+} and Fe^{2+} electrons and photon wavelength of band 6 cause strong reflectance through photon wavelength from 1.56 μm to 1.70 (Figures 5.14 and 5.15).

In ASTER data, the strong photon absorption and reflections of wavelengths of 0.520–0.600 μm, 0.630–0.690 μm and 1.600–1.700 μm are extremely entangled with electrons of iron oxide/hydroxide minerals. In other words, Fe^{3+}/Fe^{2+} iron oxides are entangled with the photon wavelengths of 0.520–0.600 μm, 0.630–0.690 μm, and 1.600–1.700 μm. Furthermore, quartz electrons are entangled with the photon wavelengths 8.125–8.475 μm, 8.475–8.825 μm, and 8.925–9.275 μm, i.e., band 10, 11, 12, respectively. Consequently, carbonate electrons are entangled with the photon wavelengths of 10.25 μm–10.95 μm and 10.95 μm–11.65 μm, respectively. Subsequently, the photon wavelengths of 8.925–9.275 μm, and 10.25–10.95 μm are entangled with mafic electrons. In this circumstance, silica-rich rocks, comprising SiO_2 electron groups, are identified at ASTER data owing to their electron entanglements with bands 10, 11, 12, 13, and 14, i.e., TIR bands (Figures 5.16 and 5.17).

Further confirmations can be also delivered by quantum cosine similarity between quantum spectral library (Chapter 4) and Marghany Quantum spectral Algorithm (MQSA) mineral images. This can be mathematically expressed by:

$$\Psi|\cos(\theta)\rangle = \frac{\rho(|\psi\rangle, C|\psi\rangle).|\Psi\rangle_{sm_r}}{\left\|\rho(|\psi\rangle, C|\psi\rangle)\right\| \left\||\Psi\rangle_{sm_r}\right\|} \tag{5.20}$$

Both Landsat-8 (Figure 5.20) and ASTER (Figure 5.21) data illustrate fluctuation in spatial similarity distribution for mineral identifications. In this regard, ASTER data have the highest similarity level of 0.83 than Landsat-8 data. This could be contributed to the advanced physical characteristics of hyperspectral data than multispectral ones. As a consequence, quantum entanglement theory forces us to be cautious in assigning physical reality to individual mineral properties in satellite data.

The quantum mineral spectral algorithms demonstrated in this chapter determine the level of the quantum entanglement properties like the angular momentum of mineral electrons interactions with photon wavelength encoded in DN of different bands. In other words, two quantum systems are entangled: photon quantum energy, and mineral quantum electron energy owing to photon-electron interactions. In this circumstance, shapes or colours in pixels using pseudo colour tools cannot be a good index for mineral occurrences.

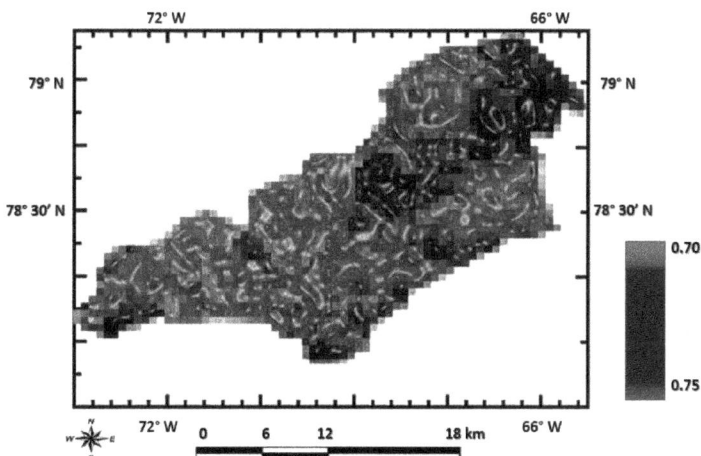

Figure 5.20. Quantum cosine similarity between the quantum spectral library and Marghany Quantum Spectral Algorithm (MQSA) for Landsat-8 data.

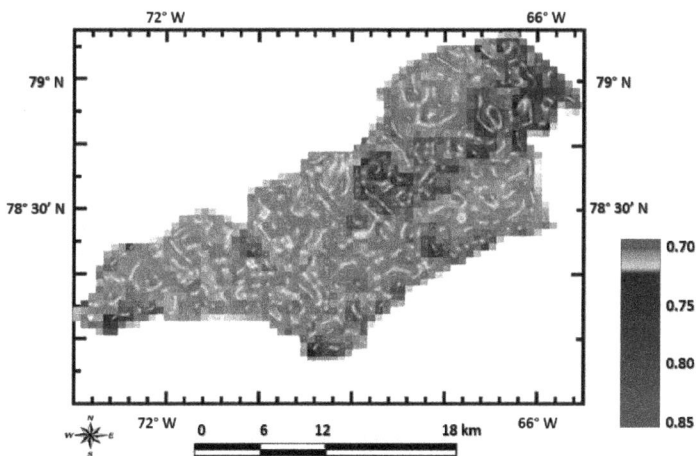

Figure 5.21. Quantum cosine similarity between the quantum spectral library and Marghany Quantum Spectral Algorithm (MQSA) for ASTER data.

In this chapter, two algorithms are developed based on quantum mechanical theory. First is "Spectral Reflectance Quantum Image Formation (SRQIF)". The second is "Marghany Quantum Spectral Algorithms (MQSA)". Both algorithms are developed and can be named as quantized Marghany's mineral spectral identifications. This algorithm delivers a promised model for accurate mineral identifications in multispectral and hyperspectral remote sensing data. In conclusion, it is advisable to implement hyperspectral bands as the peak of the observation features occur in abundance bands of short near-infrared, which are missed at multispectral data.

References

[1] Browne PR. Hydrothermal alteration in active geothermal fields. Annual Review of Earth and Planetary Sciences. 1978; 6: 229–50.

[2] Hemley JJ, Jones WR. Chemical aspects of hydrothermal alteration with emphasis on hydrogen metasomatism. Economic Geology. 1964 Jun 1; 59(4): 538–69.

[3] Reed MH. Hydrothermal alteration and its relationship to ore fluid composition. Geochemistry of Hydrothermal Ore Deposits. 1997 Jun 23; 3: 303–65.

[4] Bruhn RL, Parry WT, Yonkee WA, Thompson T. Fracturing and hydrothermal alteration in normal fault zones. Pure and Applied Geophysics. 1994 Sep 1; 142(3-4): 609–44.

[5] Crosta AP, Sabine C, Taranik JV. Hydrothermal alteration mapping at Bodie, California, using AVIRIS hyperspectral data. Remote Sensing of Environment. 1998 Sep 1; 65(3): 309–19.

[6] Kristmannsdóttir H. Alteration of basaltic rocks by hydrothermal-activity at 100–300 C. pp. 359–367. *In*: Developments in Sedimentology 1979 Jan 1 (Vol. 27). Elsevier.

[7] Mottl MJ, Holland HD. Chemical exchange during hydrothermal alteration of basalt by seawater—I. Experimental results for major and minor components of seawater. Geochimica et Cosmochimica Acta. 1978 Aug 1; 42(8): 1103–15.

[8] Ade-Hall JM, Palmer HC, Hubbard TP. The magnetic and opaque petrological response of basalts to regional hydrothermal alteration. Geophysical Journal International. 1971 Oct 1; 24(2): 137–74.

[9] Schwartz GM. Hydrothermal alteration. Economic Geology. 1959 Mar 1; 54(2): 161–83.

[10] Browne PR. Hydrothermal alteration as an aid in investigating geothermal fields. Geothermics. 1970 Jan 1; 2: 564–70.

[11] Adão T, Hruška J, Pádua L, Bessa J, Peres E, Morais R, Sousa JJ. Hyperspectral imaging: A review on UAV-based sensors, data processing and applications for agriculture and forestry. Remote Sensing. 2017 Nov; 9(11): 1110.

[12] Vasefi F, MacKinnon N, Farkas DL. Hyperspectral and multispectral imaging in dermatology. pp. 187–201. *In*: Imaging in Dermatology 2016 Jan 1. Academic Press.

[13] Paschotta R. Encyclopedia of Laser Physics and Technology. Berlin: Wiley-vch, 2008 Oct.

[14] Unninayar S, Olsen LM. Monitoring, observations, and remote sensing—global dimensions. pp. 2425–2446. *In*: Jørgensen SE and Fath BD (eds.). Encyclopedia of Ecology. Amsterdam: Elsevier Science. 2008, www.doi.org/10.1016/B978-008045405-4.00749-7.

[15] Hecker C, van Ruitenbeek FJ, van der Werff HM, Bakker WH, Hewson RD, van der Meer FD. Spectral absorption feature analysis for finding ore: a tutorial on using the method in geological remote sensing. IEEE Geoscience and Remote Sensing Magazine. 2019 Jun 17; 7(2): 51–71.

[16] Hunt GR. Spectral signatures of particulate minerals in the visible and near infrared. Geophysics. 1977 Apr; 42(3): 501–13.

[17] Salisbury JW, Walter LS, Vergo N. Availability of a library of infrared (2.1–25.0 µm) mineral spectra. American Mineralogist. 1989 Aug 1; 74(7-8): 938–9.

[18] Goetz AF, Vane G, Solomon JE, Rock BN. Imaging spectrometry for earth remote sensing. Science. 1985 Jun 7; 228(4704): 1147–53.

[19] Pearlman JS, Barry PS, Segal CC, Shepanski J, Beiso D, Carman SL. Hyperion, a space-based imaging spectrometer. IEEE Transactions on Geoscience and Remote Sensing. 2003 Aug 11; 41(6): 1160–73.

[20] Guanter L, Kaufmann H, Segl K, Foerster S, Rogass C, Chabrillat S, Kuester T, Hollstein A, Rossner G, Chlebek C, Straif C. The EnMAP spaceborne imaging spectroscopy mission for earth observation. Remote Sensing. 2015 Jul; 7(7): 8830–57.

[21] Lee CM, Cable ML, Hook SJ, Green RO, Ustin SL, Mandl DJ, Middleton EM. An introduction to the NASA Hyperspectral InfraRed Imager (HyspIRI) mission and preparatory activities. Remote Sensing of Environment. 2015 Sep 15; 167: 6–19.

[22] Markham BL, Barker JL. Spectral characterization of the Landsat Thematic Mapper sensors. International Journal of Remote Sensing. 1985 May 1; 6(5): 697–716.

[23] Li P, Jiang L, Feng Z. Cross-comparison of vegetation indices derived from Landsat-7 enhanced thematic mapper plus (ETM+) and Landsat-8 operational land imager (OLI) sensors. Remote Sensing. 2014 Jan; 6(1): 310–29.

[24] Gad S, Kusky T. Lithological mapping in the Eastern Desert of Egypt, the Barramiya area, using Landsat thematic mapper (TM). Journal of African Earth Sciences. 2006 Feb 1; 44(2): 196–202.

[25] Yamaguchi Y, Kahle AB, Tsu H, Kawakami T, Pniel M. Overview of advanced spaceborne thermal emission and reflection radiometer (ASTER). IEEE Transactions on Geoscience and Remote Sensing. 1998 Jul; 36(4): 1062–71.

[26] Steane A. Quantum computing. Reports on Progress in Physics. 1998 Feb; 61(2): 117.

[27] Cai XD, Weedbrook C, Su ZE, Chen MC, Gu M, Zhu MJ, Li L, Liu NL, Lu CY, Pan JW. Experimental quantum computing to solve systems of linear equations. Physical Review Letters. 2013 Jun 6; 110(23): 230501.

[28] Hey T. Quantum computing: an introduction. Computing & Control Engineering Journal. 1999 Jun 1; 10(3): 105–12.

[29] Preskill J. Quantum computing: pro and con. Proceedings of the Royal Society of London. Series A: Mathematical, Physical and Engineering Sciences. 1998 Jan 8; 454(1969): 469–86.

[30] Chen G, Kauffman LH, Lomonaco SJ (eds.). Mathematics of Quantum Computation and Quantum Technology. Boca Raton, FL: Chapman & Hall/CRC, 2008.

CHAPTER 6

Evolving Quantum Image Processing Tool for Lineament Automatic Detection in Optical Remote Sensing Satellite Data

◇◇

6.1 What is Meant by Lineament?

It is an interesting question that leads to understanding perfectly the economic geology, both petroleum geology and mining geology. Lineament is one of the significant geological structures that features such faults. Consequently, there are several lineament definitions. From a geological point of view, lineament is a linear (Figure 6.1) or curvilinear (Figure 6.2) feature of a surface whose chunks line up in a straight or considerably arched affiliation. In a landscape, a lineament is a linear feature, which is an articulation of a causal geological structure such as a fault (Figure 6.3). In geology, a fault is a planar fracture or discontinuity in a volume of rock across which there has been significant displacement as a result of rock-mass movements [1]. Lineaments are regularly obvious in geological or topographic maps and can show up apparent of linear features on aerial or satellite data [2].

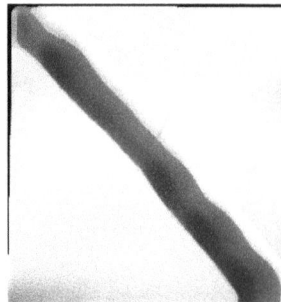

Figure 6.1. Linear feature.

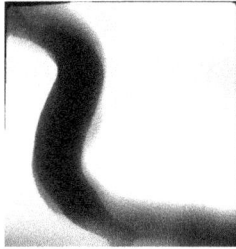

Figure 6.2. Curvilinear feature.

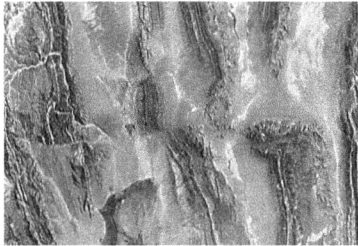

Figure 6.3. Example of fault features.

Naturally, a lineament can be seen as a fault-aligned valley (Figure 6.4), a succession of fault or fold-aligned hills (Figure 6.5), a straight coastline or definitely an amalgamation of these features. In this perspective, fracture zones, shear zones and igneous intrusions (Figure 6.6), for instance, dykes (Figure 6.7), can correspondingly be articulated as geomorphic lineaments [3].

Figure 6.4. Fault-aligned valley.

Figure 6.5. Fold-aligned hills.

Figure 6.6. Igneous intrusions.

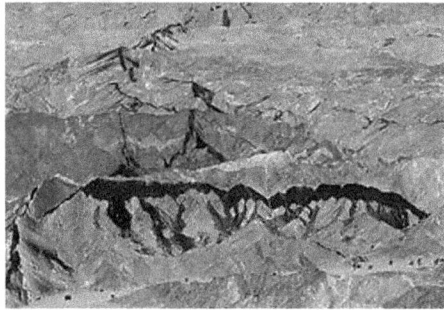

Figure 6.7. Visible dykes.

Consistent with the above perspective, lineaments commonly are demarcated merely at the provincial scale. In this understanding, they are not mesoscopic or microscopic features. Structural lineaments, therefore, intending ones that are a consequence of the localization of recognized geologic features, are characterized by structurally exact alignments of topographic features; for instance, ridges, depressions, or escarpments [4].

Moreover, the term 'megalineament' has been exploited to designate such features on a continental scale. In this regard, the trace of the San Andreas Fault (Figure 6.8) might be counted as megalineament [2]. Besides, the Trans Brazilian

Figure 6.8. San Andreas Fault.

Figure 6.9. Visible lineaments in Mars.

Lineament and the Trans-Saharan Belt, compiled together, shape conceivably the longest coherent shear zone on the Earth, ranging for approximately 4,000 km.

Needless to say that any linear feature that can be picked out as lines (appearing as such or evident because of contrasts in terrain or ground cover, tone, pattern, size, etc.) in aerial or space imagery is termed as lineaments.

On other planets and their moons, lineaments (Figure 6.9), therefore, have also been recognized. Owing to the contradictory tectonic processes tangled on other planets, lineaments' derivations may be drastically dissimilar to terrestrial lineaments.

6.2 What is the Magic of Lineament?

Lineament is keystone magic in geology mining, geology petroleum and economy geology. In this wide view, it plays a tremendous role as an index of oil and gas fields. The mapping of lineaments is also imperative in mineral resource researches since numerous ore deposits are sited along fracture zones. Moreover, along the lineament boundary, geysers, fumaroles and springs are formed. Consequently, lineaments are seen on gravity, magnetic and other geophysical data where the sedimentary patterns have fluctuated. Lithological layering can also be curved in the lineaments. The most frequently encountered feature is topographic lineaments. But lineaments are often developed by a combination of constructive geologic features and erosion. Therefore, other characteristics associated with faulting such as triangular facets, fault scarps, sag ponds, and bulges should be found along a fault lineament. Consequently, the study of true structural lineaments often provides insight into the distribution of regional structural features, ore deposits, and seismicity [4]. Subsequently, the shear joints are filled with clay material and formed layers of few millimetres up to few centimetres and the fractured zones are observed few centimetres to few meters with slickensides planes. And mostly inclined vertical to near-vertical angles with parallel to major joint sets are occurred along the fractured zones.

Lineament commonly is opened up and enlarged by erosion. Some may even become small valleys. Being zones of weak structure, they may be scoured out by glacial action and then filled by water to become elongated lakes (the Great Lakes are the prime example). Groundwater may invade and gouge the fragmented rock or seep into the joints, causing periodic dampness that we can detect optically, thermally, or by radar. Vegetation can then develop in this moisture-rich soil so that at certain

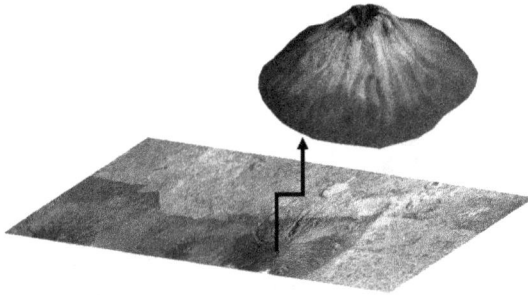

Figure 6.10. Visible volcanic lineaments.

times of year, linear features are enhanced. We can detect all of these conditions in aerial or space imagery [5–8].

Recently, it is documented that the visible volcanic lineaments (Figure 6.10) on land are parallel to seismotectonic lineaments. In this regard, lineaments are also associated with the occurrence of earthquakes. The allied fault structures must be feeble; subsequently, they are not seismogenic.

Mechanically, this permits them to discrete slices of the megathrust directly above which the crust reveals noticeably dissimilar geodynamic behaviour, proved by the distinct seismotectonic reaction of compressional buttresses and superseding extensional earthquake channels. For instance, the most recent large earthquake offshore East Japan initiated close to the prolongation of the Fukushima Lineament. In this view, the motion across the northern seismotectonic lineament reversed its sense in the days prior to the Japanese 2011 Great Earthquake [5].

Lineament and faults are key parameters that described the Earth generation or disaster mechanisms and are a significant indicator for oil explorations and groundwater storages [6]. Fortunately, the application of remote-sensing technology from space is providing geologists with the means of acquiring these synoptic data sets.

6.3 What are the Sorts of Lineaments?

Lineaments involve two primary sorts: positive and negative. In this view, positive refers to ridge trends while negative denotes the river valleys. Ridge trends as positive lineaments can be revealed as an elevated body part or structure, which can be seen through a range of hills or mountains. Besides, they can be also observed as an elongate elevation on an ocean bottom and elongate crest or a linear series of crests. An example of a ridge is the strip of mountains in the southeast area of Mt. Everest from Nepal (Figure 6.11).

A river valley is a valley formed by flowing water (Figure 6.12); entitled fluvial valley, or river valley, is usually V-shaped. The exact shape will depend on the characteristics of the stream flowing through it. Rivers with steep gradients, as in mountain ranges, produce steep walls and a bottom. Shallower slopes may produce broader and gentler valleys. However, in the lowest stretch of a river, where it approaches its base level, it begins to deposit sediment and the valley bottom becomes a floodplain.

Figure 6.11. The strip of mountains.

Figure 6.12. River valley associates with lineaments and mountains.

Perhaps the topographically negative straight lineaments signify joints, faults and shear zones. Nevertheless, the topographically positive straight lineaments are perhaps taken as dykes and dyke swarms [7–9].

The slightly curved and sub-parallel lineaments, therefore, designate foliation or bedding trends, depend on crystalline of rock sort or limestone. On the other hand, circular features may delineate ring dykes.

It can also categorize the lineament as major and minor. In this view, the major lineaments perhaps match to important shear zones, faults, rift zones and major tectonic structures or boundaries. Major lineaments, therefore, can vary from a few to hundreds of kilometres in length. On the other hand, minor lineaments or micro-lineaments perhaps correspond to fairly minor faults, or joints, fractures, bedding traces, etc. These are expressed as soil-tonal changes, vegetation alignments, springs, gaps in ridges, aligned surface sags and depressions, and impart the textural character in a larger scene.

6.4 Satellite Remote Sensing and Image Processing for Lineament Features' Detection

Lineaments are any linear features that can be picked out as lines (appearing as such or evident because of contrasts in terrain or ground cover on either side) in aerial or space imagery. If geological, these are usually faults, joints, or boundaries between stratigraphic formations. Other causes of lineaments include roads and railroads, contrast-emphasized contacts between natural or man-made geographic features (e.g., fence lines), or vague "false alarms" caused by unknown (unspecified) factors. The human eye tends to single out both genuine and spurious linear features, so that some thought to be geological may be of other origins [10].

In the early days of Landsat, perhaps the most commonly cited use of space imagery in Geology was to detect linear features (the terms "linear" or "photolinear" are also used instead of lineaments, but 'linear' is almost a slang word) that appeared as tonal discontinuities. Almost anything that showed as a roughly straight line in an image was suspected to be geological. Most of these lineaments were attributed either to faults or to fracture systems that were controlled by joints (fractures without relative offsets) [11–13].

Lineaments are well-known phenomena in the Earth's crust. Rocks exposed as surfaces or in road cuts or stream outcrops typically show innumerable fractures in different orientations, commonly spaced fractions of a meter to a few meters apart. These lineaments tend to disappear locally as individual structures, but fracture trends persist. The orientations are often systematic, meaning that in a region, joint planes may lie in spatial positions having several limited directions relative to north and horizontal fracture directions [17]. Where continuous subsurface fracture planes that extend over large distances and intersect the land surface produce linear traces (lineaments) in the Earth's crust. A linear feature, in general, can show up in an aerial photo or space images as a discontinuity that is either darker (lighter in the image) in the middle and lighter (darker in the images) on both sides, or is lighter on one side and darker on the other side. Some of these features are not geological. Instead, these could be fence lines between crop fields, roads, or variations in land use. Others may be geo-topographical, such as ridge crests, set off by shadowing. But those that are structural (joints and faults) are visible in several ways [6, 14].

Consequently, optical remote sensing techniques over more than three decades have shown a great promise for mapping geological feature variations over wide-scale [4, 10, 17]. In referring to Koike et al. [18], and Walsh et al. [19], lineament information extractions in satellite images can be divided broadly into three categories: (i) lineament enhancement and lineament extraction for characterization of geologic structure, (ii) image classification to perform geologic mapping or to locate spectrally anomalous zones attributable to mineralization [17, 20], and (iii) superposition of satellite images and multiple data such as geological, geochemical, and geophysical data in a geographical information system [4, 21]. Furthermore, remote sensing data assimilation in real-time could be a bulk tool for geological features' extraction and mapping. In this context, several investigations are currently underway on the assimilation of both passive and active remotely sensed data into automatic detection of significant geological features, i.e., lineament, curvilinear and fault.

Image processing tools that have been used for lineament feature detections are: (i) image enhancement techniques [20–22], and (ii) edge detection and segmentation [15–23]. In practice, researchers have preferred to use the spatial domain filtering techniques to get rid of the artificial lineaments and to verify disjoint lineament pixels in satellite data [20]. Further, Leech et al. [24] implemented the band-rationing, linear and Gaussian nonlinear stretching enhancement techniques to determine lineament populations. Won-In and Charusiri [11] found that High Pass Filter enhancement technique provides an accurate geological map. The High Pass filter selectively enhances the small-scale features of an image (high-frequency spatial components) while maintaining the larger-scale features (low-frequency components) that constitute most of the information in the image.

Vassilas et al. [13] and Majumdar and Bhattacharya [15], respectively, have used Haar and Hough transforms as edge detection algorithms for lineament detection in Landsat-TM satellite data. Majumdar and Bhattacharya [15] reported that the Haar transform is proper in the extraction of subtle features with finer details from satellite data. In this view, the Haar Transform delivers a transform domain in which variance energy is concerted in localized zones. Therefore, Haar Transform has low and high-frequency components and perhaps can be used for image enhancement. Vassilas et al. [13], however, reported that the Hough transform is appropriate for fault feature mapping. In this understanding, the Hough Transform is a powerful tool in edge connecting for line feature extraction, the foremost compensations being its inattentiveness to noise and its fitness to extract lines even in areas with pixel absence/pixel gaps. The transform, therefore, perceives the collinear sets of edge pixels in an image by charting these pixels into a parameter space (the Hough space) demarcated in such an approach that collinear sets of pixels in the image reveal growth to peaks in the Hough space.

Consequently, Laplacian, Sobel, and Canny are the major algorithms for lineament feature detections in remotely sensed data [4, 17, 20, 24]. Consequently, Marghany and Mazlan [10] proposed a new approach for the automatic detection of lineament features from RADARSAT-1 SAR data. This approach is based on the modification of the Lee adaptive algorithm using the convolution of the Gaussian algorithm.

6.5 How do Multispectral Remote Sensing Data Identify the Lineaments?

Lineaments are being demarcated by visual interpretation of false-colour composite, which is fused, for instance, with the 15 m Pan imageries to enhance the interpretation. The concepts of numerous bands amalgamation suggested that, for instance, LANDSAT 7 enhanced thematic mapper plus (ETM+) band 7 should be implemented, besides band 4, which was found to be the best in revealing the textural features. In this regard, the combination of bands 4 (Red band), 7 (Near-infrared) and 1 (Blue band of the visible spectrum) is an excellent tool to view clear lineament features in the false colour composite image.

The band ratio of Red band (Red), and Near-infrared (NIR) has a potential role in revealing the lineament features in such multispectral image as LANDSAT 7 enhanced thematic mapper plus (ETM+). In other words, the normalized difference vegetation index (NDVI) has a background theory whose index is based on the difference in reflectance in the near-infrared (NIR) and red bands of the electromagnetic spectrum (EMS). Its valid results fall between −1 and +1. The higher values indicate more green vegetation. This mathematically Normalized Difference Vegetation Index (NDVI) can be expressed as:

$$NDVI = \frac{(Band\,7 - Band\,4)}{(Band\,7 + Band\,4)} \tag{6.1}$$

Prior to implementing Equation 6.1, the Landsat satellite data must be corrected atmospherically, then the DN values turned into spectral reflectance. Equation 6.1 can be developed in the quantum spectral reflectance as follows:

$$\left|\Psi(\mathrm{Re}\,d)\right\rangle_{sr} = \left(\frac{\left|L(\lambda)\right\rangle_{max}}{254} - \frac{\left|L(\lambda)\right\rangle_{min}}{255}\right) \otimes \left|DN\right\rangle\left\langle DN\right| + \left|L(\lambda)\right\rangle_{min} \tag{6.2}$$

$$\left|\Psi(NIR)\right\rangle_{sr} = \left(\frac{\left|L(\lambda)\right\rangle_{max}}{254} - \frac{\left|L(\lambda)\right\rangle_{min}}{255}\right) \otimes \left|DN\right\rangle\left\langle DN\right| + \left|L(\lambda)\right\rangle_{min} \tag{6.3}$$

In the circumstance of $\left|\Psi(\mathrm{Re}\,d)\right\rangle_{sr} \notin \left|\Psi(NIR)\right\rangle_{sr}$ and $\left|\Psi(\mathrm{Re}\,d)\right\rangle_{sr} \neq \left|\Psi(NIR)\right\rangle_{sr}$, the quantum vegetation index can be expressed as:

$$\left|\Psi(\mathrm{R})\right\rangle_{sr} = \frac{\left|\Psi(NIR)\right\rangle_{sr} - \left|\Psi(\mathrm{Re}\,d)\right\rangle_{sr}}{\left|\Psi(NIR)\right\rangle_{sr} \oplus \left|\Psi(\mathrm{Re}\,d)\right\rangle_{sr}} \tag{6.4}$$

Equation 6.4 demonstrates the entanglement of electromagnetic photons of NIR and Red quantum spectral reflectance with the variation of the vegetation density variation as a function of the entanglement spectral reflectance ratio of Red and NIR photon energies $\left|\Psi(\mathrm{R})\right\rangle_{sr}$. In this understanding, the vegetation index can be represented as the wave function of the quantum photons entanglement of spectra reflectance of vegetation signature in both NIR and Red quantum spectral constrained range. The wave function $\left|\Psi(\mathrm{R})\right\rangle_{sr}$ collapses in the absence of vegetation density cover when $\left|\Psi(\mathrm{R})\right\rangle_{sr} = 0$. The wave function higher than zero indicates the existence of the vegetation covers. Therefore, the wave function $\left|\Psi(\mathrm{R})\right\rangle_{sr} = 1$ reveals the healthy vegetation covers. In this sense, the entanglement absorption between Red spectrum occurs through chlorophyll, while the entanglement reflectance in the NIR spectrum occurs by mesophyll tissue. In other words, the strong entanglement reflectance induces varieties of continuous illuminations in remote sensing sensor as a function of the particle of mesophyll tissue in vegetation leaves.

Equation 6.4, therefore, is examined in Landsat-8 imagery, which is obtained from www.earthexplorer.usgs.gov, and acquired on March 29, 2018, with path/row 117/66 along Tumpangpitu mining area, East Java (Figure 6.13). It is worth noticing that the wave function is collapsing at zero value as an index of the non-vegetation zone, which can be represented by water and urban area. The wave function approaching the value of one indicates heavy vegetation covers that is delineated by dark grey colour. This zone is mostly found in the mountain areas (Figure 6.13). This finding coincides with the study of Haeruddin et al. [31].

Geographical Location $|\Psi(R)\rangle_{sr}$

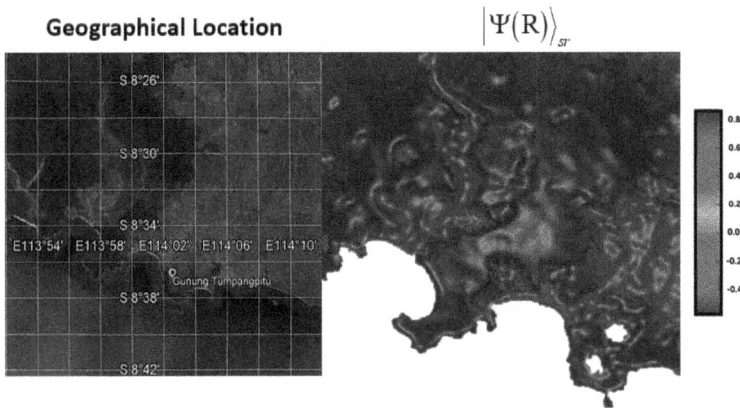

Figure 6.13. The wave function of the normalized difference vegetation index.

Equations 6.1 and 6.4 have been exploited to detect and map geologic linear features (lineaments) in hard-rock terrains as a function of tone, colour and the identified textural patterns. Besides, the presence of the lineaments can be experienced for their influences on the diversity of land cover through analysis of the vegetation index. In other words, the existence of the lineament would incessantly impact the existence of vegetation patterns around the geological structure zone. Thus, the spatial variation of vegetations is anticipated to be associated with the existence of lineaments.

Boyer and McQueen [8] proved the usefulness of NDVI in detecting fractures and faults, which can be associated with the occurrence of vegetation alignment. The mapping of such structural features will ease groundwater accretion in a rocky terrain where permeability and porosity are of negligible value [9]. In this regard, the Sobel algorithm is implemented on the wave function image $|\Psi(R)\rangle_{sr}$ to delineate the variation of lineaments as a function of NDVI as:

$$\left|\Psi\left(\frac{\partial}{\partial x}\right)\right\rangle = \left[|\Psi(R)\rangle_{sr_i}\langle|\Psi(R)\rangle_{sr_j}|\right]*\left(\left|\frac{\partial}{\partial x}G\right\rangle\right) \tag{6.5}$$

The Sobel operator relies on central variances in which it is regarded as a guess of the first Gaussian derivative. In this sense, Sobel operator is correspondent to the first derivate of Gaussian blurring $|\Psi(R)\rangle_{sr_i}\langle|\Psi(R)\rangle_{sr_j}|$ image convolutes by $(\frac{\partial}{\partial x}G)$ and attained by instigating a 3×3 mask to the image. In this circumstance, Figure 6.14 demonstrates the varying intensity of lineaments' features as a function of NDVI wave function. It is worth noticing that the lineament density in highlighted dark grey colour overprint on the heavy vegetation covers with wave function higher than 0.5 and equals 0.85. Needless to say that the lineament heavy density variations are directly proportional to the high wave function of the NDVI, which is in excellent agreement with Haeruddin et al. [31].

It is interesting to add a novel explanation why heavy vegetation covers are associated with highest lineament density variations? Owing to the quantum

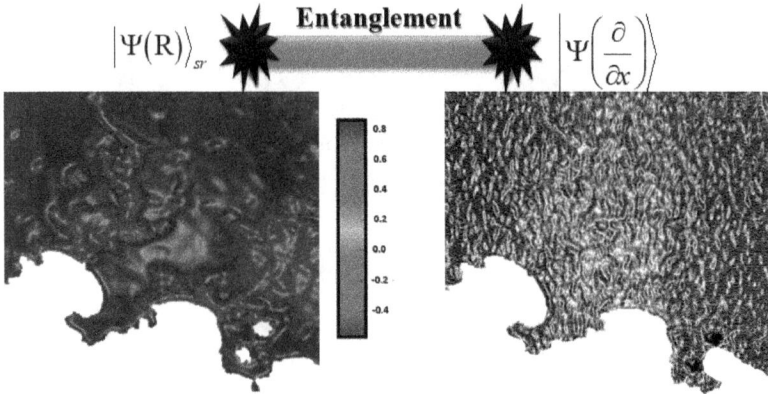

Figure 6.14. Lineament density variations as a function of the wave function of the NDVI.

entanglement between two quantum statues of vegetation covers and lineaments variation, the highest lineament densities are beneath printed vegetation. In this understanding, as well as the wave function of both quantum states being entangled, the lineaments automatically must be entangled with the level of vegetation cover.

Owing to its fairly cheap cost, truly, the archive of Landsat imagery can be downloaded free from the internet. Landsat is the utmost widespread satellite imagery in lineament mapping. On the contrary, a vantage of the dated archival imagery is that it perhaps specifies features not subsequently appear on the further current imagery, because of swelling of urbanization, etc. It should be strained, nevertheless, that where conceivable the lineament should be "ground-truthed".

6.6 Problems for Geological Features' Extraction from Remote Sensing Data

Geological studies require standard methods and procedures to acquire precise information. However, traditional methods might be difficult to use due to high earth complex topography. Regarding the previous perspective, the advantage of satellite remote sensing in its application to geology is the wide coverage over the area of interest, where much accurate and useful information such as structural patterns and spectral features can be extracted from the imagery. Yet, the abundance of geological features is not fully understood. Lineaments are considered the bulk geological features, which are still unclear, despite being useful for geological analysis in oil exploration. In this sense, lineament extraction is very important for the application of remote sensing to geology. However, the real meaning of lineament is still vague. Lineaments should be discriminated from other line features that are not due to geological structures. In this context, the lineament extraction should be carefully interpreted by geologists [24–27].

In this understanding, the identification of lineaments is prevented in remote sensing data owing to the existence of a condensed vegetation canopy, wide-ranging weathering and recent non-consolidated deposits. In this regard, fluvial terraces, pediments or alluvial sediments are deposited across fault zones creating modification in topography pattern. The deprived countenance of a lineament in remote-sensed imagery, as a consequence, does not inevitably signify that there is no joint or fault in the specific zones. Remote sensing satellite imagery, nevertheless, is commonly developing at recognizing these linear or slightly curvilinear features equated with field investigation. Consequently, the steadiness of a fault is perhaps correspondingly not visible in satellite imagery because of the "noise" instigated by the elevated aspect offered by thermal noise, stripping, and dense cloud covers. Moreover, some "lineaments" that have been described in the literature turn out to be artefacts of sunlight interacting with the ground surface, and thus do not have geological significance.

The foremost difficulties with automatic lineament extraction practices are that in approximate circumstances, the filtering algorithm creates segmented images comprising abundant false lineament pixels that would be eradicated using complex edge-linking algorithms, and these lineaments' extraction routines accomplish an arbitrary extraction of edge pixels without deliberating the topographic information inherent in remotely sensed data. Also, these procedures cannot efficiently excerpt lineaments from the low-contrast pixels and in mountain shadows, which yield short dense lineaments that are problematic to associate to tectonically noteworthy structures [25, 28].

The non-filtering technique of the Segment Tracing Algorithm (STA) is implemented to detect the actual linear features. The main principle of the STA is to distinguish a line of pixels as a vector component by inspecting domestic adjustment of the grey level in the digital image and to link reserved line constituents along with their predictable routes. In this perspective, the threshold values for the abstraction and connection of line constituents are the paths reliant on the direction of continuous linear features. Therefore, STA is performed better than traditional filtering procedures as it can locate merely incessant valleys and abstract extra lineaments analogous to the sun's azimuth and placed in shadow areas [29, 30]. Accurate automatic detection of lineaments can be achieved using the combination of STA with the Hough Transform, which is known as Segment Tracing And Rotation Transformation (START). Consequently, START traces automatically lineaments as a function of perceiving the unceasing grey-level margins [30].

6.7 Can Digital Elevation Model be Utilized in Lineament Delineation?

Digital elevation model (DEM) can be derived from either topographical maps, SRTM or aerial photos, which has also been utilized in lineament mappings. In this view, the use of DEM will certainly expand the lineament clarification owing to the outline of the 3D characteristic that can improve the interpretation of the topography, which is vital in lineament understanding. Thus, draping the satellite imagery over the DEM just delivers a stereo view, which perhaps strengthens the elucidation capability of lineament observations [26]. In general, SRTM DEM is used to determine the existence of lineament, in which the SRTM DEM must be turned into a multi-shaded relief. Subsequently, the lineament can be extracted using lineament detection algorithms such as Hough transform, Canny edge detection algorithm, STA or a combination between STA and Hough transform, for instance.

Utilizing analytical hill shading in DEM is another useful technique for delineating lineaments, which is mainly based on forming shaded topographic images of the Earth's surface elevations as a function of the reflection of artificial electromagnetic beam radiated from remote sensing sensors. In this regard, a DEM could be artificially irradiated from slight direction anticipated, which is impossible to be achieved in optical satellite imagery. Consequently, radiated beams need to be perpendicular to specific topography direction to increase any distrusted lineament. In this perspective, the impossible unseen lineaments in an aerial photo or satellite imagery can be recognized in the shaded-relief image [26–28]. In this sense, shaded-relief image merely reveals bare-ground surfaces, buried features by any vegetation or land use and can be easily identified.

Consistent with the above perspective, utilization of hill shade by DEM is successfully implemented in mapping lineament. In this section, the ASTER GDEM data is compiled with Landsat-8 that are implemented at section used to determine the lineament density analysis in the study area. The data could be downloaded at www.earthexplorer.usgs.gov, having a 30 m × 30 m (1 arc-second) resolution (Figure 6.15).

Figure 6.16 demonstrates the extraction of the lineament density in shaded-relief as a function of DEM. In this perspective, the wave function of Sobel operator entangles the DEM with shaded-relief image and variation of vegetation covers to determine the accurate lineament pixels in remote sensing data. The high density of lineament variations is entangled with highest elevation, especially high mountains in Tumpangpitu Mining Area, East Java. In other words, the lineaments' densities are directed northeast-southwest and northwest-southeast Tumpangpitu Mining Area, East Java.

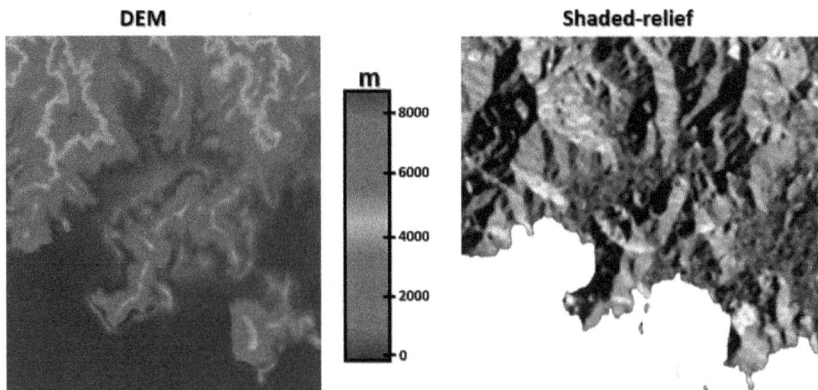

Figure 6.15. DEM and shaded-relief image of Landsat-8.

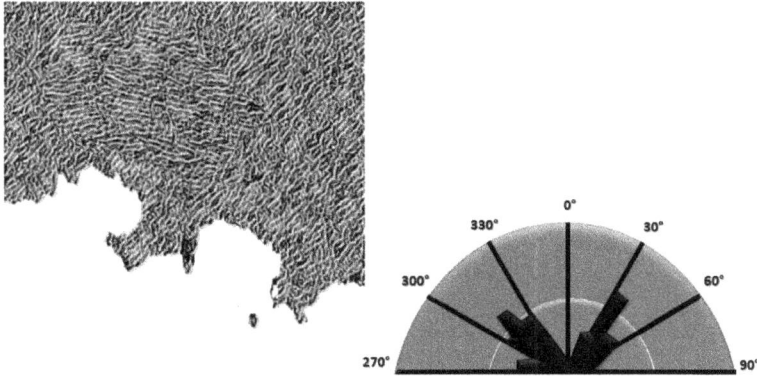

Figure 6.16. Lineament density variation derived based on the shaded-relief image.

Nevertheless, attentiveness is expected once exploiting the hill shade radiance since it can contain wide-range of the "false lineaments". Besides, landforms—which are decisive in lineament detection—are identified to perversely appear with various solar illumination directions. In this circumstance, a drumlin seen side-on will develop linear, but when observed frontal can seem similar to a circular hill [25–27].

6.8 What is the Main Question?

Concerning the above perspective, we address the question of impact of uncertainties on modelling Digital Elevation Model (DEM) for 3-D lineament visualization from multispectral satellite data without needing to include digital elevation data. This is demonstrated with Landsat-8 satellite data using a quantum fuzzy B-spline algorithm. Three hypotheses are examined:

- Lineaments can be reconstructed in Three Dimensional (3-D) visualization,
- The quantum edge detection algorithm can be used as an automatic tool to discriminate between lineaments and surrounding geological features in optical remotely sensed satellite data,
- Uncertainties of the DEM model can be solved using a quantum fuzzy B-spline algorithm to map spatial lineament variations in 3-D.

6.9 The Fuzzy B-splines Algorithm for Digital Elevation Model Reconstruction

Uncertainties associated with ASTER DEM can be solved utilizing fuzzy B-spline algorithm. In this view, the definition of the B-splines can be introduced by fuzzy B-spline. Therefore, the identification of fuzzy B-spline is established by triangle-based criteria and edge-based criteria. In this view, let us assume that the value variation of DEM is restricted by membership function $\mu_{DEM}(x) : X \rightarrow [0,1]$, which clarifies the membership degree of the DEM's element $x \in X$ to the fuzzy set *DEM*, such that $0 \leq \mu_{DEM}(x) \leq 1$. Thus, a fuzzy set *DEM* interrelated to a set $X \neq 0$ in which

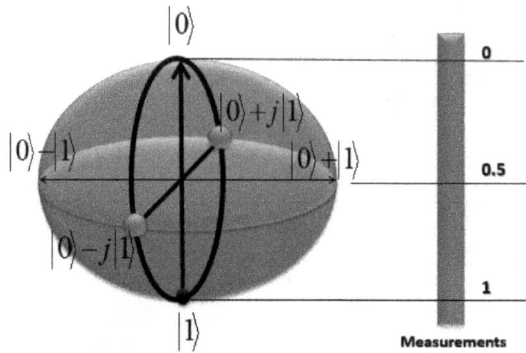

Figure 6.17. Bloch sphere.

$DEM = \{(x, \mu_{DEM}(x)) : x \in X\}$. Consequently, a fuzzy set *DEM* can be turned into quantum computing implementing a meridian on Bloch Sphere (Figure 6.17), with the North Pole $|0\rangle$ signifying 0 (false) and the South Pole $|1\rangle$ expressing 1 (truth). Therefore, fuzzy set *DEM* values along the meridian are assumed to denote the [0, 1] of the interval of fuzzy logic, as the pure quantum state (a point on the surface of the sphere) $(1/\sqrt{2})\,(|0\rangle + |1\rangle)$ embodies fuzzy value 0.5 by its measurement.

Simulation of the fuzzy quantum computing of set DEM is achieved through three keystone gates: inverter, conjunction and disjunction. In this view, in inverter gate, NOT(DEM) or ¬DEM is considered, in which Fuzzy Quantum Not $(\alpha|0\rangle + \beta|1\rangle) = \beta|0\rangle + \alpha|1\rangle$. In this circumstance, the square of the value of the general complex associated with key $|1\rangle$ is the probability of obtaining a '1' measurement of DEM reconstruction.

In conjunction gate, AND(DEM, X) or ∧(DEM, X) or DEM∩X, is presented as quantum state AND through the Toffoli gate (Figure 6.18) with one qubit $|0\rangle$. The mathematical description of the Fuzzy Quantum Conjunction Operator for DEM reconstruction is given by:

$$DEM \wedge \left(\alpha_1|0\rangle + \alpha_2|1\rangle, \beta_1|0\rangle + \beta_2|1\rangle\right) =$$
$$DEM\left(\alpha_1|0\rangle + \alpha_2|1\rangle, \beta_1|0\rangle + \beta_2|1\rangle, 0\right) = \qquad (6.6)$$
$$\left(\left(\alpha_1|0\rangle + \alpha_2|1\rangle\right) \otimes \left(\beta_1|0\rangle + \beta_2|1\rangle\right) \oplus |0\rangle\right).$$

Kronecker product of three parallel input qubits of Equation 6.6 is formulated as:

$$
\begin{bmatrix} \alpha_1 \\ \alpha_2 \end{bmatrix} \otimes \begin{bmatrix} \beta_1 \\ \beta_2 \end{bmatrix} \otimes \begin{bmatrix} 1 \\ 0 \end{bmatrix} = \begin{bmatrix} \alpha_1\beta_1 \\ \alpha_1\beta_2 \\ \alpha_2\beta_1 \\ \alpha_2\beta_2 \end{bmatrix} \otimes \begin{bmatrix} 1 \\ 0 \end{bmatrix} = \begin{bmatrix} \alpha_1\beta_1 \\ 0 \\ \alpha_1\beta_2 \\ 0 \\ \alpha_2\beta_1 \\ 0 \\ \alpha_2\beta_2 \\ 0 \end{bmatrix} \qquad (6.7)
$$

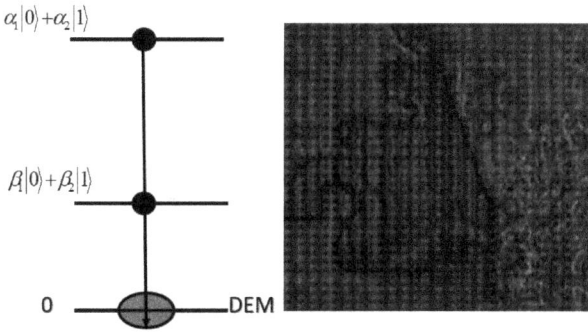

Figure 6.18. Toffoli gate simulation of initial DEM.

Therefore, Equation 6.7 is used as input to compute the Toffoli Gate as:

Toffoli Gate

$$
\begin{bmatrix}
1 & 0 & 0 & 0 & 0 & 0 & 0 & 0 \\
0 & 1 & 0 & 0 & 0 & 0 & 0 & 0 \\
0 & 0 & 1 & 0 & 0 & 0 & 0 & 0 \\
0 & 0 & 0 & 1 & 0 & 0 & 0 & 0 \\
0 & 0 & 0 & 0 & 1 & 0 & 0 & 0 \\
0 & 0 & 0 & 0 & 0 & 1 & 0 & 0 \\
0 & 0 & 0 & 0 & 0 & 0 & 0 & 1 \\
0 & 0 & 0 & 0 & 0 & 0 & 1 & 0
\end{bmatrix}
\bullet
\begin{bmatrix}
\alpha_1\beta_1 \\ 0 \\ \alpha_1\beta_1 \\ 0 \\ \alpha_1\beta_1 \\ 0 \\ \alpha_1\beta_1 \\ 0
\end{bmatrix}
=
\begin{bmatrix}
\alpha_1\beta_1 \\ 0 \\ \alpha_1\beta_1 \\ 0 \\ \alpha_1\beta_1 \\ 0 \\ 0 \\ \alpha_1\beta_1
\end{bmatrix}
\begin{matrix}
000 \\ 001 \\ 010 \\ 011 \\ 100 \\ 101 \\ 110 \\ 111
\end{matrix}
\qquad (6.8)
$$

In this regard, the solution of Equation 6.8 is illustrated by:

$$
\alpha_1\beta_1\left|000\right\rangle + \alpha_1\beta_2\left|010\right\rangle + \alpha_2\beta_1\left|100\right\rangle + \alpha_2\beta_2\left|111\right\rangle =
$$
$$
\left(\alpha_1\beta_1\left|00\right\rangle + \alpha_1\beta_2\left|01\right\rangle + \alpha_2\beta_1\left|10\right\rangle\right)\otimes\left|0\right\rangle + \left(\alpha_2\beta_2\left|11\right\rangle\right)\otimes\left|1\right\rangle
\qquad (6.9)
$$

Equation 6.9 delivers the initial simulation of the DEM based on the quantum fuzzy conjunction operator and its realization as a Toffoli gate (Figure 6.18). It is demonstrated that some visible features of the DEM of ASTER data along Tumpangpitu mining area, East Java are denoted by qubits. In this view, the qubits are begun to delineate the variations of DEM in $|0\rangle$ and $|1\rangle$. This also demonstrates the power of fuzzy quantum computing in representing such complicated data of DEM.

Subsequently, the fuzzy quantum disjunction as a function of OR(DEM, X) or $\lor$ (DEM, X) or *DEM U X* is implemented in the form of Toffoli gate to obtain:

$$
\alpha_2\beta_2\left|001\right\rangle + \alpha_2\beta_1\left|011\right\rangle + \alpha_1\beta_2\left|101\right\rangle + \alpha_1\beta_1\left|110\right\rangle =
$$
$$
\left(\alpha_2\beta_2\left|00\right\rangle + \alpha_2\beta_1\left|01\right\rangle + \alpha_1\beta_2\left|10\right\rangle\right)\otimes\left|1\right\rangle + \left(\alpha_1\beta_1\left|11\right\rangle\right)\otimes\left|0\right\rangle
\qquad (6.10)
$$

Equation 6.10 improves the determination of the DEM features than Equation 6.8 (Figure 6.19). In this circumstance, the construction of DEM features quantum fuzzy disjunction operator and its realization as a Toffoli gate delivers good

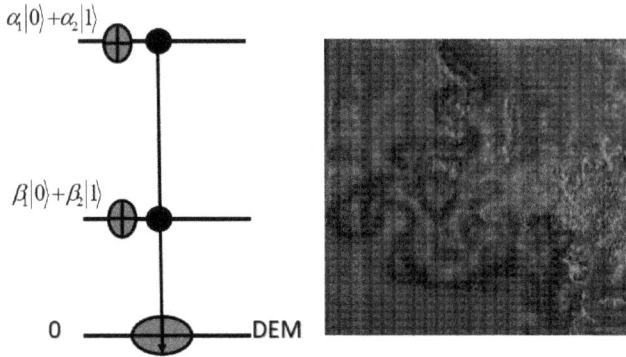

Figure 6.19. Toffoli gate simulation of initial DEM as a function of quantum fuzzy disjunction operator.

promise for DEM features' construction as the DEM is begun to assemble true X values and remove the false values in the circumstance of *DEM U X*.

6.10 Entanglement of Fuzzy Quantum for DEM Reconstruction

Let us assume the quantum fuzzy B-spline is $QS[(\alpha_1, \alpha_2), (\beta_1, \beta_2)]$, which is described as a linear combination of the basic wave functions in two topological parameters, $\Psi|\alpha\rangle$ and $\Psi|\beta\rangle$. Then, let the wave function of non-decreasing sequences of DEM real number be $\Psi|x\rangle$, which $\Psi|x\rangle \subset \Psi|\alpha\rangle$ and $\Psi|x\rangle \subset \Psi|\beta\rangle$ in the circumstance that $\Psi|\alpha\rangle \notin \Psi|\beta\rangle$ and $\Psi|\alpha\rangle \cup \Psi|x\rangle \cup \Psi|\beta\rangle$. In this understanding, the quantum polynomial function B-spline $\Psi|P_{i,1}(x)\rangle$ can be given by

$$\Psi \left| P_{i,1}(x) \right\rangle = \begin{cases} 1 & \text{If } x_i \leq x \leq x_{i+1}, \\ 0 & \text{Otherwise;} \end{cases} \tag{6.11}$$

$$\Psi \left| P_{i,p}(x) \right\rangle = \frac{x - x_i}{x_{i+p-1} - x_i} \Psi \left| P_{i,p-1}(x) \right\rangle + \frac{x_{i+p} - x}{x_{i+p} - x_{i+1}} \Psi \left| P_{i+1,p-1}(x) \right\rangle \quad \text{for } p > 1 \tag{6.12}$$

To exercise more shape controllability over the surface, and perspective invariance transformations, fuzzy B-spline is introduced. Besides having the control point as in the B-spline, fuzzy B-spline also provides a set of weight parameters $w_{i,j}$ that exert more local shape controllability to achieve projective invariance. The fuzzy B-spline surface is composed when the gate complexity is the number of 2-qubit gates, which is exploited. Formally, an algorithm with query complexity Q is gate-efficient if its gate complexity is $O(Q.\text{poly}(\log(qN)))$. Subsequently, the runtime of the quantum algorithm for obtaining precise fuzzy B-spline surface can be demonstrated implementing $O(\kappa^2 \log N\varepsilon^{-1})$, where $\varepsilon > 0$ is the error and κ is the condition number. Therefore $\kappa = \dfrac{\max_i \Psi\left(|\alpha\rangle|\beta\rangle\right)}{\min_i \Psi\left(|\alpha\rangle|\beta\rangle\right)}$ in which, if the condition number is not much larger than one, the estimated DEM from quantum B-spline surface is well-conditioned, and it means that its inverse can be computed with great precision.

$$\Psi\left(|\alpha\rangle|\beta\rangle\right) = \sum_{i=0}^{M}\sum_{j=0}^{O}\left[\left(\alpha_2\beta_2\,|00\rangle + \alpha_2\beta_1\,|01\rangle + \alpha_1\beta_2\,|10\rangle\right)\otimes|1\rangle + \left(\alpha_1\beta_1\,|11\rangle\right)\otimes|0\rangle\right]\otimes\left[C_{ij}\right]\otimes$$
$$\left[\Psi\big|P_{i,p}(x)\big\rangle\,\Psi\big\langle P_{j,p}(x)\big|\right].$$
(6.13)

Implementation of Equation 6.13 reconstructs three-dimensional shaded-relief image in the highest details than ASTER one. In comparison with ASTER shaded-relief image, the quantum B-spline reveals other geological features such as faults and drainage streams with their cautious flows in Indian Oceanside (Figure 6.20).

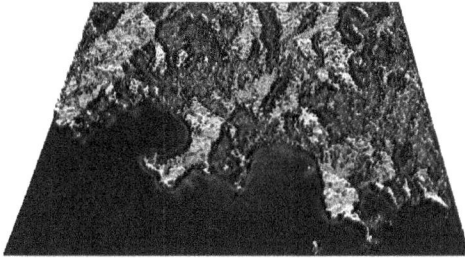

Figure 6.20. Shaded-relief image simulated using quantum fuzzy B-spline.

6.11 Quantum Edge Detection Algorithm for Lineament Mapping

In practice, edge detection algorithms count on the modelling of image gradients by dissimilar categories of filtering masks. The computational complexity of at least $O(2^n)$, therefore, is required for entirely conventional lineament edge detection algorithms, because every pixel has to be handled. In this view, a quantum algorithm promises to provide an exponential speed-up equated with standing edge mining algorithms. Nevertheless, this algorithm involves a COPY operation and a quantum black box for computing the gradients of all the pixels instantaneously. This section introduces novel procedures with a highly competent quantum algorithm that bargains the borders between several regions in satellite data in $O(1)$ time, autonomous of the image dimensions. Let us assume that the spotted couple of neighboring pixels in a satellite image column are delivered by the binary sequences $b_1.......b_{n-1}0$ and $b_1.......b_{n-1}1$, with $b_j = 0$ or 1. In this sense, their pixel values are required to be stored as the coefficients $c_{b_1}.....b_{n-1}0$ and $c_{b_1}.....b_{n-1}1$, respectively, of the matching computational foundation circumstances. In practice, let us implement a Hadamard gate H, which renovates a qubi $\dfrac{|0\rangle \to (|0\rangle + |1\rangle)}{\sqrt{2}}$ and $\dfrac{|1\rangle \to (|0\rangle - |1\rangle)}{\sqrt{2}}$ is implemented to automatically detect the lineament boundaries. In this understanding, the Hadamard transform on the preceding qubit swaps them to the new-fangled coefficients

$c_{b_1}.....b_{n-1}0 \pm c_{b_1}.....b_{n-1}1$ [33]. Therefore, H, the mathematical transformation of $\Psi(|\alpha\rangle|\beta\rangle)$, is given by:

$$\Psi(|\alpha\rangle|\beta\rangle)_{2^n-1} \otimes H : \begin{bmatrix} c_0 \\ c_1 \\ c_2 \\ c_3 \\ ... \\ c_{N-2} \\ c_{N-1} \end{bmatrix} \mapsto \frac{1}{\sqrt{2}} \begin{bmatrix} c_0 + c_1 \\ c_0 - c_1 \\ c_2 + c_3 \\ c_2 - c_3 \\ ... \\ c_{N-2} + c_{N-1} \\ c_{N-2} - c_{N-1} \end{bmatrix}. \tag{6.14}$$

In Equation 6.14, the quantum B-spline image is turned into a unity image matrix as $\Psi(|\alpha\rangle|\beta\rangle)_{2^n-1}$ represents $2^{n-1} \times 2^{n-1}$. In this regard, output image state $|g\rangle$ is delivered by:

$$|g\rangle = \left(\Psi(|\alpha\rangle|\beta\rangle)_{2^n-1} \otimes H\right)\left|\sum_{k=0}^{N-1} c_k |k\rangle\right\rangle \tag{6.15}$$

where $N = 2^n$ pixels, and to acquire the lineament boundaries between the couples of pixels using Equation 6.14, the n-qubit amplitude permutation is applied to the input image state $|g\rangle$, yielding a new image state $\left\|\sum_{k'=0}^{N'-1} c'_k |k'\rangle\right\rangle_{new}$. To this end, $\left|\sum_{k=0}^{N-1} c_k |k\rangle\right\rangle$ odd (even) elements are equivalent to the even (odd) elements of original input one. In other words, $c'_{2k} = c_{2k+1}$ and c'_{2k+2}. Consequently, the quantum amplitude transformation can be proficiently accomplished in time of $O[poly(n)]$. Subsequently, a new image state $\left\|\sum_{k'=0}^{N'-1} c'_k |k'\rangle\right\rangle_{new}$ is transferred into a single qubit using Hadamard rotation for obtaining all lineament boundary values. In this understanding, to yield complete lineament boundary values in a single step, a modified the unceasing grey-level margins of the quantum edge detection algorithm exploits a secondary qubit for encoding the $\left\|\sum_{k'=0}^{N'-1} c'_k |k'\rangle\right\rangle_{new}$ image, which yields an $(n + 1)$-qubit redundant image state as:

$$\left\|\sum_{k'=0}^{N'-1} c'_k |k'\rangle\right\rangle_{new} \otimes (|0\rangle + |1\rangle) = \sqrt{2} \otimes 2^{2-1} \begin{pmatrix} c_0, c_0, c_1, c_1, c_2, c_2, c_3, c_3, ..., c_{N-2}, \\ c_{N-2}, c_{N-1}, c_{N-1} \end{pmatrix}^T \tag{6.16}$$

By measuring the last qubit, conditioned on obtaining 1, the n-qubit state can be obtained (Figure 6.21), which is given by:

$$\left\|\sum_{k'=0}^{N'-1} c'_k |k'\rangle\right\rangle_{new} \otimes (|0\rangle + |1\rangle) = \sqrt{2} \otimes 2^{2-1} \begin{pmatrix} c_0 - c_1, c_1 - c_2, c_2 - c_3, ..., \\ c_{N-2} - c_{N-1}, c_{N-1} - c_0 \end{pmatrix}^T \tag{6.17}$$

Equation 6.17 contains the full lineament boundary information in satellite images. Consequently, Figure 6.22 reveals the 3-D high-density variations of lineaments entangled with heavy vegetation covers than urban zones. It can be confirmed that the lineament is associated with faults and it is also obvious that the

**n+1-qubit amplitude
permutation**

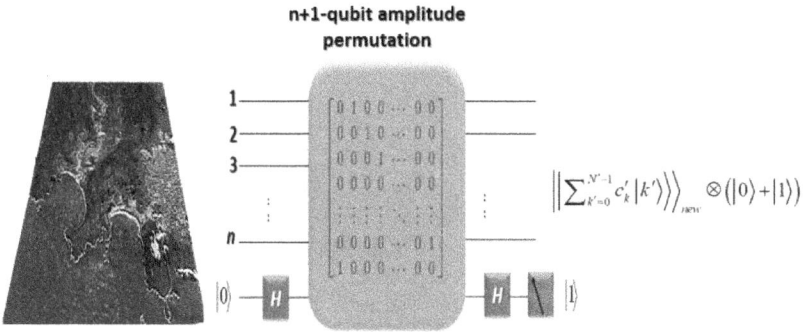

$$\left\| \sum\nolimits_{k'=0}^{N'-1} c_k' \left| k' \right\rangle \right\rangle_{new} \otimes \left(\left| 0 \right\rangle + \left| 1 \right\rangle \right)$$

Figure 6.21. Quantum edge detection algorithm based on Hadamard gate.

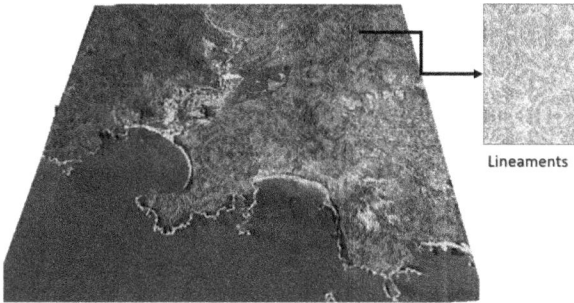

Lineaments

Figure 6.22. Automatic identification of lineament variations using 3-D quantum edge detection algorithm.

heavy capacity of lineament occurs within the mountains. This type of lineament can be named mountain lineaments.

The entanglement of fuzzy quantum B-spline and quantum edge detection algorithm retrieves the sharpest lineament and other geological features such as faults in optical remote sensing data.

In this understanding, the entanglement between quantum edge detection algorithm and quantum fuzzy B-spline delivers smoother distribution of lineaments on 3-D quantum fuzzy B-splines. When it happens that the measurement in qubit fuzzy B-spline surface gives the result 1, then the qubits α and β also collapse to value 1 (this results directly from Equation 6.13). These values are used in the next calculations of other functions that involve qubits α and β.

This chapter has demonstrated such a novel approach for automatic detection of the lineament from multispectral satellite image such as Landsat-8. The combination between quantum fuzzy B-spline and quantum edge detection can automatically delineate the density of the lineament variations. Quantum fuzzy B-spline can only reconstruct a true 3-D world of the satellite image with preserving edge boundary details. Consequently, quantum edge detection based on the Hadamard gate extracted extreme lineament details in the heavy vegetation covers.

References

[1] Lutgens FK, Tarbuck EJ, Tasa DG. Essentials of Geology. Pearson; 2017 Feb 20.

[2] Ohnaka M. The Physics of Rock Failure and Earthquakes. Cambridge University Press; 2013 Apr 11.

[3] Attoh K, Brown LD. The Neoproterozoic trans-saharan/trans-brasiliano shear zones: suggested Tibetan analogs. AGUSM. 2008 May; 2008: S51A–04.

[4] Marshak S, Van Der Pluijm BA. Earth Structure: An Introduction to Structural Geology and Tectonics. WW Norton; 2004.

[5] Lister G, Tkalčić H, Hejrani B, Koulali A, Rohling E, Forster M, McClusky S. Lineaments and earthquake ruptures on the East Japan megathrust. Lithosphere. 2018 Aug 1; 10(4): 512–22.

[6] Solomon S, Ghebreab W. Lineament characterization and their tectonic significance using Landsat TM data and field studies in the central highlands of Eritrea. Journal of African Earth Sciences. 2006 Nov 1; 46(4): 371–8.

[7] Koch M, Mather PM. Lineament mapping for groundwater resource assessment: a comparison of digital Synthetic Aperture Radar (SAR) imagery and stereoscopic Large Format Camera (LFC) photographs in the Red Sea Hills, Sudan. International Journal of Remote Sensing. 1997 May 1; 18(7): 1465–82.

[8] Boyer R, McQueen J. Comparison of mapped rock fractures and airphoto linear features. Photogrammetric Engineering and Remote Sensing. 1964; 30(4): 630–5.

[9] Mogaji KA, Aboyeji OS, Omosuyi GO. Mapping of lineaments for groundwater targeting in the basement complex region of Ondo State, Nigeria, using remote sensing and geographic information system (GIS) techniques. International Journal of Water Resources and Environmental Engineering. 2011 Aug 31; 3(7): 150–60.

[10] Marghany M, Hashim M. Lineament mapping using multispectral remote sensing satellite data. International Journal of Physical Sciences. 2010 Sep 4; 5(10): 1501–7.

[11] Won-In K, Charusiri P. Enhancement of thematic mapper satellite images for geological mapping of the Cho Dien area, Northern Vietnam. International Journal of Applied Earth Observation and Geoinformation. 2003 Jun 1; 4(3): 183–93.

[12] Robinson CA, El-Baz F, Kusky TM, Mainguet M, Dumay F, Al Suleimani Z, Al Marjeby A. Role of fluvial and structural processes in the formation of the Wahiba Sands, Oman: A remote sensing perspective. Journal of Arid Environments. 2007 Jun 1; 69(4): 676–94.

[13] Vassilas N, Perantonis S, Charou E, Tsenoglou T, Stefouli M, Varoufakis S. Delineation of lineaments from satellite data based on efficient neural network and pattern recognition techniques. pp. 355–365. *In*: Proc. Second Hellenic Conference on AI 2002 Apr.

[14] Zaineldeen UF. Paleostress reconstructions of Jabal Hafit structures, Southeast of Al Ain City, United Arab Emirates (UAE). Journal of African Earth Sciences. 2011 Feb 1; 59(2-3): 323–35.

[15] Majumdar TJ, Bhattacharya BB. Application of the Haar transform for extraction of linear and anomalous patterns over part of Cambay Basin, India. International Journal of Remote Sensing. 1988 Dec 1; 9(12): 1937–42.

[16] Mostafa ME, Bishta AZ. Significance of lineament patterns in rock unit classification and designation: a pilot study on the Gharib-Dara area, northern Eastern Desert, Egypt. International Journal of Remote Sensing. 2005 Apr 1; 26(7): 1463–75.

[17] Mostafa ME, Qari MY. An exact technique of counting lineaments. Engineering Geology. 1995 May 1; 39(1-2): 5–15.

[18] Koike K, Nagano S, Ohmi M. Lineament analysis of satellite images using a Segment Tracing Algorithm (STA). Computers & Geosciences. 1995 Nov 1; 21(9): 1091–104.

[19] Walsh GJ, Clark Jr SF. Contrasting methods of fracture trend characterization in crystalline metamorphic and igneous rocks of the Windham quadrangle, New Hampshire. Northeastern Geology and Environmental Sciences. 2000; 22(2): 109–20.

[20] Suzen ML, Toprak VE. Filtering of satellite images in geological lineament analyses: an application to a fault zone in Central Turkey. International Journal of Remote Sensing. 1998 Jan 1; 19(6): 1101–14.

[21] Novak ID, Soulakellis N. Identifying geomorphic features using LANDSAT-5/TM data processing techniques on Lesvos, Greece. Geomorphology. 2000 Aug 1; 34(1-2): 101–9.

[22] Mah A, Taylor GR, Lennox P, Balia L. Lineament analysis of Landsat Thematic Mapper images, Northern Territory, Australia. Photogrammetric Engineering and Remote Sensing. 1995; 61(6): 761–73.

[23] Marghany M. Fuzzy B-spline algorithm for 3-D lineament reconstruction. International Journal of Physical Sciences. 2012 Apr 30; 7(15): 2294–301.

[24] Leech DP, Treloar PJ, Lucas NS, Grocott J. Landsat TM analysis of fracture patterns: a case study from the Coastal Cordillera of northern Chile. International Journal of Remote Sensing. 2003 Jan 1; 24(19): 3709–26.

[25] Marghany M. Three-dimensional lineament visualization using fuzzy B-spline algorithm from multispectral satellite data. Remote Sensing Advanced Techniques and Platforms. Croatia: InTech Open Access Publisher. 2012 Jun 13: 213–32.

[26] Oguchi T, Aoki T, Matsuta N. Identification of an active fault in the Japanese Alps from DEM-based hill shading. Computers & Geosciences. 2003 Aug 1; 29(7): 885–91.

[27] Das S, Pardeshi SD, Kulkarni PP, Doke A. Extraction of lineaments from different azimuth angles using geospatial techniques: a case study of Pravara basin, Maharashtra, India. Arabian Journal of Geosciences. 2018 Apr 1; 11(8): 160.

[28] Mallast U, Gloaguen R, Geyer S, Rödiger T, Siebert C. Derivation of groundwater flow-paths based on semi-automatic extraction of lineaments from remote sensing data. Hydrology and Earth System Sciences. 2011 Aug 1; 15(8): 2665.

[29] Koike K, Nagano S, Ohmi M. Lineament analysis of satellite images using a Segment Tracing Algorithm (STA). Computers & Geosciences. 1995 Nov 1; 21(9): 1091–104.

[30] Raghavan V, Masumoto S, Koike K, Nagano S. Automatic lineament extraction from digital images using a segment tracing and rotation transformation approach. Computers & Geosciences. 1995 May 1; 21(4): 555–91.

[31] Haeruddin, Irawan JF. Identifying of the relationship between lineament density and vegetation index at Tumpangpitu Mining Area, East Java, Indonesia. In AIP Conference Proceedings 2020 Oct 26 (Vol. 2278, No. 1, p. 020006). AIP Publishing LLC.

[32] Reiser R, Lemke A, Avila A, Vieira J, Pilla M, Du Bois A. Interpretations on quantum fuzzy computing: intuitionistic fuzzy operations × quantum operators. Electronic Notes in Theoretical Computer Science. 2016 Sep 30; 324: 135–50.

[33] Yao XW, Wang H, Liao Z, Chen MC, Pan J, Li J, Zhang K, Lin X, Wang Z, Luo Z, Zheng W. Quantum image processing and its application to edge detection: theory and experiment. Physical Review X. 2017 Sep 11; 7(3): 031041.

CHAPTER 7

Quantum Support Vector Machine in Retrieving Clay Mineral Saturation in Multispectral Sentinel-2 Satellite Data

7.1 Salinity, Soil and Geological Minerals

The core stone question is: what is the relationship between salinity and soil? The salinity in soils is perhaps owing to the dissolved mineral salts. The foremost cations are sodium, calcium, magnesium, and potassium; the foremost anions are chloride, sulfate, bicarbonate, carbonate, and nitrate. In this view, soil salinity is well-defined as the highest meditation of solute salts counting Na^+, Ca^{2+}, and Mg^{2+} in soils, instigating further than 4 dS/m for soil electric conductivity. In this regard, soil electric conductivity is equivalent to 0.2 MPa of osmotic potential formed by 40 mM sodium chloride (NaCl) in the solution. Consequently, electric conductivity is the electric current transferred in the water and soil capacities, which is computed in the unit of ds/m. Subsequently, it is an excellent index of the total ions and dissolved solids' concentrations in both soil and water [1–3].

Soil salinity, therefore, determines the sort of soil mineral composition. The soil minerals can generally describe parts of the soil volume. Needless to say that clay minerals in the soil framework are "inferior inorganic compounds of $< 2\,\mu m$ size" counting Fe, Al, and Mn oxides (hydroxides and oxyhydroxides), besides noncrystalline phases. Accordingly, clays stimulus soil are a function of both soil bulk characteristics and connotations with clay massive accumulations across outer/inner surfaces, for instance, positive ion interchanges clay bulk characteristics. Since the physical characteristics of the clay minerals are absolutely shaped by the temperature and quantity of rainfall, they are commonly exploited as climatic indicators. For example, climate circumstances swing from cool/dry to warm/moist, and the foremost clay minerals change from chlorite/illite to vermiculite, which swaps into montmorillonite then kaolinite [3–5].

The clay minerals incorporate extraordinary construction blocks. Some of them are tetrahedral, which is a pyramid structure polyhedron made of four triangular faces, three of which they encounter at every vertex. These have silicon in the middle, and 4 oxygen at every corner. These shape in layer sheets with an octahedron, which has eight exceptional faces.

The extraordinary cations in octahedron is what makes every clay unique, and is surrounded with the aid of 6 distinctive oxygen or hydroxide molecules. This sounds very complex; however, these one-of-a-kind layers assemble collectively like layers of a cake and decide the very essential properties of soils.

Consequently, soil carbonates percolate from the surface with time and amass in the subsoil at a specific depth. In this regard, the existence of carbonate is extensively exploited as a fundamental physical soil characteristic to designate soil sorts and compute soil erosion [1–4].

7.2 Mineral Soil Classifications

Minerals are divided into eight classes depending upon their chemical composition, and specifically their dominant anion or anionic group. The classes comprise silicates, sulfates, chromates, molybdates, carbonates, nitrates, borates, oxides and hydroxides, native elements, and sulfides.

Soil minerals involve primary and secondary minerals. In this regard, primary minerals are not to be subjected to substantial chemical or mechanical alteration because of their deposition in sedimentary rocks or their crystallization within igneous or metamorphic rocks [5–7]. They are left from the parent material of magma or lava, and habitually originate in the silt fraction of soils and also in the sands. Mutual keystone minerals in soil environments contain silicates, oxides of iron (Fe), zircon (Zr) and titanium (Ti), and phosphates (P). Consequently, secondary minerals are altered consequences of the chemical cessation or alteration of primary minerals under ambient circumstances and/or re-crystallized. Since primary minerals regularly have particle size reductions during weathering, secondary minerals mostly originate in the clay and fine-silt fractions. Conventional secondary minerals originating in soils comprise alumino-silicates, oxides and hydroxides, amorphous minerals, carbonates, and sulfates [1, 5, 7].

In soil environments, mineral occurrences are the consequence of heritage from parent materials, precipitation from soil solution, which is termed as neoformation. Consequently, alteration of prevailing minerals into new phases is well-known as diagenesis. The mineralogical structure of a specific soil, therefore, is definited generally by the composition of the parent material(s), and the strength and period of the weathering schedule. In this sense, silt and sand fractions have a tendency to be compiled primarily of inherited keystone minerals. Clay fractions, however, are subjugated by secondary minerals, either transported from other environments or *in-situ*. Nevertheless, it is not rare for carbonates, sulfates, secondary silica minerals, as well as Fe- and Al-hydroxides to be meaningful components of the coarse soil fraction. Therefore, quartz, feldspars, and mica minerals are found in clay fractions.

The variety of global enduring environments, yet, creates approximately dissimilar soil mineral spreading inclinations [5, 8].

7.3 Remote Sensing of Mineral Soils

In previous Chapters 4 and 5, we have discussed the reason behind spectral library uncertainties. In other words, the precise spectral libraries are not consistent over all the Earth. In this view, clay minerals are commonly exploited as climatic gauges because their nature is precisely inspired by weather conditions; for instance, temperature fluctuations and quantity of precipitation at the location throughout pedogenesis. In this circumstance, the dominant clay minerals fluctuate from chlorite/ illite to vermiculite, which changes to montmorillonite and finally transforms into kaolinite. These alterations are a function of climate fluctuations as they fluctuate from cool/dry to warm/moist. In this understanding, the foremost clays in soils indicate diagnostic absorptions in the SWIR spectrum domain, which are formed by vibrational transitions and regularly reveal sharp and narrow characteristics. The analytical bands are chiefly constrained on ~ 1.4 µm (implications instigated by OH), ~ 1.9 µm (implications triggered by molecular water), and ~ 2.2 µm (amalgamation tones instigated by Al-OH). Moreover, roughly weak absorption bands in the 2.3–2.5 µm region are interrelated to the occurrence of Fe-OH and/or Mg-OH in the clay minerals.

Chlorites are demonstrated as a set of clay minerals encompassing precise octahedral positive ion; for instance, Fe, Mg, and Al. Their reflectance spectra reveal a weak absorption band at roughly 1.4 µm and triple absorption features near 2.3 µm. The bands at 2.25 and 2.35 µm are interrelated to Fe-OH and Mg-OH, correspondingly. Therefore, illite is considered by three protuberant absorptions at approximately of 1.4, 1.9, and 2.2 µm. Two secondary analytical Al-OH absorption peaks almost 2.3 and 2.4 µm are improved by Fe and Mg cation swap. Vermiculite has dual broad absorptions at 1.4 and 1.9 µm and dual weak absorptions near 2.2 and 2.3 µm. Montmorillonite has three strong and sharp absorption bands at approximately 1.4, 1.9, and 2.2 µm, which are analogous to but commonly more robust than illite [6]. Consequently, kaolinite is featured by dual spectral doublets: one is approximately 1.4 µm, and the second is close to 2.2 µm.

At a specific depth in the clay, carbonates are seeped from the surface and amass with a period in the soil [5]. Widely, the existence of carbonate is exploited as a keystone of the soil characteristics to identify soil sorts and enumerate soil erosion. Numerous absorptions of carbonates in the VNIR domain are induced by implications and amalgamations of vital vibrations of the CO_3^{2-} ion. In this view, in the NIR region, carbonates have a strong absorption band at approximately 2.35 µm. However, carbonates have three weak absorption bands at approximately 1.9, 2.0, and 2.16 µm. Therefore, carbonates also have double-band structures at approximately 2.35 µm absorption [6, 9].

Generally, water, soil and various biochemical compounds on Earth contain three basic elements as nutrients such as N, P and K. Nitrogen is intimately related to the various ecological and biochemical process. The significant absorption features of N are found in the visible, short wave near-infrared (SWNIR) and near-infrared

(NIR) regions of the spectrum. N absorption bands are located at 2.35, 2.3, 2.18, 2.06 and 1.51 μm wavelengths. The spectral reflectance properties for many biochemical compounds such as Chls, proteins and plant leaves include the N fingerprint. The significant absorption features of P are across the visible, SWNIR and NIR regions. Phosphorous indirectly affects the spectral characteristics of diverse biochemical compounds.

Potassium concentration in leaves has an insignificant effect on needle morphology. The presence of K affects the NIR reflectance. It thickens the sclerenchyma cell walls in leaves and the high concentration of K increases the NIR reflectance. The evaluation of the nutrient concentration is very much guided by the image data and spectral measurements. Several analyses protocols and methods such as derivative analysis, spectral matching and spectral index analysis are developed for accurate assessment of nutrients contents. An efficient model for water quality prediction and evaluation must contain all the nutrients and salinity factors.

7.4 Can Marshlands be Indicator for Mineral Occurrences?

At present, there is no study to investigate the mineral occurrences in marshlands utilizing remote sensing data. Marshlands are rich in soil minerals that drain very slowly. In this view, the clay minerals are formed owing to the accumulations and sedimentation of soil minerals. One of the richest marshlands is the Mesopotamian marshlands of southeastern Iraq shielded over 15,000 km². In the Middle East and western Eurasia, therefore, it is the largest wetland ecosystem. Needless to say that these marshes are developed because of the convergence of the Tigris and Euphrates rivers. Accordingly, they are treasured water reserves, countrywide heritage and conservation habitat of the Middle East. This chapter is developed to investigate the occurrence of the clay minerals in the central marshlands of southern Iraq, which are situated in a triangular zone between the Tigris and Euphrates Rivers (Figure 7.1). One of the well-known marshlands is Al-Hawizeh marshland, which is located in the east of Tigris River. Therefore, Al-Hawizeh marshland is the foremost one as it has the coverage area of exactly 2,500–3,000 km² [10–14].

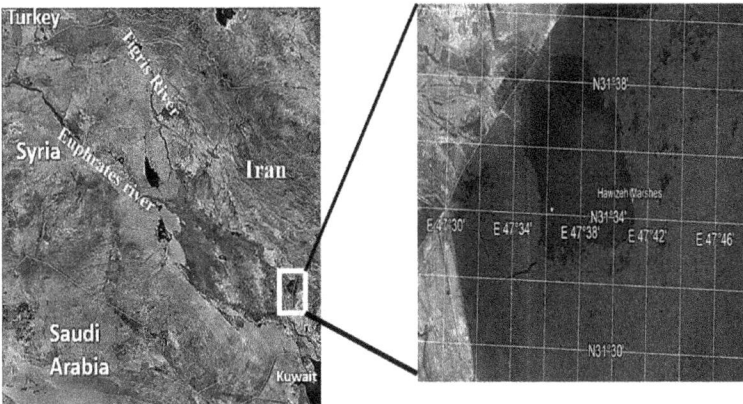

Figure 7.1. Satellite image of the Al-Hawizeh marshland.

Geographically, Al-Hawizeh marshland is sited at longitude 47° 32″–47° 45″ E and latitude 31° 30″–31° 42″ N (Figure 7.2), which is located on the east side of Tigris River between Amarah and Basrah region, which receives the water from Tigris. Al-Hawizeh marsh has a length of 80 km and a width of 30 km, while its depth is approximately ranged from 1.5 to 4 m [15].

Total of 54 *in-situ* field locations are used to collect various samples of the soil and carried to the laboratory to determine different sort of minerals (Figure 7.3). The collected soil samples are homogeneously mixed and put to accomplish equilibrium with air for 2 hours in the trays/paper dishes. Subsequently, drying the soil, clods are creased softly and grounded with the assistance of wooden pestle and mortar. Gravel, soft chalk, limestone, stones and concretions should be removed from the samples.

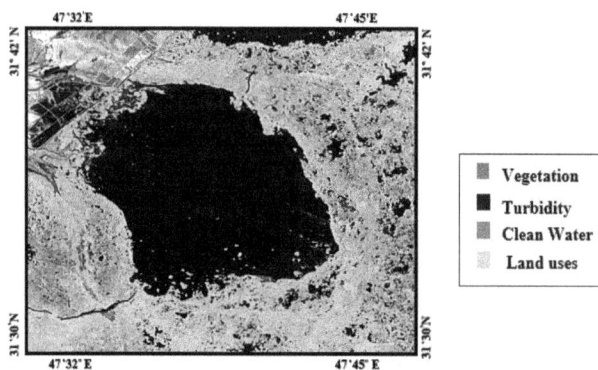

Figure 7.2. Geographical locations of Al-Hawizeh marshland.

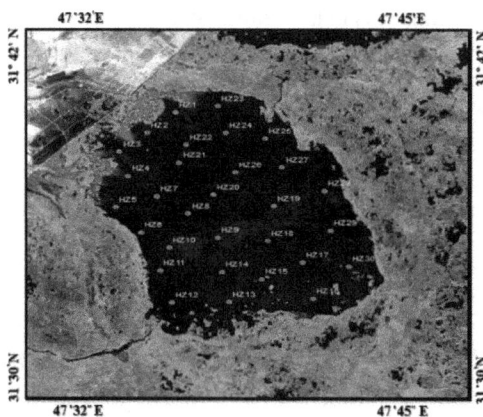

Figure 7.3. *In-situ* soil sample collections.

7.5 How to Compute Cation Exchange Capacity in Laboratory?

The Cation Exchange Capacity (CEC) is the capacity of soil to hold exchangeable cations. The higher the CEC of soil, the more cations it can retain. Soil differ in their capacities to hold exchangeable K^+ and other cations. The cations' exchange capacity depends on the amount and kind of clay and organic matter present. High clay soil can hold more exchangeable cations than low clay soils. CEC also increases as organic matter increases. Clay mineral usually ranges from 10 to 150 meq./ 100 g. In CEC values, organic matter ranges from 200 to 400 meq/100 g. So the kind and amount of clay and organic matter content greatly influence the CEC of soil. Cation exchange is an important reaction in soil fertility in causing and correcting soil acidity and basicity [16, 18].

The cation exchange capacity (CEC) resolve comprises gauging the entire quantity of negative charges per unit weight of the soil which are neutralized by the exchangeable cations. It is demarcated as the capacity of the soils to absorb the total of exchangeable cations articulated as milliequivalents per 100.0 gm of soil or C mol (P^+)/kg of soil on the oven-dry basis. In other words, the sum of cations equals base cations plus acid cations $(Ca + Mg + K + Na) + (H + AI)$, for instance. If the basic cation concentration is low and acidic cation concentration is high, this indicates high degree of weathering and leaching, and the soil is likely to be acidic [17].

Let us assume that the Total Exchangeable Bases is TEB, which equals to exchangeable $Ca + Mg + Na + K$. Thus TEB = $Ca + Mg + Na + K$ (mmoles$_c$/kg). Therefore, the "Total Exchangeable Acids" is TEA, which equals Exch H plus Al, i.e., CEC = TEA + TEB. In this regard,

Percentage Base Saturation (%BS) = TEB/CEC × 100   (7.1)

Percentage Acid Saturation (%AS) = TEA/CEC × 100   (7.2)

The percentage saturation of clay ion $\%C_I$ is mathematically computed as:

$$\%C_I = \frac{C}{CEC \times 100} \tag{7.3}$$

where C is the exchangeable concentration of mineral clay ions, which should be used later to determine any type of clay minerals in optical remote sensing data [16–18].

In this regard, under certain conditions, extremely acidic, with pH < 4.5, hydrous oxides of Fe and Al and some aluminosilicate clays can develop a positive charge. Consequently, charges develop from protonation to the OH group at the exposed crystal edge; for instance, $Al—OH + H^+$ ---------- $Al—OH_2^+$. When positive charges develop on clays, the clays then adsorb anions (anion adsorption). For instance, Cl^-, SO_4^{2-} and $H_2PO_4^-$ can be held this way.

Consequently, clays are frequently negatively charged and hence adsorb cation in solution, which is referred to as clay colloids or micelles. A colloid is formed when one phase is divided within another phase, i.e., a solid (clay) exists within a liquid phase (soil solution). In fact, the clays are small and finely dispersed in the soil solution. Figure 7.4 demonstrates that different cations are adsorbed around a clay colloid. In this sense, cations in the soil are from rock weathering and minerals, fertilisers and the breakdown of organic matter. Subsequently, preferential order of adsorption is $Al^{3+} > Ca^{2+} > Mg^{2+} > K^+ > Na^+ > H^+$. Indeed, clay will absorb Al^{3+} in preference to Ca^{2+} if the equal concentration of each cation is present. In this

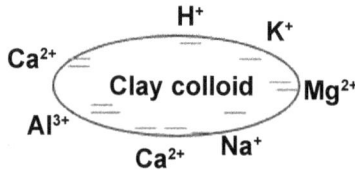

Figure 7.4. A concept of clay colloid.

understanding, cations are adsorbed, meaning they are loosely held at the clay surfaces and thus can be exchanged with other cations in the soil solution [17, 19].

If a clay micelle with calcium ions is exposed to high concentrations of K ions, the Ca ions will be exchanged for the K ions, which is known as cation exchange (Figure 7.5). Needless to say that the reaction is balanced to satisfy all the negative charges. Furthermore, the cation exchange reaction shows the effect of adding lime (Figure 7.6).

Clay particles and organic matter carry negative charges over their surface owing to which they adsorb positively charged particles (cations). There are dissimilar sorts of cations, e.g., H^+, Ca^{++}, K^+, NH_4^+, Mg^{++}, etc. These cations can swap each other relying upon their concentration (mass action) and replacing power. The replacement of cations by one another is known as cation exchange. The order of replacing the power of some cations is given as $H^+ > Ca^{++}. Mg^{++} > K^+ > NH_4^+ > Na^+ > Si^{++++}$ [17–19].

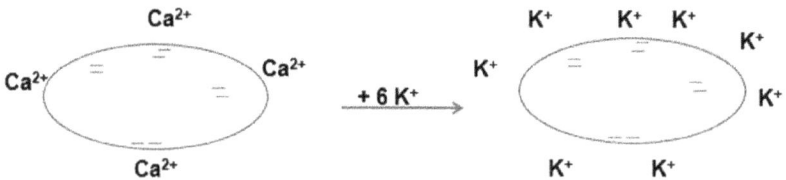

Figure 7.5. Concept of cation exchange.

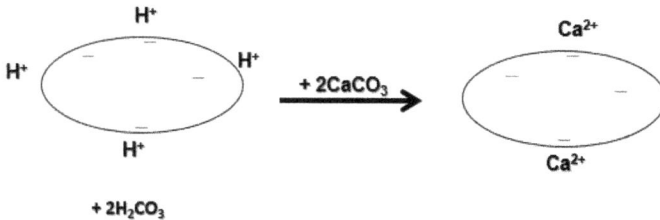

Figure 7.6. Lime adding to cation reaction.

7.6 Sentinel-2 Satellite Data

Sentinel-2 (S2) multi-spectral instrument (MSI) collects multispectral data with 13 bands in the visible/near-infrared (VNIR), and shortwave infrared (SWIR) regions of the spectrum, every 5 days or less. These data can be obtained free of charge from https://glovis.usgs.gov/app. It is different from traditional multispectral satellite sensors such as Landsat, by including three vegetation red edge bands.

Blue (0.490 μm), Green (0.560 μm), Red (0.665 μm) and NIR (0.842 μm) bands are collected at 10 m spatial resolution, while the three Vegetation Red Edge bands (0.705 μm, 0.740 μm, and 0.783 μm), Narrow NIR band (0.865 μm), and two shortwave infrared bands (1.610 μm and 2.190 μm) collect data at 20 m spatial resolution.

We used S2 MSI Level 1C data representing TOA (top of atmosphere) reflectance scaled by 10,000 available in GEE. The S2-TOA imagery was used because of the higher data accessibility compared to processed S2-Surface Reflectance (SR) data that was not available for the 2017 timeframe. All 10 m bands of S2 images were resampled to 20 m to be consistent with the 20 m bands of Sentinel-2 images [20].

The MSI measures reflected radiance through the atmosphere within 13 spectral bands. The spatial resolution is dependent on the particular spectral band:

- 4 bands at 10 meter: blue (0.490 μm), green (0.560 μm), red (0.665 μm), and near-infrared (0.842 μm).

- 6 bands at 20 meter: 4 narrow bands for vegetation characterization (0.705 μm, 0.740 μm, 0.783 μm, and 0.865 μm) and 2 larger SWIR bands (1.610 μm and 2.190 μm) for applications such as snow/ice/cloud detection or vegetation moisture stress assessment.

- 3 bands at 60 meter: mainly for cloud screening and atmospheric corrections (443 nm for aerosols, 0.945 μm for water vapour, and 1.375 μm for cirrus detection).

Table 7.1. Characteristics of Sentinel-2 satellite data.

Band number	Central wavelength (μm)	Bandwidth (μm)	Spatial resolution (m)
1	0.433–0.453	0.020	60
2	0.458–0.523	0.065	10
3	0.543–0.578	0.035	10
4	0.650–0.680	0.030	10
5	0.698–0.713	0.015	20
6	0.733–0.748	0.015	20
7	0.773–0.793	0.020	20
8	0.785–0.900	0.115	10
8a	0.855–0.875	0.020	20
9	0.935–0.955	0.020	60
10	1.365–1.385	0.030	60
11 (SWIR-1)	1.565–1.655	0.090	20
12 (SWIR-2)	2.100–2.280	0.180	20
TCI*	RGB	Composite	10

7.7 How to Retrieve Clay Potential Percentage in Remote Sensing Data?

This section developed a novel algorithm to determine clay saturation in optical remote sensing data. Both multispectral and hyperspectral images have dual bands of short-wave infrared (SWIR), SWIR-1 and SWIR-2 bands, respectively. In Sentinel-2, SWIR-1 ranges between 1.565 and 1.655 µm, while SWIR-2 ranges between 2.100 and 2.280 µm. Retrieving percentage saturation of clay in SWIR-1 and SWIR-2 bands is required to derive a new algorithm. To this end, let us assume that the percentage saturation of clay C_I is embedded in both SWIR-1 and SWIR-2 bands as $C_I \in SWIR_{1,2}$ and probability $p \le 0.5$. From the point of view of the quantum mechanics [21–23], let $\Psi|C_I\rangle \alpha \Psi|SWIR_1\rangle|SWIR_2\rangle^{-1}$. In this understanding, both SWIR-1 and SWIR-2 must be in the form of the spectral reflectance $|S\rangle_r \sim |SWIR_1\rangle|SWIR_2\rangle^{-1}$. Following this expectation, the quantum state of the percentage saturation of clay probability occurrences $\Psi|C_I\rangle$ is proportional directly with the quantum state of spectral reflectance of $|S\rangle_r$, i.e., $\Psi|C_I\rangle \alpha \Psi|S\rangle_r$. Both circumstances can be expressed as:

$$\Psi|C_I\rangle \alpha \Psi|SWIR_1\rangle|SWIR_2\rangle^{-1}$$

or   (7.4)

$$\Psi|C_I\rangle \alpha \Psi|S_r\rangle$$

Equation 7.4 can be solved in linear form by implanting constant values $|\hat{b}\rangle$ for the right side of the equation as follows:

$$\Psi|C_I\rangle = |\hat{b}\rangle + \sum_{i=1}^{k}|S_r\rangle_i \otimes \left[\frac{\partial|C_I\rangle_i}{\partial|S_r\rangle_i}\right]$$ (7.5)

Given a Hermitian $K \times k$ matrix H and a unit vector $\vec{h}$ and $\vec{x}$ is satisfying:

$$H\hat{b} = \vec{h}$$ (7.6)

where

$$H = \begin{bmatrix} n & \sum S_{2j} & .... & \sum S_{kj} \\ \sum S_{2j} & \sum S_{2j}^2 & .... & \sum S_{2j}^2 S_{kj} \\ .... & .... & .... & .... \\ \sum S_{kj} & \sum S_{kj}S_{2j} & .... & \sum S_{kj}^2 \end{bmatrix}$$ (7.6.1)

$$\hat{b} = \begin{bmatrix} \hat{b}_1 \\ \hat{b}_2 \\ : \\ : \\ \hat{b}_k \end{bmatrix}$$ (7.6.2)

and

$$\vec{h} = \begin{bmatrix} \sum \Psi|C_I\rangle_j \\ \sum \Psi|C_I\rangle_j |S_r\rangle_{2j} \\ \vdots \\ \sum \Psi|C_I\rangle_j |S_r\rangle_{kj} \end{bmatrix} \tag{7.6.3}$$

Thus, H is presented the estimated matrix $k \times k$ estimated for both spectral radiance $|S_r\rangle$ of SWIR-1 and SWIR-2 data, respectively, that is used to find out the percentage saturation of clay C_I. Consequently, both $\vec{b}$ and $\vec{h}$ are $k \times 1$ column vectors. The solution to the least-squares normal equation is

$$\hat{b} = H^{-1}\vec{h} \tag{7.7}$$

Given a solution of least-squares normal equations, the retrieval $\Psi|C_I\rangle$ is computed using the fitted multiple regression model as:

$$\Psi|C_I\rangle = \left[\sum_{i=1}^{k} |S_r\rangle_i \otimes \left[\frac{\sum_{i=1}^{k}(|S_r\rangle_i - |\bar{S}_r\rangle) \otimes (|C_I\rangle_i - |\bar{C}_I\rangle)}{\sum_{i=1}^{k}(|S_r\rangle_i - |\bar{S}_r\rangle)^2} \right] \right] \oplus \left[|\bar{C}_I\rangle - |\bar{S}_r\rangle \otimes \left(\frac{\sum_{i=1}^{k}(|S_r\rangle_i - |\bar{S}_r\rangle) \otimes (|C_I\rangle_i - |\bar{C}_I\rangle)}{\sum_{i=1}^{k}(|S_r\rangle_i - |\bar{S}_r\rangle)^2} \right) \right] \pm$$

$$\left[\frac{\sum_{i=1}^{k}\sqrt{\left[\Psi|C_I\rangle - \Psi|C_{lab}\rangle\right]^2}}{k} \right] \tag{7.8}$$

where $|S_r\rangle$ is computed by:

$$|S_r\rangle = \frac{\sum_{i=1}^{k} \left(\frac{|S(\lambda_i)\rangle_{max}}{254} - \frac{|S(\lambda_i)\rangle_{min}}{255} \right) \otimes |DN\rangle_i + |S(\lambda_i)\rangle_{min}}{\sum_{j=1}^{k} \left(\frac{|S(\lambda_j)\rangle_{max}}{254} - \frac{|S(\lambda_j)\rangle_{min}}{255} \right) \otimes |DN\rangle_j + |S(\lambda_j)\rangle_{min}} \tag{7.8.1}$$

Equation 7.8 is the general derived equation to estimate the percentage saturation of clay C_I in any multispectral or hyperspectral data. A linear hyperbolic quantum algorithm is considered to determine the variability of percentage saturation of clay C_I in optical remote sensing data. This quantum algorithm aims at uncertainty reduction in computing percentage saturation of clay C_I. This algorithm can be titled as "quantized Marghany clay saturation algorithm".

7.8 Quantized Marghany Clay Saturation Algorithm in Al-Hawizeh Marsh

Figure 7.7 demonstrates the clay quantum spectral reflectance as part of computing the percentage of clay saturation in SWIR-1 and SWIR-2 of Sentinel-2 satellite data.

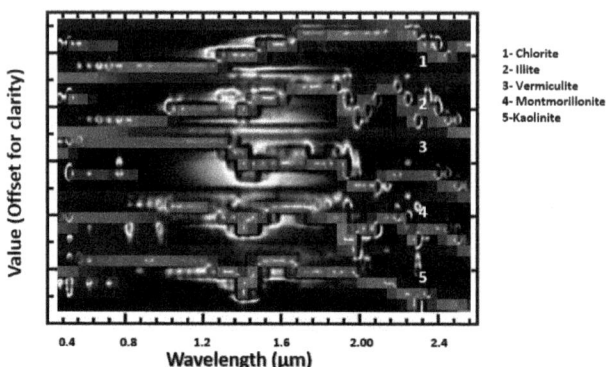

Figure 7.7. Quantized Marghany clay saturation algorithm along Al-Hawizeh marshland.

In this sense, the foremost clays in soils indicate indicative absorptions in the SWIR spectral domain. Therefore, these absorption spectral bands are created owing to vibrational transitions and frequently reveal piercing and fine absorption features (Figure 7.7). Subsequently, the diagnostic spectral reflectance bands are primarily absorbed on approximately 1.400 μm, which are generally formed by OH. Moreover, water molecules forms spectral reflectance at approximately 1.900 μm, while AI-OH causes spectral reflectance at approximately 2.200 μm.

The existence of Fe-OH and/or Mg-OH in the clay minerals is causing weak absorption spectral bands in approximately the domain of 2.300 to 2.500 μm. In the circumstance of water occurrences in Al-Hawizeh marsh, weak reflectance is shown near 1.468 μm and 1.970 μm, which is an excellent indicator for montmorillonite spectra. On top of that, the dominated dual spectral of Kaolinite is shown near 1.400 μm and 2.200 μm, respectively. In this view, chlorites are a cluster of clay minerals comprising precise octahedral cations, for instance Fe, Mg, and Al, which reveal a weak absorption band at approximately 1.400 μm. On the other hand, at the spectral domain of 2.250 and 2.350 μm, respectively, triple absorption features occur owing to Fe-OH and Mg-OH, correspondingly. Besides, at 1.400 μm, 1.900 μm, and 2.200 μm, respectively, three bulging absorptions occur owing to the existence of illite. However, at 1.400 μm, 1.900 μm, and 2.200 μm, montmorillonite has sharper and stronger absorption peaks than illite. Therefore, near 2.200 and 2.300 μm, vermiculite has two weak absorptions, while at 1.400 and 1.900 μm, it has broad absorptions. Consequently, Fe and Mg modify dual secondary diagnostic Al-OH absorption peaks almost 2.344 and 2.445 μm.

Figure 7.8 reveals the percentage rate of the saturation of clay in the Al-Hawizeh marsh on June 30th, 2019. The surrounding marsh closed to the Al-Hawizeh marsh is dominated by the maximum percentage rate of the saturation of clay of approximately 60%. Al-Hawizeh marsh water pool shows zero percentage rate of the saturation of clay at the centre of the pool. Indeed, water molecules cause weak reflectance near 1.468 μm and 1.970 μm as shown in Figure 7.7. Urban zone at the top end left corner of Figure 7.8 is also free of clay accumulations. The 60% of the clay rate saturation is also found in shallow pool water owing to clay discharges from vegetation covers into the Al-Hawizeh water pool.

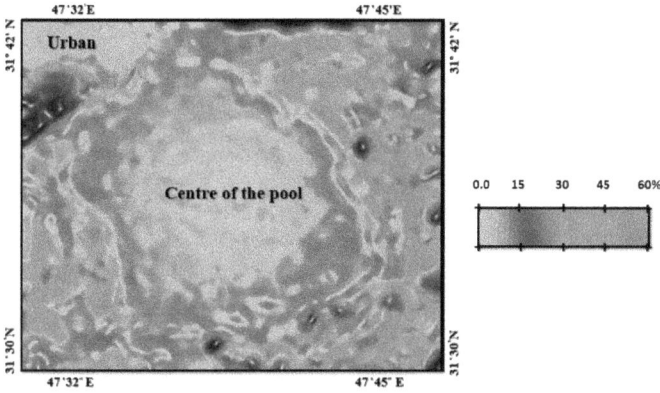

Figure 7.8. The percentage rate of the saturation of clay in the Al-Hawizeh marsh on June 30th, 2019.

Consequently, the 100% rate of the saturation of clay in the Al-Hawizeh marsh is observed during winter time of December 22nd, 2019 (Figure 7.9). Considerable influences of the high rate of precipitation during the winter period assist in accumulating high rate of clay than the summer season. On top of that, urban zone is also dominated by the high rate of saturation of the clay, which is approximately 80%.

Validation of the quantized Marghany clay saturation algorithm is achieved using the root mean square error (RMSE), which is formulated as:

$$RMSE = \sqrt{n^{-1}\sum_{i=1}^{n}\left[\%C_I - \Psi\left|\%C_I\right\rangle_{M\,\text{arg}\,hany}\right]^2} \tag{7.9}$$

where n presents the number of the tested samples. $\%C_I$ is laboratory estimated one and $\Psi\left|\%C_I\right\rangle_{M\,\text{arg}\,hany}$ is quantized Marghany clay saturation algorithm. In this view, the quantized Marghany clay saturation algorithm can determine the accurate percentage rate of the saturation of clay from the Sentinel-2 satellite image, which is proved by r^2 of 0.92, $p < 0.00005$ and root mean square error (RMSE) of ±0.32% (Figure 7.10).

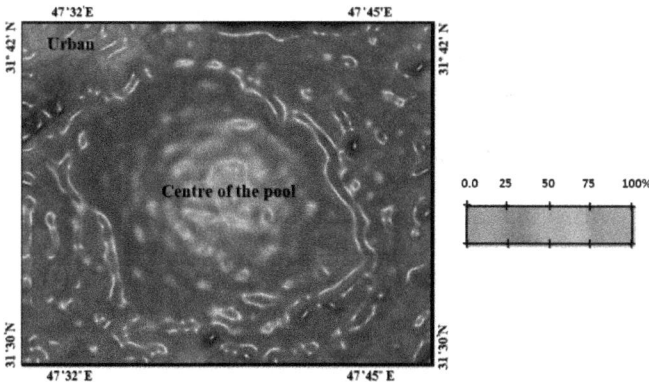

Figure 7.9. The percentage rate of the saturation of clay in the Al-Hawizeh marsh on December 22nd, 2019.

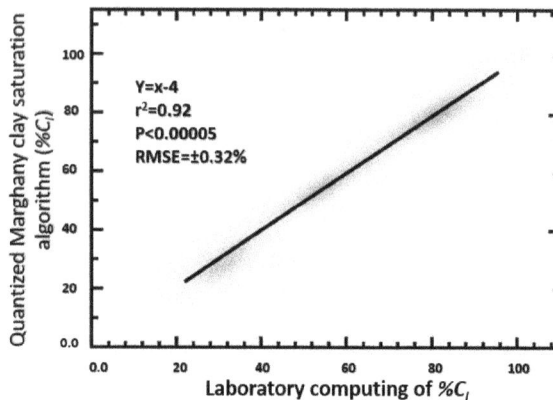

Figure 7.10. Accuracy of quantized Marghany clay saturation algorithm.

The quantized Marghany clay saturation algorithm reduces the error in computing the rate of the clay saturation in such multispectral data of the Sentinel-2 satellite image. In this view, quantized Marghany clay saturation algorithm entangled quantized spectral reflectance of Sentinel-2 data with laboratory computing framework of clay saturation in the *in-situ* collected samples. In this circumstance, the quantized Marghany algorithm behaves as accurately matched algorithms with fewer errors in which the algorithm itself involves the reduction of the error rate as demonstrated in Equation 7.8. In other words, the quantized Marghany algorithms entangle two quantum state of clay occurrences: one computed as based on the laboratory sample analysis and the other where the quantum state is encoded as spectral clay absorption in SWIR-1 and SWIR-2 domains.

7.9 Support Vector Machines

Now the question is: what is meant by support vector machines (SVM)? A support vector machine (SVM) is a kind of deep learning algorithm that executes supervised learning for cataloguing or regression of data clusters. In this view, an SVM makes up a learning model that allocates new illustrations to one cluster or another. In these operations, SVMs are well-known as a non-probabilistic, binary linear classifier. Consequently, SVMs can exploit approaches, for instance, Platt Scaling through probabilistic classification settings [24–26].

Corresponding to other supervised learning machines, an SVM necessitates categorized data to be accomplished. Clusters of training data are considered for classification. Training data for SVMs are catalogued discretely in dissimilar themes in space and systematized into evidently detached clusters. Post handing out abundant training samples, SVMs can behave as an unsupervised learning algorithm. In this understanding, the accurate separation of training data with the frontier about the hyperplane being exploited and smooth between both edge sides in SVMs of selected trained data. However, the hyperplane that separates the dataset with the maximum margin is considered the best one because it increases generalization to the new dataset. SVM is sensitive to noise (mislabeled) data, thus the accuracy is lowered for non-labeled data with the maximum margin (Figure 7.11). The separating hyperplane is defined as:

$$w^T \phi(x) + b = 0 \tag{7.10}$$

where w is the decision hyperplane normal vector, X_i is a vector of the dataset mapped to a high dimensional space with i = 1, 2,....n, yi is the class label of data point i (+1 or −1), ξ is the classification error, and ρ is the margin separator called the width of separation between support vectors of classes given by $\rho = \dfrac{2}{\|w\|}$.

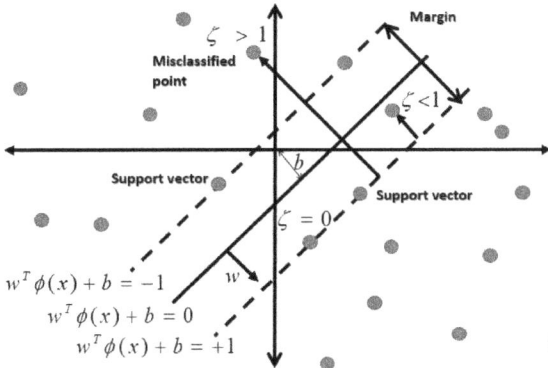

Figure 7.11. SVM classifications with support vectors' critical point close to the decision boundary.

Now, how does SVM work? It exploits a procedure termed the kernel trick to transform input training data and then grounded on these transformations, it acquires an optimal boundary between the probable outputs. A support vector machine is also known as a support vector network (SVN). This method of image classification is accurate and efficient for remote sensing. SVM needs small amounts of training data that is located in the area of feature space near interclass boundaries. Although SVM classification needs the only sample per class, the sample must be close to the boundary of the class. Moreover, maximum likelihood (ML) classification requires a large training sample size, especially when the data occurs in high dimensional feature space [25, 28].

Figures 7.12 and 7.13 demonstrate the percentage rate of clay saturations that are delivered by SVM from Sentinel-2 satellite data in Al-Hawizeh marsh. Like the quantized Marghany clay saturation algorithm, the spatial variation of the percentage rate of the clay saturations shows high concentration during wintertime (Figure 7.13). On the other hand, SVM in comparison with laboratory work shows highest RMSE of 1.5% than the quantized Marghany clay saturation algorithm and lower r^2 value of 0.75 (Figure 7.14). Indeed, the SVM requires to specify a kernel function as well as the slow relation development of multiclass. Unlike the quantized Marghany algorithm, SVM is unable to reduce the standard errors owing to the increment of training samples through classification procedures. However, the major advantage of SVM is that it will not consume time for collecting data during the training.

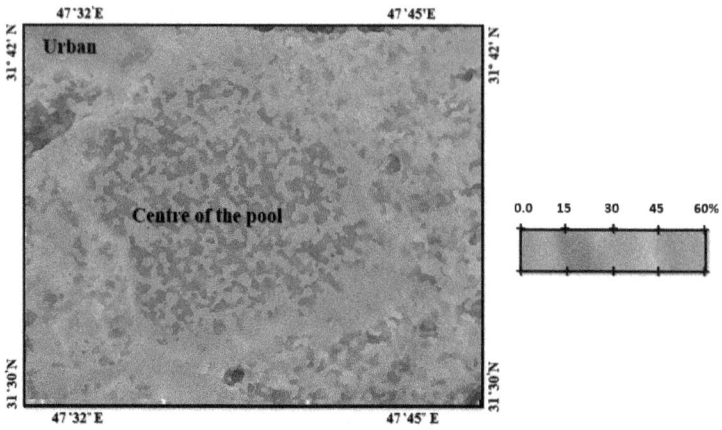

Figure 7.12. SVM for computing percentage rate of the saturation of clay in the Al-Hawizeh marsh on June 30th, 2019.

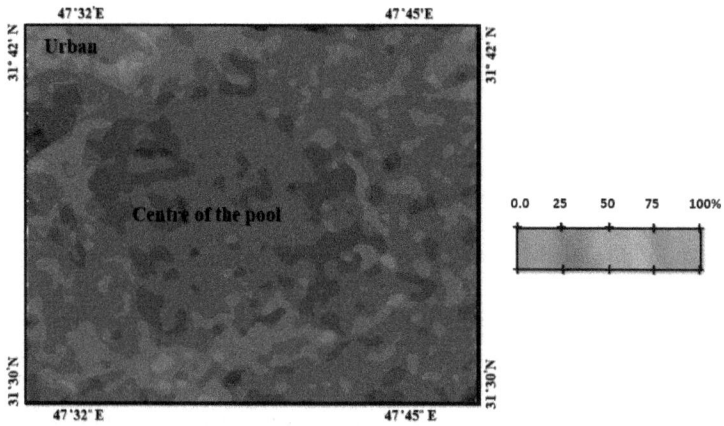

Figure 7.13. SVM for computing percentage rate of the saturation of clay in the Al-Hawizeh marsh on December 22nd , 2019.

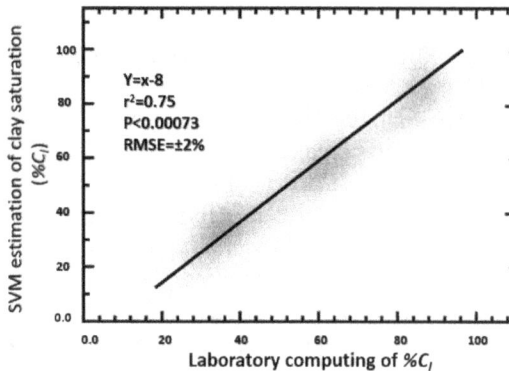

Figure 7.14. Accuracy of SVM.

7.10 Quantum Support Vector Machines

Let us assume the training data for the percentage rate of quantized Marghany's clay saturation algorithm are $\Psi|\%C_I\rangle_{M \text{ arg } hany}$, $\{(\%\vec{C}_{1_j}, \Psi|\%C_I\rangle_{M \text{ arg } hany_j}): \%\vec{C}_{1_j} \in \mathbb{R}^d$, $\Psi|\%C_I\rangle_{M \text{ arg } hany_j} = \pm 1\}$, and a kernel k. In this view, identification of the classified training data of quantized Marghany's clay saturation can be formulated as:

$$\Psi|\%\vec{C}_I\rangle_{M \text{ arg } hany} = \text{sgn}\left(\sum_{j=1}^{m} \alpha_j k\left(\left|\%\vec{C}_I\right\rangle_{M \text{ arg } hany_j}, \left|\%\vec{C}_I\right\rangle_{M \text{ arg } hany}\right) + b\right). \tag{7.11}$$

In the foundation of training set labels of $\Psi|\%\vec{C}_I\rangle_{M \text{ arg } hany}$, the anticipated SVM parameters are encoded in the amplitudes of the ultimate state-run:

$$|b, \vec{\alpha}\rangle = \left[b^2 + \sum_{k=1}^{m} \alpha_k^2\right]^{-0.5}\left(b|0\rangle + \sum_{k=1}^{m} \alpha_k|k\rangle\right), \tag{7.12}$$

Then there exists a quantum algorithm that solves the training problem of the least-square support vector machine within the accuracy ε in the time $O(\log(md)\kappa^2\varepsilon^{-3})$. Determination of the new classified clay features can be expressed mathematically as:

$$\Psi|\%\vec{C}_I\rangle_{M \text{ arg } hany} |0\rangle|0\rangle = \sum_{j=1}^{m+1} \langle u_j|\%C_{M \text{ arg } hany}\rangle|0\rangle|0\rangle$$

$$\xrightarrow[result=1]{Compute} \sum_{j=1}^{m+1} \frac{\langle u_j|\%C_{M \text{ arg } hany}\rangle}{\tilde{\lambda}_j}|u_j\rangle \tag{7.13}$$

Equation 7.13 says that $\tilde{\lambda}_j$ is the eigenstate of the Hamiltonian matrix, and unitary u_j can be given by:

$$|u\rangle = \left[b^2 + \sum_{k=1}^{m} \alpha_k^2|\%C_{I_k}|^2\right]^{-0.5}\left(b|0\rangle|0\rangle + \sum_{k=1}^{m} \alpha_k|\%\vec{C}_I||k\rangle|\%\vec{C}_{I_k}\rangle\right) \tag{7.14}$$

$$|\%C_{M \text{ arg } hany}\rangle = \left[M|\%C_{M \text{ arg } hany}|^{-2}\right]^{-0.5}\left(|0\rangle|0\rangle + \sum_{k=1}^{m}|\%\vec{C}_I||k\rangle|\%\vec{C}_I\rangle\right) \tag{7.15}$$

In Equation 7.15, the Hermitian matrix M can be expressed as $\langle\%C_{M \text{ arg } hany}| M |\%C_{M \text{ arg } hany}\rangle$. If we calculate only the kernel matrix by quantum means, we have a complexity of $O(M^2(M + \varepsilon^{-1} \log N))$. There is more to gain: exponential speedup in the number of training examples is possible when using the least-squares formulation of support vector machines. In this circumstance, obtaining the normalized kernel matrix can assist in an exponential speedup of QSVM as:

$$\hat{k} = \frac{1}{tr(k)}\sum_{i,j=1}^{m}\langle\%C_{M \text{ arg } hany_i}|\%C_{M \text{ arg } hany_j}\rangle\|\%C_{M \text{ arg } hany_i}\|\%C_{M \text{ arg } hany_j}\||i\rangle\langle j|. \tag{7.16}$$

$\hat{k}$ is a normalized Hermitian matrix, which makes it a prime candidate for quantum self-analysis. The exponentiation, therefore, is completed procedure of $O(\log N)$.

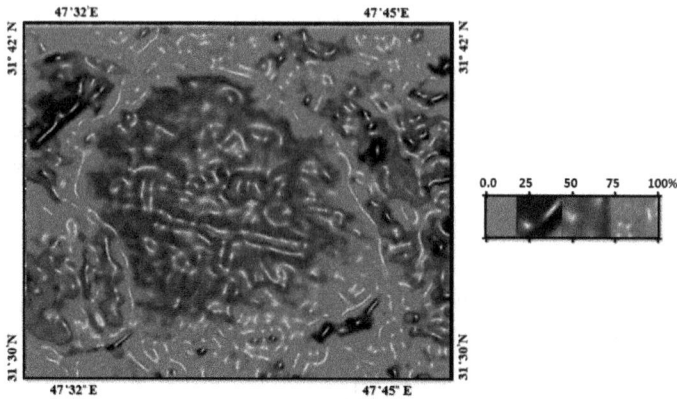

Figure 7.15. QSVM clay saturation results.

Figure 7.15 demonstrates the determination of the quantized Marghany's clay saturation percentage rate involving QSVM. It is worth noting that QSVM improved the cluster spatial variations of the quantized Marghany algorithms in which the boundaries between different clusters of clay saturation percentage rate are extremely identified in the Sentinel-2 SWIR-1 and SWIR-2 spectral domains. In other words, the sharpest distinguished cluster boundaries are well identified. Like quantized Marghany's clay saturation algorithm, QSVM delivers a maximum clay saturation rate of 100% in the central pool during December 22nd, 2019. However, QSVM improves the quantized Marghany's clay saturation algorithm by identifying clear clay concentration pattern on the centre of the pool owing to clay discharging from land into the water pool of Al-Hawizeh.

7.11   Why Does QSVM Entangle Quantized Marghany's Clay Saturation Algorithm?

QSVM demonstrates higher accuracy than SVM with less RMSE of ±0.25% (Figure 7.16). On the other hand, QSVM improves the accuracy of the quantized Marghany's clay saturation algorithm in identifying clear cluster features owing to an error reduction difference of ±0.07%. In this sense, QSVM reveals extreme agreement with the quantized Marghany's clay saturation algorithm with r^2 value of 0.98 and RMSE ± 0.089% (Figure 7.17), which is approximately close to the differences error estimation between both QSVM and the quantized Marghany's clay saturation algorithm against laboratory clay saturation percentage rate estimation.

Consequently, QSVM involves Grover's search to replace sequential minimum optimization in a discretized search space. Also, involving least-squares support vector machines translate an optimization problem into a set of linear equations, which requires the quick computation of the kernel matrix—this is one source of the speedup in the quantum version. Relying on this formulation, and assuming quantum input and output space, quantum support vector machines can achieve an exponential speedup over their classical counterparts [28–30].

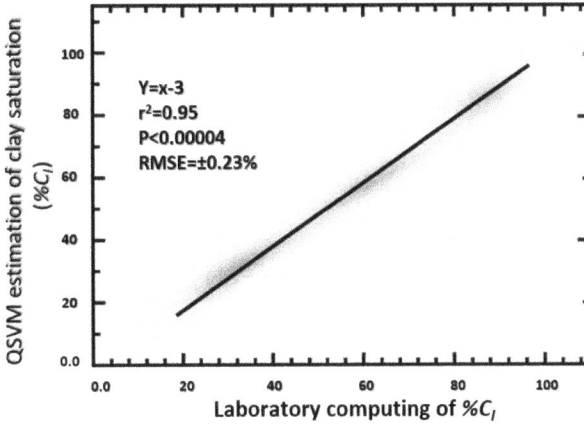

Figure 7.16. Accuracy of QSVM algorithm.

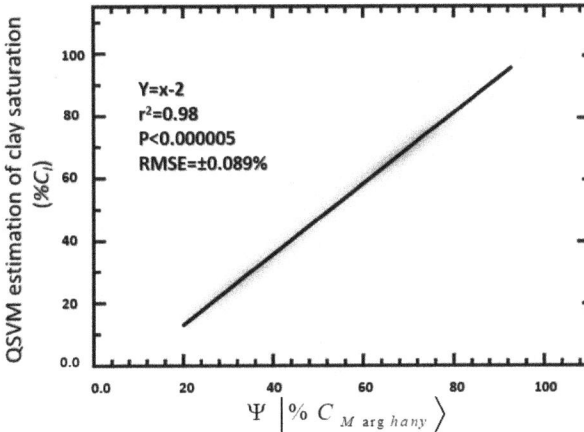

Figure 7.17. Correlation between QSVM and the quantized Marghany's clay saturation algorithm $\Psi|\%C_{M\,arg\,hany}\rangle$.

Particularly, QSVM is entangled with quantized Marghany's clay saturation algorithm and then utilized to classify a new-coming clay saturation percentage rate. In this understanding, this effort has established the possibility of performing QSVM and quantized Marghany's clay saturation algorithm in a near-term quantum computer. However, further efforts are required to improve quantum hardware and error corrections to accomplish a huge-gauge testing satellite data.

To this end, this chapter demonstrates the novel quantum algorithm in identifying clay saturation percentage rate as a function of SWIR-1 and SWIR-2 of Sentinel-2. This novel quantum algorithm is named "quantized Marghany clay saturation algorithm". The highest accurate achievement of the quantized Marghany clay saturation algorithm is proved by an RMSE of $\pm0.32\%$. Needless to say that QSVM entangles with quantized Marghany clay saturation algorithm with r^2 value of 0.98. The next chapter will introduce a novel quantum computing algorithm for oil seep automatic detection.

References

[1] Schulze DG. An introduction to soil mineralogy. Soil Mineralogy with Environmental Applications. 2002 Jan 1; 7: 1–35.

[2] Dixon JB, Schulze DG. Soil mineralogy with environmental applications. Soil Science Society of America Inc.; 2002.

[3] Ketterings QM, Bigham JM, Laperche V. Changes in soil mineralogy and texture caused by slash-and-burn fires in Sumatra, Indonesia. Soil Science Society of America Journal. 2000 May; 64(3): 1108–17.

[4] Amonette JE, Zelazny LW. Quantitative Methods in Soil Mineralogy. Madison, WI: Soil Science Society of America.

[5] Foth HD. Fundamentals of soil science. Soil Science. 1978 Apr 1; 125(4): 272.

[6] Fang Q, Hong H, Zhao L, Kukolich S, Yin K, Wang C. Visible and near-infrared reflectance spectroscopy for investigating soil mineralogy: A review. Journal of Spectroscopy. 2018 Jan 1; 2018.

[7] Karathanasis AD. Subsurface migration of copper and zinc mediated by soil colloids. Soil Science Society of America Journal. 1999 Jul; 63(4): 830–8.

[8] Karathanasis AD. Mineral equilibria in environmental soil systems. Soil Mineralogy with Environmental Applications. 2002 Jan 1; 7: 109–51.

[9] Hunt GR, Ashley RP. Spectra of altered rocks in the visible and near infrared. Economic Geology. 1979 Nov 1; 74(7): 1613–29.

[10] Green EA. Hydropolitics in the Middle East and US policy. NAVAL WAR COLL NEWPORT RI; 1993 Jun.

[11] Nicholson E, Clark P. Iraqi Marshlands. Politico's Pub.; 2003.

[12] Al-Yamani FY, Bishop JM, Al-Rifaie K, Ismail W. The effects of the river diversion, Mesopotamian Marsh drainage and restoration, and river damming on the marine environment of the northwestern Arabian Gulf. Aquatic Ecosystem Health & Management. 2007 Sep 14; 10(3): 277–89.

[13] Zhang H, Abed FH. Development of a GIS database for Iraqi marshlands ecosystem studies. p. 749241. In International Symposium on Spatial Analysis, Spatial-Temporal Data Modeling, and Data Mining 2009 Oct 15 (Vol. 7492). International Society for Optics and Photonics.

[14] Grabe NH. A review of the water quality of the Mesopotamian (southern Iraq) marshes prior to the massive desiccation of the early 1990s. Marsh Bulletin. 2009; 4(2): 98–120.

[15] Al-Handal A, Hu C. MODIS observations of human-induced changes in the Mesopotamian Marshes in Iraq. Wetlands. 2015 Feb 1; 35(1): 31–40.

[16] Aprile F, Lorandi R. Evaluation of cation exchange capacity (CEC) in tropical soils using four different analytical methods. Journal of Agricultural Science. 2012 Jun 1; 4(6): 278.

[17] Lorenz PM, Meier L, Kahr G. Determination of the cation exchange capacity (CEC) of clay minerals using the complexes of copper (II) ion with triethylenetetramine and tetraethylenepentamine. Clays and Clay Minerals. 1999; 47(3): 386–8.

[18] Hendershot WH, Lalande H, Duquette M. Ion exchange and exchangeable cations. Soil Sampling and Methods of Analysis. 1993; 19: 167–76.

[19] Rhoades JD. Cation exchange capacity. Methods of Soil Analysis: Part 2 Chemical and Microbiological Properties. 1983 Feb 1; 9: 149–57.

[20] Liu X, Fatoyinbo TE, Thomas NM, Guan W, Zhan Y, Mondal P, Lagomasino D, Simard M, Trettin CC, Deo R, Barenblitt A. Large-scale high-resolution coastal mangrove forests mapping across West Africa with machine learning ensemble and satellite big data. Frontiers in Earth Science. 2020; 8: 677.

[21] Wang G. Quantum algorithm for linear regression. Physical Review A. 2017 Jul 31; 96(1): 012335.

[22] Schuld M, Sinayskiy I, Petruccione F. Prediction by linear regression on a quantum computer. Physical Review A. 2016 Aug 30; 94(2): 022342.

[23] Gilyén A, Song Z, Tang E. An improved quantum-inspired algorithm for linear regression. arXiv preprint arXiv:2009.07268. 2020 Sep 15.

[24] Balabin RM, Lomakina EI. Support vector machine regression (LS-SVM)—an alternative to artificial neural networks (ANNs) for the analysis of quantum chemistry data? Physical Chemistry Chemical Physics. 2011; 13(24): 11710–8.

[25] Noble WS. What is a support vector machine? Nature Biotechnology. 2006 Dec; 24(12): 1565–7.

[26] Wang L (ed.). Support Vector Machines: Theory and Applications. Springer Science & Business Media; 2005 Jun 21.

[27] Ding SF, Qi BJ, Tan HY. An overview on theory and algorithm of support vector machines. Journal of University of Electronic Science and Technology of China. 2011 Jan; 40(1): 2–10.

[28] Ding C, Bao TY, Huang HL. Quantum-inspired support vector machine. arXiv preprint arXiv:1906.08902. 2019 Jun 21.

[29] Zhu X, Xiong J, Liang Q. Fault diagnosis of rotation machinery based on support vector machine optimized by quantum genetic algorithm. IEEE Access. 2018 Jan 17; 6: 33583–8.

[30] Ezawa M. Variational Quantum Support Vector Machine based on Γ matrix expansion and Variational Universal-Quantum-State Generator. arXiv preprint arXiv:2101.07966. 2021 Jan 20.

CHAPTER 8

Automatic Detection of Oil Seeps in Synthetic Aperture Radar Using Quantum Immune Fast Spectral Clustering

8.1 What are Oil Seeps?

An oil seep is a natural leak of crude oil [1] (Figure 8.1). Therefore, what is the difference between the oil spill and natural oil seep? Bear this in mind, an oil spill is not caused just by accidents or as a consequence of extreme weather events disrupting infrastructures; the ocean's floor also naturally releases oil. Active seepage takes place when subsurface hydrocarbons penetrate shallow sediments and the overlying water column (Figure 8.2); this happens in basins where hydrocarbons are actively generating or contain migration pathways [1]. Oil naturally migrates through cracks from deep deposits below the ocean floor, reaching in some cases surface waters, generating oil slicks [2].

Over time, natural seeps discharge oil gradually, tolerating ecosystems to acclimate. However, an oil spill is the release of a liquid petroleum hydrocarbon into the environment, especially the marine ecosystem, due to human activity,

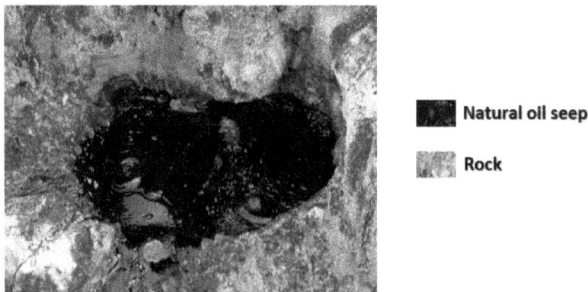

■ Natural oil seep

▒ Rock

Figure 8.1. Natural oil seep.

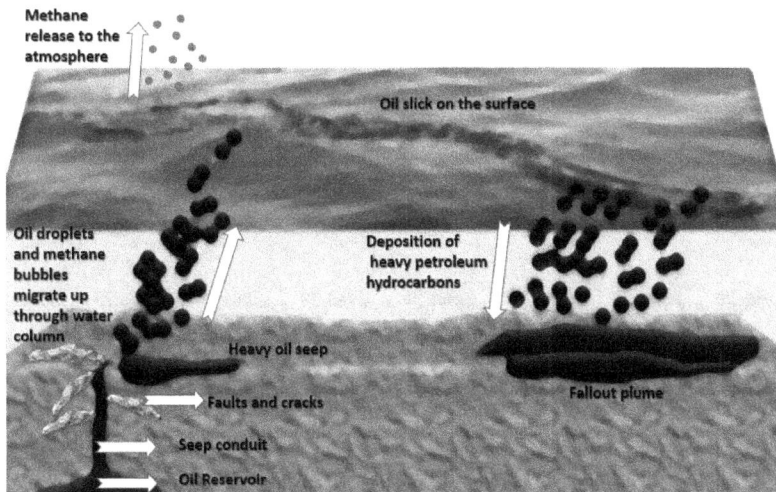

Figure 8.2. Formation of the natural oil seeps in the seafloor.

Figure 8.3. Oil spill on the sea surface.

and is a form of pollution. The term is usually given to marine oil spills (Figure 8.3), where oil is discharged into the ocean or coastal waters, but spills perhaps correspondingly occur on land.

In locations where seeps are found, oil flows slowly up through networks of cracks, forming springs of hydrocarbons similar to the La Brae tar pits on land. Lighter compounds rise buoyantly to the water's surface and evaporate or become entrained in ocean currents; others fall to the seafloor and collect over hundreds or thousands of years.

Seeps are often observed in locations where oil and gas extraction activities are additionally sited. As a consequence, numerous surface slicks and tar balls induced through seeps are frequently accredited to discharges from oil and gas platforms. The vital question arises then: if oil takes place naturally in the ocean and if seeps are the largest single source, why is there concern about the occasional unintentional spill? The answer lies in the nature and charges of oil inputs through these one-of-a-kind sources.

Generally, seeps are extremely old and move at a tremendously low rate. In this regard, elements that move out are still incredibly regularly toxic. However, organisms that live nearby are adapted to conditions in and around seeps. An insufficient extremely rare species of animals is even able to exploit the hydrocarbons and other chemicals discharged at seeps as a mine of metabolic energy. Besides, rather than being made up entirely of crude oil, the material flowing from seeps is frequently immensely biodegraded by microbial accomplishment deep below the seafloor.

In contrast, the production, transportation, and consumption of oil with the aid of humans commonly results in tremendously short, high-volume inputs of oil and sophisticated hydrocarbon products in sites that have by no means experienced considerable publicity to these chemical substances and so do no longer have many natural defences to them. As a result, seeps are regularly looked upon as a residing laboratory for scientists to learn about how natural processes have an effect on the destiny of discharged oil or how individual species or communities of plants and animals are successfully addressing the responsibility of occurrences of the other case of toxic chemicals. In this view, healthy environments can be determined through the occurrences of oil spills or oil seeps as they have a dangerous damages of ecosystems [1–3].

8.2 Behaviour of Oil and Gas Jets and Plumes Below the Sea Water Surface

On the continental margins, natural seeps discharge hydrocarbons in the forms natural gas and liquid oil. Consequently, hydrocarbons perhaps also are escaped from sunken vessels and damaged pipelines. The question now is: how do the properties of hydrocarbon liquids and gases modify post-contact with seawater? The majority of chemical compounds in oil and gas do not dissolve in water so, once in contact with seawater, they do not mix but rather break up into droplets and bubbles [4].

In deep seawater, where there are circumstances of intense pressure and low temperature, the delivery of hydrocarbon mixtures can generate natural gas hydrates. These are solid deposits with an ice-like crystalline assembly that breaks the solubility and bioavailability of the hydrated composites.

The dimensions of droplets and bubbles rely on the equilibrium between the disparaging force of mixing energy in numerous stalls of the environment, mainly the discharge spot (vent, pipe, etc.) and the resistive force primarily because of the interfacial tension between the oil and water or gas and water (the interfacial tension force is the physical factor behind the statement "oil and water do no mix") [5].

Yet, the vital question is: how do natural seeps behave as they release across the water column? A huge volume of hydrocarbon generates oil seeps below seawater plume. For example, the plume discharges, it attunes water in it, and ultimately, every individual oil droplet and gas bubble prolongs expanding at its "terminal" speed, which results from an equilibrium between its drag and its buoyancy.

Consequently, the significant question is: how far-off and rapidly can they plunge into the marine environment? Gas bubbles naturally escalate at speeds between 10 and 30 cms^{-1}, with larger velocities allied with huge bubbles. In this understanding, huge (millimeter-scale) oil droplets arise at comparable velocities of up to 20 cms^{-1}. However, minor oil droplets can spread incredibly gently, while micro-scale oil droplets have barely stimulated upwards at all [3–5].

8.3 Onshore Seep Occurrences

The keystone question is: what are the main characteristics of the onshore seeps? Onshore seeps tend to be tiny and it is not necessary of a detectable volume of gas and oil depleting the reservoir. The occurrence of onshore seeps is owing to the existence of onshore fold and thrust belts as accretions have either been broken or restructured to tertiary cons. In this sense, the connection between surface seeps and the leaking traps is extremely complicated. Nevertheless, such a sophisticated geology structural is hardly confronted in offshore basins, which does not commonly expand. Consequently, validation of the occurrence of seeps, particularly in offshore basins, is optimistic. However, in the vast majority of circumstances, offshore seep occurrence does not reveal of ruptured or drained traps.

8.4 Offshore Seep Occurrences

Naturally, seeps migrate from the seafloor vent up to the surface as oil-coated gas bubbles. In this regard, offshore seep oil and gas can be detected easier than onshore ones. Consequently, the gas bubble bursts and the seeps access the sea surface forming a thin oil films. In the circumstance of the calm sea surface, thin oil film appears as attractive, shimmering circular shapes; naturally, their diameters range from 0.5 m to 1 m in diameter, which is identified as 'oil pancakes'.

The most important example of offshore seep is the ones located in the Gulf of Mexico. Consequently, in the Gulf of Mexico, there are more than 600 natural oil seeps that leak between one and 5 million barrels of oil per year, equivalent to roughly 80,000 to 200,000 tonnes. When petroleum seeps accumulate across the water column it may additionally shape an extraordinary sort of volcano recognized as an asphalt volcano.

8.5 Sort of Seeps

There are two sorts of seep that can occur, relying on the level of overpressure. Capillary failure can take place in reasonable overpressure conditions, causing low-intensity of seepage till the overpressure equalizes and resealing occurs [6]. In some cases, the moderate overpressure cannot be equalized due to the fact that the pores in the rock are small, so the displacement pressure, the pressure required to break the seal, is very high. If the overpressure continues to enlarge to the point that it overcomes the rock's minimum stress and its tensile power earlier than overcoming the displacement pressure, then the rock will fracture, inflicting local and excessive intensity seepage till the pressure equalizes and the fractures close [7].

8.6 How Does Remote Sensing Technology Identify Natural Oil and Gas Seeps?

It is a challenging question to be answered. In fact, natural oil and gas seeps originated on the seafloor. However, there is a great possibility for remote sensing technology to accomplish information about potential marine zone of oil and gas explorations based on occurrence of oil seeps on sea surface. Natural seeps, therefore, escape from its source on the seafloor into the water column toward the sea surface forming a thin downstream layer which is controlled by the wind and current movements. In this understanding, the signature of the natural seeps through the water column is illustrated in the bubble stream. The bubble streams and slicks have been denoted as "plumes", which can be misleading because unlike a real plume trajectory movements across the sea surface. In this context, a bubbling stream or a surface slick cannot be identified as boundary current and further diffuse as it drifts from its original reservoir source.

Therefore, floating of the natural seeps on the sea surface is known as the slick origin or surfacing footprint, which is the leading end of the slicks on the sea surface. In this view, the perimeter represents the range (in km) of oil seep in sea surface. Consequently, the velocity and direction of the water column currents, besides on the depth of the source, have a great impact on the surfacing perimeter.

Consistent with the above perspective, natural seeps have a unique surface signature, which could be detectable in remote sensing data. The cornerstone question now is: how do remote sensing data monitor natural seeps?

The visible spectral bands, for instance, in the LANDSAT TM sensor do not deliver excellent discrimination between sea surface water, and natural seeps, but request further image processing such as image enhancement to yield good consequences. However, the sharp visible spectral signature of natural seeping can also be improved in the circumstances of wind and wave impacts, which is known as sunlight (Figure 8.4). In this view, sunlight is a phenomenon that occurs when sunlight reflects off the surface of the ocean at the same angle that a satellite or other sensor is viewing the surface. In the affected area of the image, smooth ocean water becomes a silvery mirror, while rougher surface waters appear dark. Sometimes the sunlight

Slick

Figure 8.4. Example of sunlight phenomenon in MODIS data.

region of satellite images reveals interesting ocean or atmospheric features that the sensor does not typically record. Figure 8.4 reveals sunlight in MODIS data covering the slick zone south of Indonesia and north of Australia. In this circumstance, the possible seepage detection can be delivered by optical satellite data such as MODIS.

The thermal spectral bands can deliver excellent seep detection in optical remote sensing data. Indeed, seeps reveal several thermal properties; for instance, thermal inertia, thermal capacity, and thermal conductivity. Therefore, seeps have lower thermal inertia than seawater, especially during high surface temperature at day time. In this circumstance, seeps can be detected clearly by the thermal sensors during the afternoon and owing to its high surface temperature. As it can act as a black body in which absorbing a high amount of heat and developing warmer boundary than the surrounding seawater can be detected in the thermal sensors. On the other hand, slick loses heat more rapidly than surrounding seawater during night time which becomes cooler than seawater. In this regard, thin seeps and surrounding sea surface are not thermally detectable in thermal sensors during night time. Needless to say, seep thermal properties fluctuate diurnally owing to the mixing of seeps with seawater.

Consequently, the thermal capacity of seawater is twice of the covered seeps. In other words, seawater can store heat than seeps. However, the thermal conductivity of oil is greater than water since the water regularly has low thermal conductivity. In this understanding, the high rate of heat passing through the seeps is larger than the seawater.

Overcoming the disadvantages of both visible and thermal spectral bands in seep detection, ultraviolet spectrum band can achieve this task. In fact, the tin seep films of approximately 1.5 μm can be detected using ultraviolet spectrum owing to their high reflectivity. Besides the high level of solar radiation, the phenomenon of fluorescence in the ultraviolet spectra is accomplished in existing heavy natural seeps. However, this requires clear weather during the day time as it is difficult to function during the night time. Consequently, the atmospheric scattering can impact the quality of acquired ultraviolet data. The acquired image must be captured at an elevation less than 1 km. Indeed, scattering of ultraviolet spectra towards the sensor at a high latitude of more than 1 km can distort the signal. In this circumstance, the seeps are also detectable in ultraviolet laser spectra that are generated between 0.3 to 0.355 μm. Indeed, seeps usually absorb this range of laser light that is radiated from the fluorosensors [8–10].

Also, it emits radiation in the visible range of 0.4 to 0.65 μm with a sharp peak at 0.48 μm [8–10].

In this understanding, by comparing the fluorescent response of the oil with traces of various known oils, an active ultraviolet sensor assists to discriminate between light refined crude and heavy crude. Moreover, a thin thickness of seeps between 0.15 μm to 10 mm can be detected using fluorescence only in clear weather.

8.7 Why Do Microwave Data Have Advantages on Top of Optical Data in Seep Monitoring?

This question is the keystone in understanding microwave remote sensing in seep monitoring. The answer behind this question has been addressed widely in Marghany's novel theory "Synthetic Aperature Radar Automatic Detection Algorithms (SARADA)". SAR data have advantages on the top of the optical satellite systems, such as Landsat TM and airborne systems in that SAR satellite in fixed polar orbit observes night and day and penetrates the heavy cloud cover. In this sense, SAR sensors form images of the sea surface itemizing its physical morphology. Therefore, SAR images represent slicks as smooth patches of the surface that can be interrelated by investigation to petroleum seepage.

Besides, SAR (Synthetic Aperture Radar) can image surface oil seeps remotely with wide swath coverage (typically 100 × 100 km scenes for ERS and 165 × 165 kms for Radarsat Wide 1) and at low cost. Likewise, SAR is boundless skies and is being endlessly attained, hence delivering multi-temporal satellite data over any area of the Earth. Such repeat seeps offer the site for continuous surface sampling from which vital geochemical information on the reservoir's oil can be attained prematurely of the drill [8].

8.8 Offshore Seep Imagine in SAR Data

Damping the capillary surface wave by seep bubbles ahead of the oil reservoir can be revealed in the form of the slick as seeps mix with sea surface water. This mechanism is illustrated on SAR data as dark patches stuck to the sea surface. SAR imaging mechanisms for seeps is controlled by wind speed that it does not exceed 5 ms^{-1} and is not less than 2 ms^{-1}. In this circumstance, the dark curved and circled form of natural seeps can be easily recognized in SAR data. Consequently, a thicker oil spill can be imaged in SAR data although wind speed is approximately 10 ms^{-1}. In this wind circumstance, growth of the millimetre to centimetre capillary and gravity waves take place in the sea surface and can be damped by slick. At wind speed less than 2 ms^{-1}, the sea surface is extremely smooth because the capillary wave has not developed yet and seepage-slicks can be "camouflaged" by dense natural slicks or obscured.

Consequently, the capillary and gravity waves are destroyed at a wind speed of 5 ms^{-1}. In this circumstance, the slicks are dispersed mechanically and are not detected by SAR sensors. Therefore, how does swell growth influence the seep imagine in the SAR data? Seeps can develop on swell at the lower local wind shear in which dual SAR images are needed for seep or slick observation.

8.9 What are the Physical Seep Parameters Identified in SAR Data?

Developing automatic slick detection is a function of knowing the following physical seep parameters: size, backscattering, context, repetition, ocean features, streaming direction, and edge characteristics. The minimum resolvable seepage slick size is ranged from 100 to 150 m long, reliant on sea shape and sensor. In this circumstance, slicks smaller than this will not be resolved by SAR. Consequently, backscatter reduction is an accurate index in identifying slicks in SAR data. In this view, seeps or slicks commonly have superior distinction and sharpness edge enhancement than natural films have, but this characteristic is a function of wind speed as explained in the previous Section 8.8. Backscatter reduction can be associated with sites over plausible migration conduits or geographical sites within shipping lanes or in areas of oil production, which is termed as context. The most significant physical seep parameter is repetition, which is demonstrated on successive SAR images. However, seep repetition is constrained to the leakiest basin sorts as episodic seepage is a parameter of utmost seepage and in furthermost offshore basins. Consequently, ocean features such as eddies, upwelling, and oil slick are required to be investigated and described in the slick examination [11].

Moreover, in the streaming path, seepage slicks will coincide with dominant wind and current directions, which is modulated by tidal effects. In this view, fresh pollution slicks, especially from ships, are characterized by narrow line and have a dissimilar featheredge where older slick material is driven by wind. Besides, sharp seep edge is a function of wave facets, current, and wind. In these circumstances, the slick thickness is constrained to dynamic ocean components such as wave, wind and current pattern [9–11].

8.10 SAR Polarization Signals

A SAR antenna can originate from shoot and receive electromagnetic waves with well-defined polarization. In short, polarization denotes the route of propagation of an electromagnetic wave vector's tip: vertical (up and down), horizontal (left to right), or circular (rotating in a constant plane left or right). The direction of polarization is well-defined by the orientation of the wave's electric field, which is permanently $90°$, or perpendicular, to its magnetic field.

By fluctuating the polarization of the sending signal and receiving numerous dissimilar polarized images from the similar sequence of pulses, SAR systems can acquire full information on the polarimetric characteristics of the experimental surface target, which can expose the assembly, orientation and environmental circumstances of the surface targets. Linearly well-oriented structures, for instance, ripples in the sand or building incline towards reflecting signal and maintain the coherence pattern (a similar linear route) of the incident polarimetric signal. Therefore, polarimetric signals are randomly slanted towards structures, for instance, tree leaves scatter and depolarize the signal as it bounces several times towards radar receiver. In this regard, multiple polarizations and wavelength combinations deliver dissimilar or matching surface information.

In other words, imaging radars have dissimilar polarization configurations. In this view, a single-polarization system, or "single-pol," transmits and receives a single polarization, typically the same direction, resulting in a vertical-vertical (VV) image (Figure 8.5) or horizontal-horizontal (HH) image (Figure 8.6). Subsequently, a dual-polarization system, or "dual-pol," might transmit in one polarization but receive in two, resulting in either HH and HV (Figure 8.7) or VH and VV imagery. In this view, dual-polarization delivers additional detail about surface features over the unlike and harmonizing echoes. Moreover, a quad-pol system would alternate between transmitting H and V waves and would receive both H and V, resulting in HH, HV, VH and VV imagery. To operate in quad-pol mode, however, the radar must pulse at twice the rate of a single- or dual-pol system since the transmited polarization

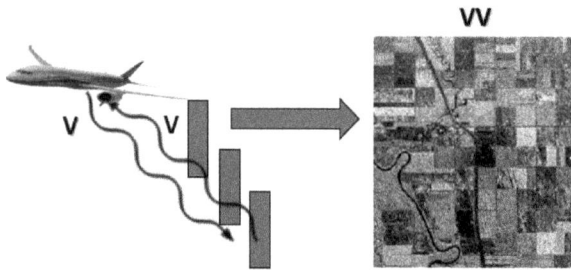

Figure 8.5. VV polarization.

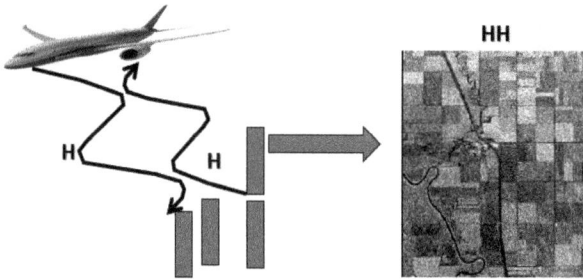

Figure 8.6. HH polarization.

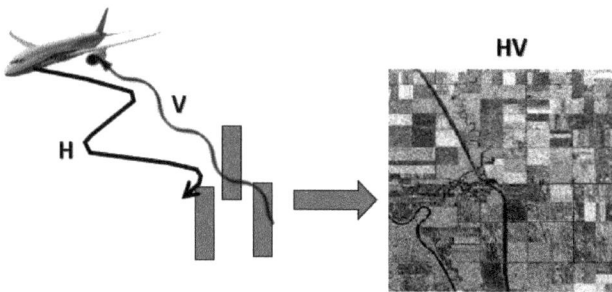

Figure 8.7. VH polarization.

signal has to alternate between H and V pulse by pulse. As this type of operation can cause interference between the received echoes, a variant of quad-pol known as quasi-quad-pol can be used. Quasi-quad-pol mode operates two dual-pol modes simultaneously: an HH/HV mode in the lower bounds of the transmit frequency band and a VH/VV mode in the upper portion. Because the frequencies are different, the two modes don't interfere with each other, but for this same reason, the observed data are mutually incoherent or have no phase relationship with each other [12].

While most spaceborne radar structures are linearly polarized, it is additionally viable to create a signal that's circularly polarized on transmission [12]. This is commonly accomplished through concurrently transmitting H and V indicators that are phase-shifted through 90°. The ensuing wave's electric-powered field vector tip attracts a circular route as it rotates between the offset amplitudes. Various mixtures of right-circular and left-circular polarization transmit and receive configurations of synthesizing signals in single-, dual-, and quad-pol modes of observations. A hybrid model that transmits a circularly polarized wave (R or L) and receives H and V is recognised as compact-pol. Compact-pol combines the appropriate properties of dual-pol, e.g., discrimination between oriented and random surfaces, whilst better balancing the power between the received channels [13].

Depolarization, for instance, is usually explained in terms of volume scatter or multiple reflections. At short wavelengths, when the terrain can be considered very rough, like- and cross-polarized images are almost identical. At longer wavelengths, however, when the terrain is considered relatively smooth, noticeable differences stand out. This has resulted in the limited geological application in identifying rock types [12].

In practice, VV ocean backscatters are extremely appropriate for measuring ocean wind speeds than HH yields. VV ocean backscatter is larger than HH returns. The alternations between VV and HH growth with growing incident angle is high at C-band. HH is more sensitive than VV to the changes in the local incident angle produced by variations in the long-scale sea surface slope. The change in the normalized radar cross-section (NRCS) backscatter from the sea surface owing to the changing local incident angle is known as tilt modulation [14]. In this view HH has larger tilt modulation than VV which makes HH more sensitive to the local incident angle. At low wind speed, nevertheless, HH is more sensitive to wind speed than VV and the discrimination between both polarizations is peaking. Therefore, HH is extremely sensitive to wave steepness and whitecaps than VV polarization [12–14].

8.11 Quantum Fully-polarized SAR Image Processing

The first step for QImP is the encoding of the 2-D image data into a quantum mechanical system (i.e., QImR). The QImR model substantively determines the types of processing tasks and how well they can be performed. Our present work is based on a QImR where the image is encoded in a pure quantum state, i.e., encoding the pixel values in its probability amplitudes and the pixel positions in the computational basic states of the Hilbert space.

Consider the fully-polarized SAR image S_{xy}, in which natural oil seep pixel backscattering properties can be identified by a backscattering matrix as:

$$S = \begin{pmatrix} S_{HH} & S_{HV} \\ S_{HV} & S_{VV} \end{pmatrix} \tag{8.1}$$

where S_{xy} is the complex backscattering efficiency, with x denoting the received wave polarization and y indicating the transmitted wave polarization. Therefore, the radar transmits only linear combination of horizontal and vertical ($\pi/4$) or circularly (CTLR, DCP) polarized signal and receives both horizontal and vertical polarizations in the compact polarimetric SAR modes. In this view, the 2-D quantity vector $\vec{K}$ is the projection of the complete backscattering matrix on the transmited polarization state. Hence, the quantity vector $\vec{K}$ of three foremost compact polarimetric SAR modes can be delineated as:

$$\vec{k}_{\pi/4} = [S_{HH} + S_{HV} \quad S_{VV} + S_{HV}]^{T}/\sqrt{2} \tag{8.2}$$

$$\vec{k}_{CTLR} = [S_{HH} - iS_{HV} \quad -iS_{VV} + S_{HV}]^{T}/\sqrt{2} \tag{8.3}$$

$$\vec{k}_{DCP} = [S_{HH} - S_{VV} + i2S_{HV} \quad i(S_{HH} + S_{VV})]^{T}/2 \tag{8.4}$$

Following Souyris et al. [15], the magnitude of linear coherence and the cross-polarization ratio can be correlated with parameter γ as:

$$\gamma = \frac{1 - |\rho_{HHVV}|}{\langle |S_{HV}|^2 \rangle} \times \langle |S_{HH}|^2 \rangle + \langle |S_{VV}|^2 \rangle \tag{8.5}$$

Equation 8.5 demonstrates that the probability of polarimetric SAR oil-spill sorting leans on great dissimilarity of the polarimetric mechanisms for clean water surface and oil-seep-covered sea surface [16]. In this context, based on different polarimetric scattering behaviours, mineral oil, and biogenic slicks can be better distinguished: for the oil-seep-covered area, Bragg scattering mechanism is largely suppressed and high polarimetric entropy can be witnessed, while in the case of a biogenic slick, Bragg scattering is still dominant, but with a lower intensity. In this understanding, the cross and co-polarization ratio, which is the power ratio between HH and VV or HV and HH/VV channels, can be obtained through dual-pol systems. In the tilted Bragg scattering model adopted by Minchew et al. [16], cross and co-polarization ratio are only a function of dielectric constant and incidence angle. Nonetheless, definite circumstances are continuously beyond obscured and further investigations are deeply required. Co-polarization ratio was also proved to be possible to discriminate natural oil seeps from look-alike features associated with low-wind conditions and surface current effects [17].

Polarimetric SAR decomposition parameters entropy (E) and average alpha angle ($\bar{\alpha}$) can be used to investigate the scattering patterns of a natural oil seep surpassed to the slick-free ocean. In this regard, E measures the randomness of the scattering mechanisms and ($\bar{\alpha}$) portrays the scattering mechanism. For oil-seep-covered areas, damping is strong, the correlation between co-polarization channels is low and thus E and ($\bar{\alpha}$) are high. In this view, the prevailing non-Bragg scattering is considered. Nonetheless, for the clean sea surfaces, the relationship between co-polarization channels is high, and E and ($\bar{\alpha}$) are low, indicating that surface Bragg scattering dominates [18]. Let consider that $E, \bar{\alpha} \in S_{xy}$, where $E \notin \bar{\alpha}$ or $E \neq \bar{\alpha}$. In this sense, the full polarimetric SAR image describes $I = (S_{xy})_{P \times L}$ as P and L present pixel and lines, while i and j are the positions of the natural-oil-seep pixel value in row and column, respectively. In this regard, $i = 1, ..., P$ and $j = 1, ..., L$ (Figure 8.8). In other words, I and O are the input and output images, respectively. On the classical computer, $P \times L$ SAR image can be signified as a matrix and encoded with at least 2^n bits, i.e., $[n = [log_2 [PL]]$. Consequently, the SAR image can be embodied as a quantum state $|Q\rangle$ and encoded in n qubits (Figure 8.9).

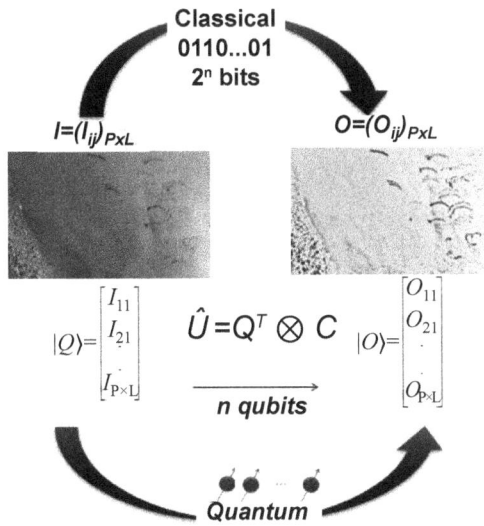

Figure 8.8. Quantum procedures of image processing.

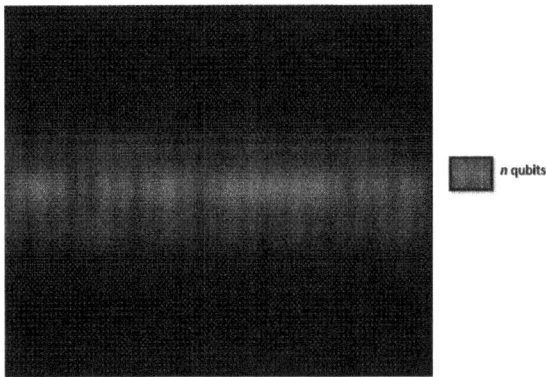

Figure 8.9. SAR image encoded as n qubits.

Hence, the quantum image transformation is achieved by unitary evolution $\hat{U}$ under a suitable Hamiltonian. In the quantum context, the linear transformation $|T_{lin}\rangle$ (Figure 8.10) can be represented as:

$$|T_{lin}\rangle = \hat{U}|Q\rangle \qquad (8.6)$$

The mathematical expression of unitary evolution can be given as:

$$\hat{U} = Q^{T_{lin}} \times C \qquad (8.7)$$

where $Q^{T_{lin}}$ and C are unitary operators equivalent to the classical operations. That is, the corresponding unitary operations of n qubits can be signified as a direct product of dual independent operations, with one acting on the first l-$log_2 L$ and qubits and the other on the last p-$log_2 P$ qubits.

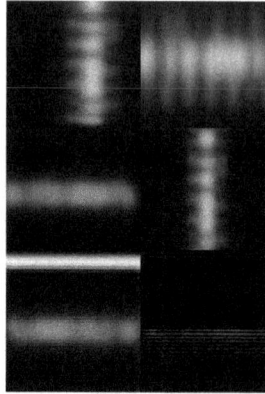

Figure 8.10. The procedure of linear transformation of quantum SAR image.

Finally, the wanted information about natural oil seep-covered can be extracted using QImP. To obtain entirely the constituents of the SAR state $|Q\rangle$ would demand $O'(2^2)$ operation. In other words, when the prerequisite oil seep information is, e.g., a binary consequence, for pattern matching and recognition, the number of involved operations could be significantly smaller.

Conversely, the well-known approaches of SAR data segmentation can be approximately allocated to region-based approaches and edge-based techniques, for instance, morphological methods, threshold segmentation, clustering techniques [14], and procedures of the random field. The following section introduces a new approach for automatic detection of a natural oil seep in SAR image using the quantum immune fast spectral clustering.

8.12 Quantum Immune Fast Spectral Clustering

The quantum immune fast spectral clustering depicts the cluster centres as quantum antibodies, whose tiniest information unit is a Q-bit, which can be formulated as:

$$|Q\rangle = P|0\rangle + P|1\rangle \tag{8.8}$$

where P is the probability amplitude of both $|0\rangle$ and $|1\rangle$, respectively. Consider the antibody population size N, representative point number m, and encoding length L_{en}. In this regard, the random numbers of N rows and m × L_{en} columns in $[-1,1]$ are preliminary created the antibody population. For the SAR image oil seep segmentation, let us implement the affinity function with space information. Foremost, perform the boundary expansion on the segmented oil seep image to construct a kernel window of 7 × 7 spatial neighborhood of m representative points. Succeeding, the number of samples, which fits into the major oil seep class in each neighbourhood, is signified as A. Furthermore, the divergence within the oil seep class is initiated to designate the scattering of sample oil spill pixels in the SAR image, which is expressed as:

$$\beta = m^{-1} \sum \min \left\| c_i - c_j \right\|^2, \ i, j = 1, \dots, m \tag{8.9}$$

Here c_i and c_j are the representative natural oil seep pixels. Subsequently, the affinity function for automatic detection of oil seep in SAR image can be expressed as:

$$F = 0.01^* A + m^{-1} \sum \min \left\| c_i - c_j \right\|^2, \tag{8.10}$$

To implement the quantum immune fast spectral clustering, the quantum immune operation procedures have to be considered.

8.13 Quantum Immune Operation

The quantum immune operation involves (i) quantum immune cloning operation, (ii) quantum updating operation, and (iii) quantum immune selection operation. Let us consider that $A = \{a_1, a_2, ..., a_N\}$ be the quantum antibody population (Figure 8.11). Therefore, the quantum immune cloning operation $A_c(t)$ can be identified as

$$A_c(t) = I_{ij} \times a_{ij}(t)[(A(t))] \tag{8.11}$$

Equation 8.11 indicates that the dimension row vector resembles all elements, which is one and can be the scale of antibody a_{ij} post cloning, which can be expressed as:

$$q_{ij} = Int \left(n_c \times F(a_{ij}) \left(\sum_{j=1}^{n} F(a_{ij}) \right)^{-1} \right) \tag{8.12}$$

Equation 8.12 demonstrates the operator of the transferred numerical value into an integer and Int ($\cdot$). Besides, Equation 8.12 involves the affinity value $F(a_{ij})$ of the antibody element a_{ij}. In other words, Equations 8.11 and 8.12 achieved the procedure of quantum immune cloning operation. After the quantum immune cloning operation, the quantum updating operation is considered. In this context, the quantum rotation gate G_r [12] is exploited to embody the updating of the quantum population N (Figure 8.12), which can be exploited as:

$$G_r(\theta) = \begin{bmatrix} \cos(\theta) & \sin(\theta) \\ \sin(\theta) & \cos(\theta) \end{bmatrix}. \tag{8.13}$$

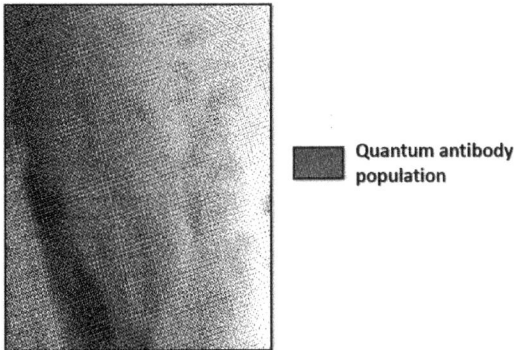

Quantum antibody population

Figure 8.11. Quantum antibody population for SAR data generation.

N=20 x10⁴   **N=20 x10³**

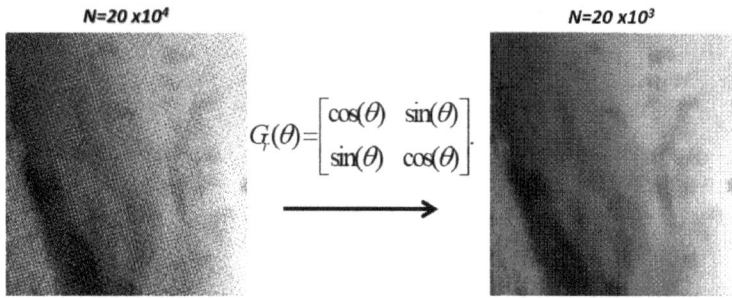

$$G_r(\theta) = \begin{bmatrix} \cos(\theta) & \sin(\theta) \\ \sin(\theta) & \cos(\theta) \end{bmatrix}.$$

Figure 8.12. Quantum rotating gate result.

The impact of G_r can improve the searching of antibody element a_{ij} that belongs to the oil-seep-covered pixels (Figure 8.12). It is interesting to notice that updating the quantum population leads to discrimination clusters in the SAR image. In this regard, the probability amplitude of ith Q-bit spins to ξ_i and η_i as:

$$\begin{bmatrix} \xi_i' \\ \eta_i' \end{bmatrix} = G_r(\theta_i) \times \begin{bmatrix} P|0\rangle \\ P|1\rangle \end{bmatrix} \tag{8.14}$$

where θ is the rotation angle which organizes the convergence rate, which is defined as $\theta = \delta \times s(P|0\rangle, P|1\rangle)$. Therefore, δ is a coefficient which regulates the speed of convergence and its observed quantities are 0.005π to 0.1π. $s(P|0\rangle, P|1\rangle)$ that regulates the path of oil seep pixel exploring (Figure 8.13), whose values are given in the lookup table (Table 8.1).

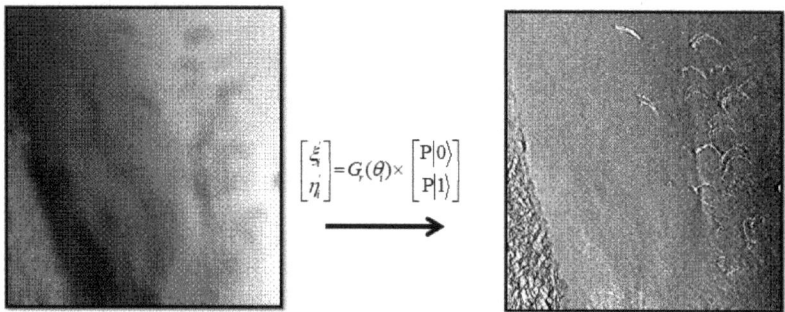

$$\begin{bmatrix} \xi \\ \eta \end{bmatrix} = G_r(\theta) \times \begin{bmatrix} P|0\rangle \\ P|1\rangle \end{bmatrix}$$

Figure 8.13. Quantum probability of determining oil spill pixels.

Table 8.1 indicates that b_i is the ith bit of the existing antibody, $best_i$ is the ith bit of the existing best antibody, and $F(b_i)$ is the affinity of the present antibody. In this circumstance, the dark patch pixels sign as natural seep clusters, while oil spill, look-alikes and surrounding sea surface pixels sign as false clusters. Finally, the quantum immune selection operation is performed to pick out the unsettled individual pixels of the oil seeps from the subpopulation post cloning to generate a new population (Figure 8.14). For the best antibody $b_i(t)$ in each subpopulation, if $f(a_{ij} < f(b_{ij}))$, then a_{ij} is replaced by $b_{ij}(k)$.

Table 8.1. θ look-up table.

b_i	$best_i$	$F(b_i) \geq$ $F(best)$	θ	$s(\text{P}\vert 0\rangle, \text{P}\vert 1\rangle)$			
				$\text{P}\vert 0\rangle, \text{P}\vert 1\rangle > 0$	$\text{P}\vert 0\rangle, \text{P}\vert 1\rangle > 0$	$\text{P}\vert 0\rangle = 0$	$\text{P}\vert 1\rangle = 0$
0	0	false	0	0	0	0	0
0	0	true	0	0	0	0	0
0	1	false	0	0	0	0	0
0	1	true	Δ	-1	1	± 1	0
1	0	false	Δ	1	-1	0	± 1
1	1	true	Δ	1	-1	0	± 1
1	1	false	Δ	1	-1	0	± 1

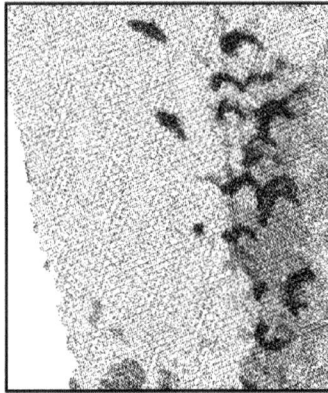

Figure 8.14. Quantum immune selection operation.

8.14 Spectral Embedding

Spectral embedding is achieved once the representative SAR pixels m have been selected, i.e., $\mathbb{R} = \{r_i\}$, $i = 1,2,3,....,m$, which can be implemented as spectral embedding. In this view, consider that the sampling pixels $a_1, a_2,....., a_n$, $K \in R^{n \times n}$ present the similarity matrix, the diagonal matrix is presented by $S \in R^{n \times n}$, and the elements $\vert C_k \vert$ of the diagonal matrix $d \in R^{m \times m}$ ($\vert C_k \vert$) as C_k is the cluster dimension equivalent to k^{th} representative pixels. In this view, the mathematical explanation of the quantization submatrix W can be given by:

$$W = K(r_s, r_t)\vert C_t \vert \tag{8.15}$$

where s,t = 1,2,3,...,m. In this regard, the extrapolation matrix $\tilde{E}$ is given by:

$$\tilde{E} = K(a_i, r_j)\vert C_j \vert, \quad i = 1,..., n, \tag{8.16}$$

Consistent with above Equations 8.15 and 8.16, the diagonal matrices of W and $\tilde{E}$

$$d_{\mathbb{R}} = \sum_{k=1}^{m} \vert C_K \vert K(r_s, r_t) \tag{8.17}$$

The eigen decomposition of $d_{\mathbb{R}}$ is formulated as:

$$\frac{d_R^{-0.5}}{W d_R^{-0.5} v^{-1}} = \lambda v \tag{8.18}$$

Equation 8.18 is performed and extrapolated to

$$U = \frac{d_X^{-0.5}}{\tilde{E} d_R^{-0.5} v^{-1} \lambda} \tag{8.19}$$

where U is unorthogonal eigenvectors, while the orthogonal ones may have better performance in practice [11–14]. Indeed, oil seep cluster pixels can be identified as depicted in Figure 8.15.

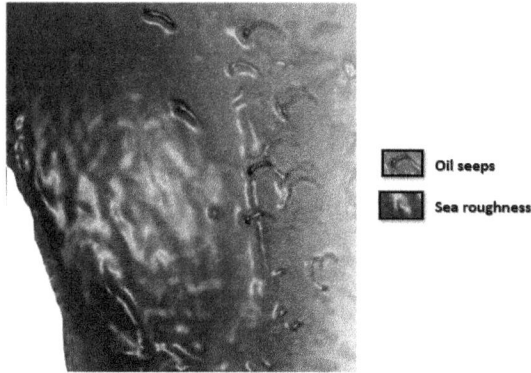

Figure 8.15. Orthogonal identification of oil seeps from surrounding sea roughness.

8.15 Automatic Detection of Oil Seep in Full Polarimetric SAR

The SAR images were acquired by UAVSAR (uninhabited aerial vehicle synthetic aperture radar) throughout the DWH (Deepwater Horizon) oil spill disaster in the Gulf of Mexico in 2010. UAVSAR is a fully-polarimetric L-band SAR. They have a centre frequency of 1.2575 GHz. Moreover, they have dual multi-look, i.e., 3 and 12 looks in the range and azimuth directions, respectively. UAVSAR data have a 5 m slant range resolution and 7.2 m azimuth resolution. In this section, UAVSAR data were acquired on June 23, 2010, shows the southern Louisiana coastline, covering the area around Grande Isle and the entrance to Barataria Bay (Figure 8.16). The oil spill pixels in both images have a damping of normalized radar cross-section NRSC of −35 dB along the oil spill pixels. It can assume that NRSC values range from −30 to −25 dB. In this understanding, natural oil seeps are dominated by higher NRSC than oil spill-covered sea surface owing to its thin shapes.

UAVSAR instrument has a lower noise floor range between −35 dB to −53 dB, which allows for better oil spill and natural oil seep imagines. In fact, the instrument noise floor is grasped only at the far edge of the swath for the HV backscatter from oil-covered or natural seep-covered sea surface. This can assist to quantify the radar cross-section of water with an L-band radar, even with natural oil seeps damping the surface waves.

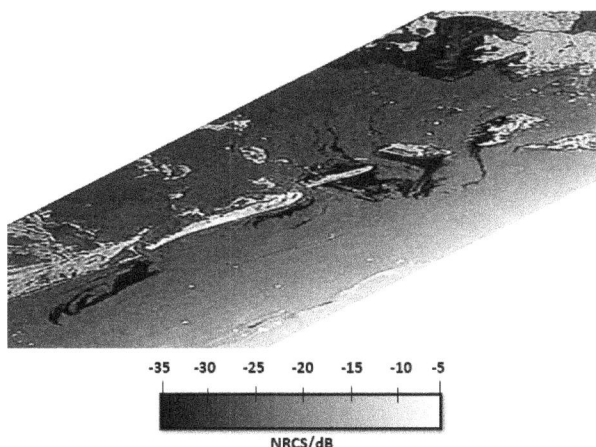

Figure 8.16. Normalized radar cross-section of UAVSAR image was acquired on June 23, 2010.

Specific quantities are operated for parameter settings, i.e., quantum antibody population, encoding length, cloning scale, and maximum iteration number. In this regard, we set N equals 100, encoding length L equals 20, and cloning scale S equals 50 under the circumstance of 100 iteration number to ensure an accurate result. Figure 8.17 delivers the results, which are obtained by quantum immune fast spectral clustering (QIFSC). QIFSC can discriminate automatically between Bragg scattering and non-Bragg scattering. The dynamic fluctuation between both Bragg and non-Bragg scattering is accurately determined. Moreover, QIFSC can deliver several clusters in UAVSAR, for instance, surrounding sea surface, natural oil seeps and thick oil spill (Figure 8.17). Subsequently, if we reset the N equals 200, encoding length L equals 100, and cloning scale S equals 90 and increase the iteration number to 400, we acquire sharp oil seeps morphology features without oil spill. Besides, the clear discrimination of oil spill from natural oil seeps is shown in Figure 8.18.

The increase of the number of parameters leads to the accurate isolated clusters of the oil slick and natural oil seeps. In other words, these oil spill and natural oil seep clusters have revealed a variety of the oil seep characteristics, such as area complexity of oil-covered sea surface, thick curved and circular spreading patterns (Figure 8.19). In fact, UAVSAR characterizes an oil spill and natural seeps by

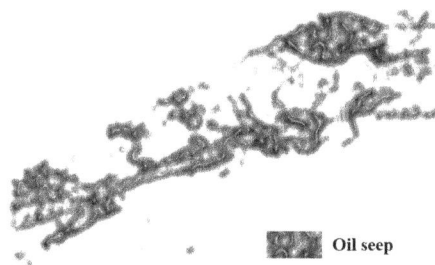

Figure 8.17. Oil seep discrimination in UAVSAR data using QIFSC.

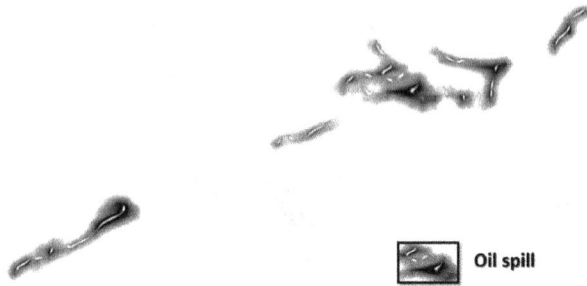

Figure 8.18. Oil spill discriminations in UAVSAR image using QIFSC.

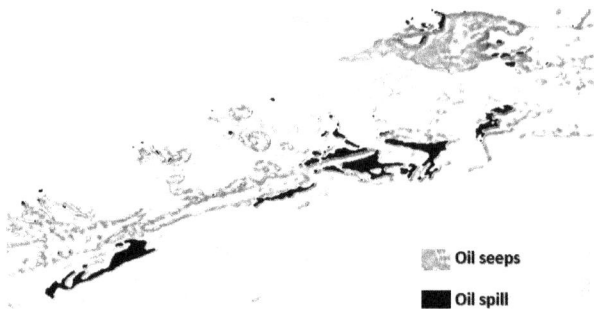

Figure 8.19. Natural oil seep and oil spill clusters using QIFSC.

detecting variations in the roughness of its surface and, for thick slicks, changes in the electrical conductivity of its surface layer. In this regard, UAVSAR "sees" an oil spill and seeps at sea as a smoother (radar-dark) area against the rougher (radar-bright) ocean surface because most of the radar energy that hits the smoother surface is deflected away from the radar antenna. UAVSAR's high sensitivity and other capabilities enabled the team to separate thick and thin oil and seeps for the first time using a radar system.

The advantages of global search in quantum immune and spectral embedding in UAVSAR data are combined in QIFSC. QIFSC can find the optimal solutions more frequently with the maximum clustering accuracies in 300 runs. QIFSC is operated in both optimal solution and stability [14].

It appears that the QIFSC output result is superior, which reveals more detail of the oil clusters and less false alarms than the raw UAVSAR data. It is easy to realize an unbroken oil slick in the QIFSC image.

8.16 Applications of QIFSC to Other Satellite Polarimetric SAR Sensors

QIFSC is implemented also on satellite polarimetric SAR sensors, which involved Sentinel-1A quad-polarization (VV and VH) SAR image of natural oil seeps in the Black Sea on October 10, 2014 (Figure 8.20). In the Black Sea of the coastal waters of Georgia, the slick patterns are strangely dissimilar in structure from ship spills,

Figure 8.20. Sentinel-1A data along with Georgia coastal waters.

Figure 8.21. Automatic detection of oil seeps using QIFSC.

and widespread slicks of biological origin are frequently detected at the sea surface. It is worth noticing that the slick is dominated by NRCS of damping range from –30 to –20 dB.

Figure 8.21 depicts the results revealed by QIFSC for quad-polarization data. It is noticed that QIFSC can automatically detect oil seeps from its surrounding area. Besides, it delivers two clusters of oil seeps from thick to thinner [20].

Subsequently, QIFSC can also detect only the thick oil spill from coherency of quad-polarization (Figure 8.21). In fact, the magnitude of the correlation coefficient between HH + VV and HH − VV are useful for suppressing lookalikes [19]. Moreover, quad-polarization SAR images are susceptible to noise. Pauli decomposition has the advantages of anti-interference and general high adaptability [21]. In general, the Pauli decomposition images are clearer than the original quad-polarization SAR images. The image preprocessing stages are as follows: (i) the original quad-polarization SAR data are decomposed by Pauli, and (ii) the obtained Pauli decomposition images are filtered by 3 × 3 Boxcar filtering. However, QIFSC is able to optimize the impact of noise, in addition to providing automatic clustering and automatic detection of oil seeps in quad-polarization SAR data and also determine their physical morphology, i.e., thickness.

Generally, the QIFSC can deliver the same results with single Sentinel-1A data, which was acquired on February 19, 2016, along the coastal waters of Turkey (Figure 8.22a). In this regard, Figure 8.22b shows the variety of clusters of oil seeps, which are illustrated as thick and thin seeps. This confirms that QIFSC can deliver the same results for both single SAR polarization and quad-polarization data.

In this regard, QIFSC is also implemented to TerraSAR-X strip map (HH/VV) data over East coast of Malaysia, Kuala Terengganu (Dungun) on November 9 2015 (Figure 8.23a). It is interesting to notice that QIFSC produces three clusters, which

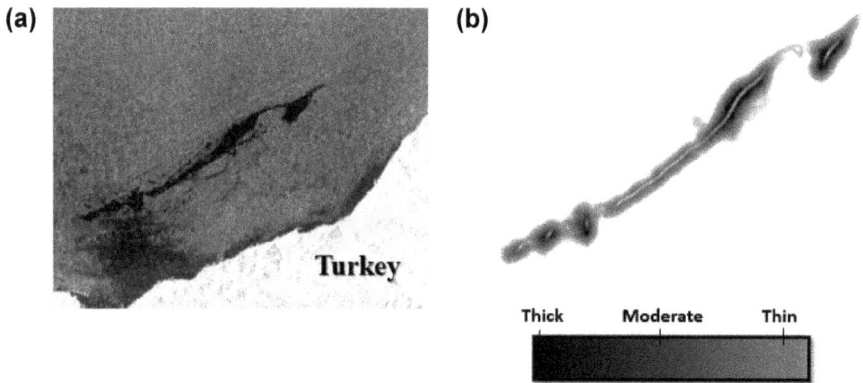

Figure 8.22. Investigation on (a) single Sentinel-1A using (b) QIFSC clustering of oil seeps.

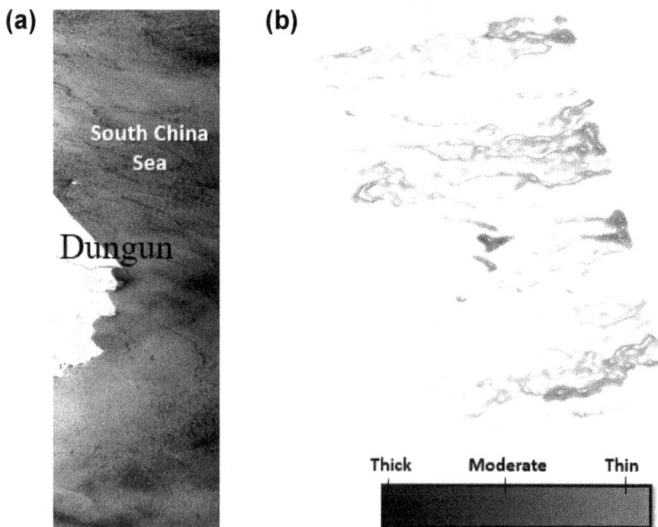

Figure 8.23. Oil seep clusters, which are obtained from (a) TerraSAR-X strip map by (b) QIFSC.

are thick oil seeps, medium oil seeps and light oil seeps too (Figure 8.23b). This indicates the level of natural oil seeps spreading under the dynamic action of both wind and sea surface hydrodynamic impacts in the South China Sea, Malaysia.

QIFSC provided unexpected precious clusters of oil seep characteristics, which are related to its thickness, from both polarimetric SAR data and dual-polarization data. The achievement accuracy of QIFSC is presented as a higher true positive rate (TPR) in ROC curves of different sensors. In terms of the ROC area, an oil seep can be clustered by 99% among the different SAR sensors by using QIFSC (Figure 8.24).

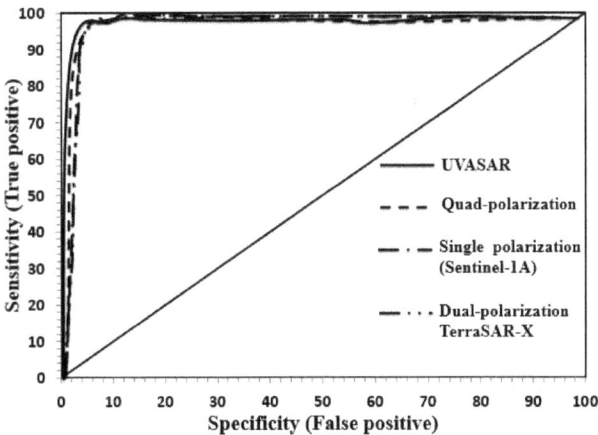

Figure 8.24. ROC for QIFSC performance.

8.17 Why Can QIFSC Precisely Cluster Different Kinds of Oil Seep?

The oil-seep-covered water is considered as a decoherence zone while the surrounding sea surface is coherent. In quantum mechanics, quantum decoherence is the loss of coherence or assembling of the phase angles between the constituents of oil-seep-covered water and the surrounding sea surface in a quantum superposition. One significance of this dephasing is conventional or probabilistically additive behaviour. Quantum decoherence offers the attendance of wave function collapse, i.e., Bragg scattering and non-Bragg scattering. In other words, quantum decoherence is considered the decline of the physical characteristics of sea roughness backscatter into a single possibility of oil-seep-covered water, which is imagined as decoherence pixels in SAR images. In this understanding, decoherence speculation of oil-seep-covered water in SAR data is the mechanism, which determines the location of the quantum-classical boundary, i.e., Bragg scattering. In this view, decoherence befalls when oil-seep-covered water interacts with its surrounding environment in a delectrical irreversible approach. In this circumstance, thick oil-seep-covered water with a lower delectrical than its surrounding prevents Bragg and non-Bragg scattering in the quantum superposition of the entire SAR scene's wavefunction from interfering with each other. This is why the dumping of backscatter occurs in the oil-covered area than its surroundings. However, decoherence does not spawn definite wave function collapse. It only delivers a description of the accomplishment of wave function collapse, as the quantum nature of the oil seeps in the SAR data "disintegrates" from its surrounding environment. That is, components of the wavefunction are isolated from a coherent sea surface and attain phases from their instantaneous surroundings. In this circumstance, a total superposition of the global or universal wavefunction immobile exists (and remains coherent at the global level of the sea roughness); nonetheless, its vital providence remains as a logical global interpretation concern along the oil-seep-covered water. In this regard, QIFSC can play a vital role to determine the lost information in the oil spill pixels. Besides, QIFSC can solve the computational gridlock problems of conventional spectral clustering algorithms on large-scale SAR image segmentation. In fact, QIFSC signifies the oil seeps' representative pixels and boundaries through an operative encoding technique. In this understanding, the quantum immune operation optimizes the selection of representative oil seep pixels, while the calculated affinity function for the clustering oil seeps and its surrounding environments diminishes the computation quantity to a huge SAR data used. Subsequently, QIFSC can count the huge amount of eigenvalues with a value of one, which leads to the precise determination of cluster number across the SAR data. In other words, the attendance of spectral distribution in this algorithm allows for excellent separation between different clusters. The boundaries between different oil seep clusters are well identified. Indeed, the largest eigenvalue is set as one, and the other eigenvalues diminish steadily, while the breach between the second and third eigenvalues is prevalent, which reveals that the three or two-classes of the oil spill can be easily separated.

According to Marghany [14], QIFSC is accomplished better than that genetic algorithm, particularly in clustering the oil spill pattern, while NSGA-II can preserve the morphology of oil spill footprint boundaries, i.e., spreading. Moreover, the conventional segmentation approaches, especially for full polarimetry SAR, for instance Entropy E, Anisotropy A, Alpha angle α and Co-Polarized Phase Difference (CPD) are not able to cluster the degree of oil spill variations from thick to thin because of hydrodynamic diffusion. However, co- and cross-polarization ratio, the degree of polarization, entropy H, Alpha angle α, and anisotropy A [17–19] under the control of certain incidence angle range, can still distinguish oil seep from its surrounding environment. In this sense, QIFSC is accurately performed better than the co-polarized phase difference, conformity coefficient, and co-pol correlation coefficient [21–24] for oil seeps' automatic detection and clustering.

In this view, the immune clonal selection algorithm takes both local examination and global search into account, which could handle the optimal solution with higher precision compared to conventional evolutionary algorithms [23]. In this context, the quantum immune clonal algorithm can be used for function optimization and multiuser detection of code division multiple access [24–26]. Quantum computing and the developed algorithm can be implemented with a genetic algorithm to improve its accuracy [25]. Finally, quantum image processing is required for advanced SAR image segmentation and processing.

This chapter demonstrated the novel technique for automatic detection of the natural oil seeps in full polarimetric SAR data and single polarization ones. To this end, quantum immune fast spectral clustering reveals excellent tools for natural oil seep automatic detection in complex radar data. Quantum immune fast spectral clustering can automatically cluster the natural oil seeps that are covered by a thick oil spill. The main physical characteristics of natural oil seeps are that they occur in the forms of irregular and circular patterns and seem smaller and thinner than oil spills. This chapter also for the first time detected natural oil seeps along the East Coast of Malaysia, Terengganu, Dungun.

References

[1] Kvenvolden KA, Cooper CK. Natural seepage of crude oil into the marine environment. Geo-marine Letters. 2003 Dec; 23(3): 140–6.

[2] Zatyagalova VV, Ivanov AY, Golubov BN. Application of Envisat SAR imagery for mapping and estimation of natural oil seeps in the South Caspian Sea. In Proceedings of the Envisat Symposium, Montreux, Switzerland (ESA SP-636 2007 Apr 23).

[3] Pesch S, Jaeger P, Jaggi A, Malone K, Hoffmann M, Krause D, Oldenburg TB, Schlüter M. Rise velocity of live-oil droplets in deep-sea oil spills. Environmental Engineering Science. 2018 Apr 1; 35(4): 289–99.

[4] Lardner R, Zodiatis G. Modelling oil plumes from subsurface spills. Marine Pollution Bulletin. 2017 Nov 15; 124(1): 94–101.

[5] Dong P, Lu B, Gong S, Cheng D. Experimental study of submerged gas jets in liquid crossflow. Experimental Thermal and Fluid Science. 2020 Apr 1; 112: 109998.

[6] North CP. Gluyas, J, Swarbrick, R. 2004 Petroleum Geoscience, Blackwell Science Ltd. Teaching in Earth Sciences. 2003; 28: 41–2.

[7] Ortoleva PJ. Basin Compartments and Seals, AAPG Mem. 1995; 61.

[8] Mazumder S, Saha KK. Detection of oil seepages in oceans by remote sensing. In Proceedings of the 6th International Conference & Exposition on Petroleum Geophysics 2006.

[9] Ivanov AY. Remote sensing of oil films in the context of global changes. Remote Sensing of the Changing Oceans. 2011: 169–91.

[10] Lu Y, Tian Q, Wang X, Zheng G, Li X. Determining oil slick thickness using hyperspectral remote sensing in the Bohai Sea of China. International Journal of Digital Earth. 2013 Jan 1; 6(1): 76–93.

[11] Williams A, Lawrence G. AAPG Studies in Geology No. 48/SEG Geophysical References Series No. 11, Chapter 12: The Role of Satellite Seep Detection in Exploring the South Atlantic's Ultradeep Water; 2002.

[12] Moreira A, Prats-Iraola P, Younis M, Krieger G, Hajnsek I, Papathanassiou KP. A tutorial on synthetic aperture radar. IEEE Geoscience and Remote Sensing Magazine. 2013 Apr 18; 1(1): 6–43.

[13] Tomiyasu K. Tutorial review of synthetic-aperture radar (SAR) with applications to imaging of the ocean surface. Proceedings of the IEEE. 1978 May; 66(5): 563–83.

[14] Marghany M. Automatic Detection Algorithms of Oil Spill in Radar Images. CRC Press; 2019 Oct 8.

[15] Souyris JC, Imbo P, Fjortoft R, Mingot S, Lee JS. Compact polarimetry based on symmetry properties of geophysical media: The/spl pi//4 mode. IEEE Transactions on Geoscience and Remote Sensing. 2005 Mar; 43(3): 634–46.

[16] Migliaccio M, Nunziata F, Brown CE, Holt B, Li X, Pichel W, Shimada M. Polarimetric synthetic aperture radar utilized to track oil spills. Eos, Transactions American Geophysical Union. 2012 Apr 17; 93(16): 161–2.

[17] Kudryavtsev VN, Chapron B, Myasoedov AG, Collard F, Johannessen JA. On dual co-polarized SAR measurements of the ocean surface. IEEE Geoscience and Remote Sensing Letters. 2013 Jul; 10(4): 761–5.

[18] Zhang B, Perrie W, Li X, Pichel WG. Mapping sea surface oil slicks using RADARSAT-2 quad-polarization SAR image. Geophysical Research Letters. 2011 May 1; 38(10).

[19] Zheng H, Khenchaf A, Wang Y, Ghanmi H, Zhang Y, Zhao C. Sea surface monostatic and bistatic EM scattering using SSA-1 and UAVSAR data: Numerical evaluation and comparison using different sea spectra. Remote Sensing. 2018 Jul; 10(7): 1084.

[20] Mityagina M, Lavrova O. Oil slicks from natural hydrocarbon seeps in the Southeastern Black Sea, their drift and fate as observed via remote sensing. pp. 7926–7929. In IGARSS 2018–2018 IEEE International Geoscience and Remote Sensing Symposium 2018 Jul 22. IEEE.

[21] Salberg AB, Rudjord Ø, Solberg AH. Oil spill detection in hybrid-polarimetric SAR images. IEEE Transactions on Geoscience and Remote Sensing. 2014 Oct; 52(10): 6521–33.

[22] Nunziata F, Gambardella A, Migliaccio M. On the degree of polarization for SAR sea oil slick observation. ISPRS Journal of Photogrammetry and Remote Sensing. 2013 Apr 1; 78: 41–9.

[23] De Castro LN, Von Zuben FJ. The clonal selection algorithm with engineering applications. pp. 36–39. In Proceedings of GECCO 2000 Jul 8 (Vol. 2000).

[24] Han KH, Kim JH. Quantum-inspired evolutionary algorithms with a new termination criterion, H/sub/spl epsi//gate, and two-phase scheme. IEEE Transactions on Evolutionary Computation. 2004 Apr; 8(2): 156–69.

[25] Jiao L, Li Y, Gong M, Zhang X. Quantum-inspired immune clonal algorithm for global optimization. IEEE Transactions on Systems, Man, and Cybernetics, Part B (Cybernetics). 2008 Oct; 38(5): 1234–53.

[26] Tung F, Wong A, Clausi DA. Enabling scalable spectral clustering for image segmentation. Pattern Recognition. 2010 Dec 1; 43(12): 4069–76.

CHAPTER 9

Quantum Interferometry Radar for Oil and Gas Explorations

9.1 What is Reservoir Geomechanics?

The question is now: what is meant by Geomechanics? Geomechanics (from the Greek prefix geo meaning "Earth", and "mechanics") includes the geologic learning of the performance of soil and rock. Consequently, several features of geomechanics coincide with fragments of geotechnical engineering, engineering geology and geological engineering. In this regard, modern technology growths associate with seismology, and mechanical fields reveal precisely information about reservoir geomechanics phenomena [1].

The hypothetical and practical learning of mechanical behaviour of geological material is referred to as geomechanics. In this perspective, oil and gas exploration and production activities can cause the mechanical failure of the reservoir, over, side and under burden, formations to be optimized and reduced by geomechanics. In this view, mechanical failures involve, but are not restricted to, drilling of oil and gas wells, hydraulic fracturing, water/gas flooding, and depletion. Therefore, a geology structural of reservoir can collapse due to the stress is caused by drilling of oil and gas wells [2]. Subsequently, the role of geomechanics is to predict when failure would occur, assess its risks and opportunities and recommend mitigation plan(s).

Every geological disposition in the subsurface (crust) is imperilled to stresses instigated by numerous geological evolution and transformation factors; for instance, deposition (gravitational loading), tectonics, uplift, pressure inflation or deflation, stress relaxation, and thermal impacts. In these perspectives, stresses in the subsurface are not static through time nor can these be embodied in a similar manner all over the place. Correspondingly, creation's intensity characteristics are exposed to fluctuations through the geological time from deposition to the present day.

Any formation/reservoir geomechanical assessment starts with characterizing the stresses, strength and pressure profiles. Hence, understanding the geological history of the formation of interest is crucial to its reliable geomechanical characterization.

9.2 What is the Role of Reservoir Geomechanics in Oil and Gas Explorations?

Within the general reservoir management framework, the forecasting of the subsurface compaction in creating hydrocarbon reservoirs is a vital concern. Consequently, casing deformation and failure of wellbore are great risk impacts, which must be disallowed to guarantee the great safety of the drilling operations and diminish vital economical risks. Besides, the prediction of land subsidence instigated by the compaction of the rock creation can be of foremost implication.

In this sense, the platform sinking can be caused by anthropogenic settlements; for instance, they can cause sinking of the oil platform, the Ekofisk field in the North Sea. Moreover, such a pipeline excessive deformations and large environmental impact especially in coastal areas can be a great issue in oil and gas explorations.

Now, how the oil and gas reservoirs can cause land subsidence? Anticline reservoir geological blocks the hydrocarbon upward movement, hence enabling the accretion on the top (Figure 9.1). Mainly, hydrocarbon reserves, therefore, are located in sedimentary rocks of clastic and chemical origin.

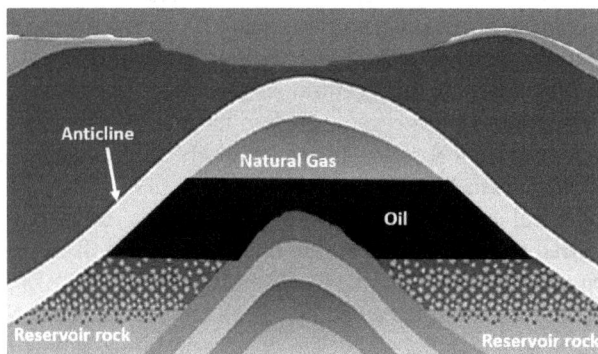

Figure 9.1. An anticline reservoir's typical structure.

The misuse of underground resources can instigate fluctuations in the geological structure, which reveals dissimilar rates of surface deformation in the oil exploitation zone (Figure 9.2). In this regard, oil exploitation will diminish reserves, creating a conforming reduction in pressure in the reservoir. In this view, shrinking of the reservoir can cause land subsidence due to an effective upsurge stress, which is formed by the reduction in pore pressure.

Post injection of the CO_2 or liquid, the generation correspondingly experiences stress variations that spread from the deep chunk of the reservoir to the surface. Simultaneously, the frequency of surface movement or deformation is relevant to geological circumstances, mining devices, and other influences. For instance, asymmetrical deformation of the ground triggered by human accomplishments critically impacts the protection of oil and gas manufacture instruments and the sustainability of oil and gas fabrications.

The vital question now is: what is the best remote sensing technology that delivers accurate information on oil and gas explorations? Interferometric synthetic

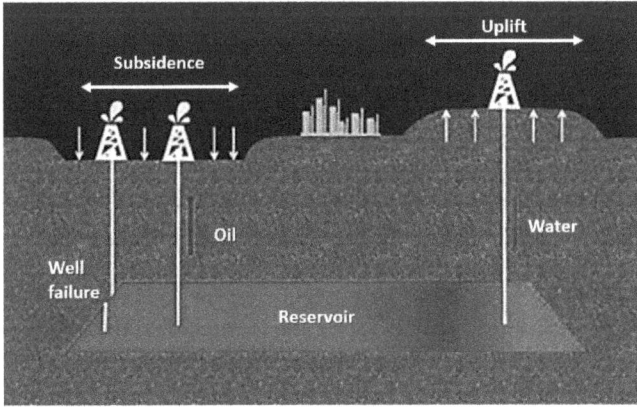

Figure 9.2. Land deformation owing to oil and gas explorations.

aperture radar (InSAR) technology, which can deliver distinguished-precision, large-scale, surface-based distortion scrutinizing consequences, has widespread implementations in surface distortion scrutinizing. Exploiting InSAR technology for oil and gas exploration deformation monitoring will notably contribute to the training of oil field difficulties.

9.3 Physics of Interferometry

Interferometry is an intimate technique in which electromagnetic waves are superimposed, instigating the phenomenon of interference, which is exploited to extract information [3]. In this regard, Young's slits is the fundamental concept of creating interferometry technique (Figure 9.3). Briefly, Thomas Young exploited this experiment to 'verify' that the light beam was a wave at a time when the light was thought to be a particle. The light is departing through dual slits interferes and creates a pattern that is straightforward to explain wave-particle dualities, but which cannot be clarified if the light turns only like particles.

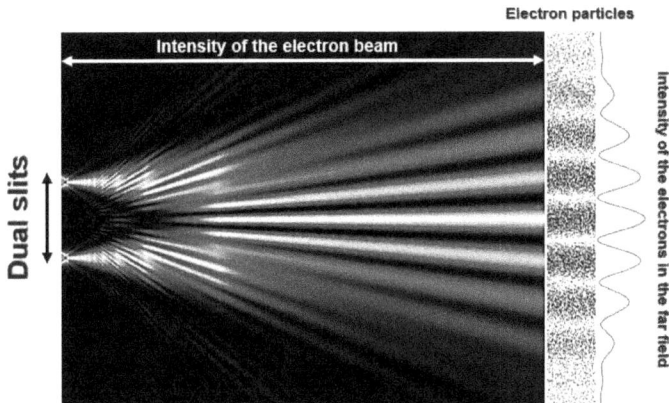

Figure 9.3. Young's slits experiment.

The phenomenon of interference between electromagnetic waves is primarily based on this idea. When dual or greater waves traverse the identical space, the remaining amplitude at every point is the sum of the amplitudes of the individual waves. In some cases, such as in a line array, the summed variant will have a greater amplitude than any of the components individually; this is known as constructive interference (Figure 9.4).

In different cases such as in noise-canceling, the summed variant has a smaller amplitude than the aspect variations; this is referred to as destructive interference (Figure 9.5).

Wave-particle duality: If the light beams are just a wave, then the electrons would absorb some energy, no matter what the frequency. If the light beams are just a wave, then the emission of an electron would take longer when a lower intensity light is used, not instantaneously (Figure 9.4).

Nonetheless, light is not just a wave. It can also behave as though it was made of tiny energy packets or particles. We call these particles photons (Figure 9.6). It is one of these photons that will hit one electron on the plate, the electron will absorb

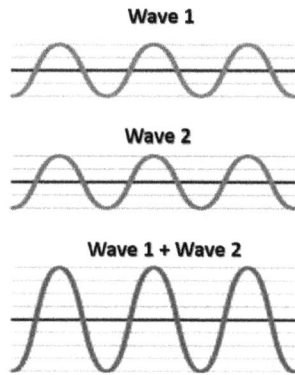

Wave 1

Wave 2

Wave 1 + Wave 2

Figure 9.4. Constructive interference.

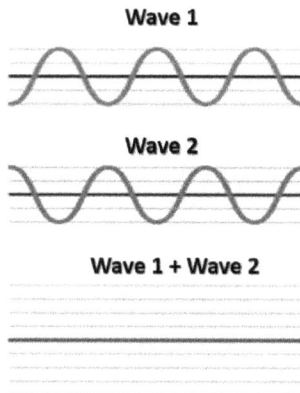

Wave 1

Wave 2

Wave 1 + Wave 2

Figure 9.5. Destructive interference.

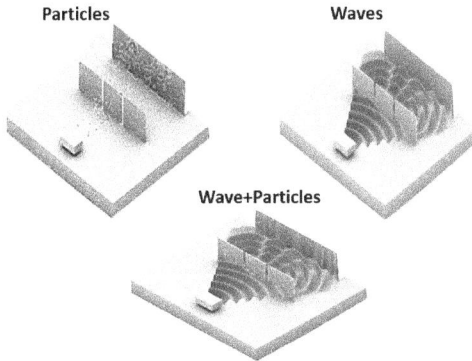

Figure 9.6. Sketch of the wave-particle duality of microwave signals.

the energy and it will fly off the plate. So if the intensity is greater, i.e., there are more photons, then more electrons can be knocked off.

9.4 What is Synthetic Aperture Interferometry?

Interferometric Synthetic Aperture Radar (InSAR) is a technique that maps millimetre-scale deformations of the Earth's surface with radar satellite measurements. Specified the continuous change of the Earth's crust the ability to yield measurements at night and throughout any weather condition makes this technique extremely valuable [3]. In this regard, let us assume that two complex SAR images S_1 and S_2 are separated by the baseline B (Figure 9.7). The radar phase difference ϕ_1 for a common transmitter is mathematically given by:

$$\phi_1 = \frac{4\pi}{\lambda}\left(\Delta R + \zeta\right) \tag{9.1}$$

where ΔR is the slant range difference from satellite to target, respectively, at different times, and λ is the SAR fine mode wavelength. For the surface displacement measurement, the zero-baseline InSAR configuration is ideal as $\Delta R = 0$, so that

$$\phi = \phi_d = \frac{4\pi}{\lambda}\left(\zeta\right) \tag{9.2}$$

Equation 9.2 demonstrates that the phase consists of the records of the distance between the antenna and the ground surface and is acquired as a fraction (remainder) of the distance (actually, twice the distance due to the spherical-trip) when divided by using the wavelength of the radio wave. Nonetheless, if nothing is done, it is challenging to use the phase as significant data owing to the a fractional portion of phase no longer contain the all wave magnitude wide-ranging of an integer portion.

Zero-baseline, repeat-pass InSAR configuration is hardly achievable for either spaceborne or airborne SAR. Therefore, a method to remove the topographic phase, as well as the system geometric phase in a non-zero baseline interferogram, is needed. If the interferometric phase from the InSAR geometry and topography can strip off from the interferogram, the remnant phase would be the phase from

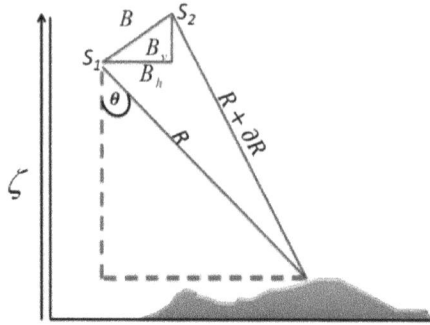

Figure 9.7. Principle of InSAR geometry.

block surface movement, providing the surface maintains high coherence [3]. Zebker et al. [5] exploited the three-pass method to remove the topographic phase from the interferogram. This method requires a reference interferogram, which is promised to contain the topographic phase only. The three-pass approach has the advantage in that all data is kept within the SAR data geometry, while the DEM method can produce errors by misregistration between SAR data and cartographic DEM [5]. The three-pass approach is restricted by data availability. The three-passes DInSAR technique uses another InSAR pair as a reference interferogram that does not contain any surface movement event as:

$$\phi' = \frac{4\pi}{\lambda} \Delta R' \tag{9.3}$$

Incorporating Equations 9.2 and 9.3 delivers the phase difference, only from the surface displacement as

$$\phi_d = \phi - \frac{\Delta R}{\Delta R'} \phi' = \frac{4\pi}{\lambda} \zeta \tag{9.4}$$

For an exceptional case where $\frac{\Delta R}{\Delta R'}$ in Equation 9.4 is a positive integer number, phase unwrapping may not be necessary [3]. However, this is not practical and it is difficult to achieve from the system design for a repeat-pass interferometer. From Equation 9.4, the displacement sensitivity of DInSAR is given as:

$$\frac{\partial \phi_d}{\partial \zeta} = \frac{4\pi}{\lambda} \tag{9.5}$$

The important dimension in InSAR is the component of the surface topography changes in the slant range direction. The resulting range displacement ΔR is therefore given by:

$$\Delta R = |m| \sin(\theta_i - \zeta) \tag{9.6}$$

where m is the surface topography deformation vectors that are describing the direction and magnitude of the surface changes, θ_i is the SAR data incident angle, and ζ the surface topography's elevation. Using Equation 9.4 into 9.6 then, the sea surface topography deformation is estimated by

$$|m| = \frac{\lambda\phi_d}{2\pi\sin(\theta_i - \zeta)} \tag{9.7}$$

9.5 Interferograms

Interferograms are the result of subtracting the phases of two SAR images, a process called Differential-InSAR (D-InSAR). The phase difference is depicted by the colour of the pixels in an interferogram. The interferogram shows repeating colour cycles. Because the same phase value is repeated in every cycle of the radar wave, the colours repeat. In this understanding, the colour of fringes in an interferogram reveals the phase difference resulting from the difference in distance between two SAR data at the point. For instance, an area where the phase difference is 0 degree is showed in light-blue colour, 60 degrees showed in blue colour, 180 degrees showed in red colour, and so on. If the phase difference is 360 degrees, the distance that a radio wave travels back and forth between the radar and the ground changes by just one wavelength [6–9].

For instance, for SAR wavelength of 23.6 cm, the phase difference of 360 degrees is equivalent to the movement of 11.8 cm (one-way distance is half of round-trip one). Similarly, the point whose difference in phase is 60 degrees shifts 2.0 cm $23.6 \times \left(\dfrac{60}{360}\right) 0.5 \approx 2.0$ *cm.* In this regard, the phase difference is equivalent to the quantity of deformation. As the colour of fringes represents the magnitude of deformation at the point that contains phase difference (Figure 9.8).

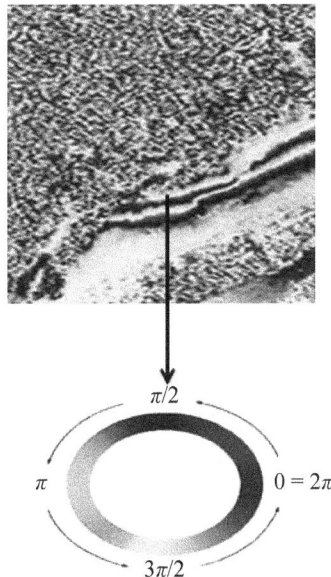

Figure 9.8. Wave cycles in an interferogram.

It can be realized that the equivalent colour cycle repeats three times, in the direction of the increasing phase. This means that the western part has subsided three times over a radar full wavelength concerning the eastern part between the two image acquisitions that were used to produce the interferogram. Because we know the exact length of a full-wave cycle, the deformation can be derived [7, 11–3]. This process is called unwrapping. Nonetheless, if there was a strong atmospheric activity during one or both of the image acquisitions used for producing the interferogram, the deformation signal can be polluted with atmospheric noise. To resolve deformation from the atmosphere, a time series of interferograms can be used to model the atmospheric signals [4, 10].

9.6 Phase Unwrapping

The phase acquired by InSAR is only measured in values between 0 and 360 degrees. Therefore, the magnitude of deformation cannot be determined from it without modification. The phase that originally takes on a wide range of value is folded (wrapped) in the range between 0 and 360 degrees. The work of unwrapping and turning the phase back to the magnitude of deformation is termed phase unwrapping. This precious technique is crucial to the 3-pass and 4-pass processes and can be smeared to the automatization and the precision enhancement of the baseline approximation, and the quantitative analysis by smearing it to the baseline estimation and conclusive results [14–18].

In the process mentioned above, if the atmosphere delay affects the whole of the image, the image often has a large margin of error. Therefore, it is essential to perform the atmospheric compensation and calculate the time average of the interferograms formed at distinct times.

In addition to comprehending how precise the InSAR data is, it is significant to distinguish how consistent it is. To conduct deformation measurements, the phase recorded is used by the satellite. This is the fraction of a complete wave cycle that is in addition to the number of full wavelengths the signal has travelled. Phase measurements can be thought of as the time on a clock. There is a finite number of times that can be demonstrated on a clock and every twelve hours those times repeats. Similarly, there are a finite number of phase measurements that can be recorded within one wave cycle (0 to 2π). Once a wave cycle has completed, the phase measurements then repeat, in which the phase measurements are shown in a circle, like a clock (Figure 9.9).

Nonetheless, the values have a new sense as soon as a new revolution has taken place. For a clock, a new revolution indicates a new 12-hour cycle has begun. To entirely comprehend what time it is, one requests to distinguish the date and/or which part of the day it currently is. Likewise, a new revolution of the phase cycle shows that a full-wave cycle has been accomplished. A fully understanding of phase calculations is required accurate discrimination in the value of full-wave cycles associates with the phase measurements. Resolving this to determine the actual deformation is called 'unwrapping' [18–22].

To originate the genuine deformation from sequential phase measurements in time, correlations in time and/or space are exploited to 'unwrap' the measurements.

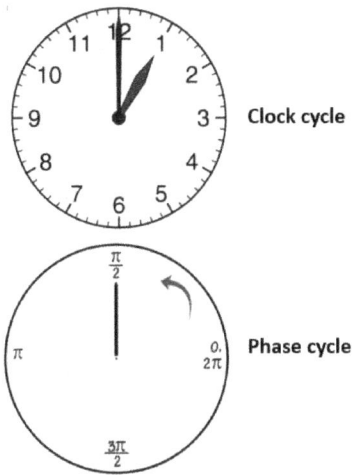

Figure 9.9. Both clock time and radar phase are ambiguous.

For instance, the occurred deformation can be assumed as linear during (a part of) the measurement time-span or that the deformation rate alteration is smoothly in space. If a wrong assumption is made, this may lead to supposed 'unwrapping errors'.

9.7 How to Understand SAR Interferograms?

InSAR measures in the direction along a straight line between a SAR antenna and the ground (the direction of the satellite's line of sight(LOS)). As it is mentioned in previous chapters, SAR satellite transmits and receives a radio wave obliquely downward (obliquely upward from the ground). Besides, the track on which the satellite observes the ground from the east or the west is often used. The ground movement is three dimensional (east and west, north and south, up and down) but InSAR can measure it in only a single direction, in the LOS direction. Consequently, the direction the ground moved in is difficult to be determined.

The "LOS distance becomes longer" indicates that the ground moved in the direction away from the satellite. In other words, if the satellite observes the ground from the east, the ground moved down or west. At this time, we cannot distinguish the ground movement whether the ground relocated in the north-south direction. The ground might move down, west, or down-west. Then, the ground might change in the direction away from the satellite since the ground moved down-east but the downward component countered the east component. Conversely, these differences cannot be detected [22–24].

The result of InSAR expresses the ground deformation in the change of colours. The gentle deformation of the ground causes the gentle change of colour, the steep deformation of the ground does the steep change of colour (Figure 9.10).

The difference in displacement between the places in the same colour is one of the integral multiples of the half SAR wavelength (11.8 cm). In this understanding, the existence displacement is just 11.8 cm if the colour turns back to the same colour (Figure 9.11).

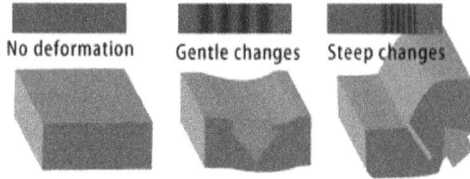

Figure 9.10. The scenario of ground deformation as based on changing of colour cycle.

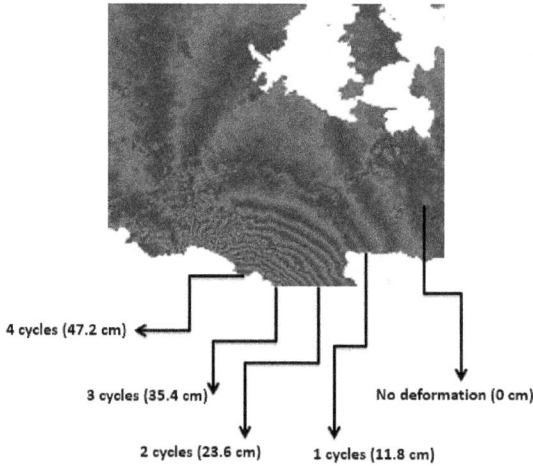

4 cycles (47.2 cm)

3 cycles (35.4 cm)

No deformation (0 cm)

2 cycles (23.6 cm)   1 cycles (11.8 cm)

Figure 9.11. Example of deformation magnitude from interferogram pattern.

The interferometric phase involves the signature owing to topography φ_{topo}, displacement φ_{disp}, atmospheric effect φ_{atmo}, baseline error $\varphi_{baseline}$, and noise φ_{noise}, that is mathematically expressed as:

$$\phi = \varphi_{topo} + \varphi_{disp} + \varphi_{atmo} + \varphi_{baseline} + \varphi_{noise} \tag{9.8}$$

Equation 9.8 reveals that the phase concerning surface warp can be assimilated by eliminating other constituents. The topographic phase can be modeled using a digital elevation model (DEM). The atmospheric effects, therefore, can host spatially correlated artifacts of a few centimeters, and the baseline error can be simulated and eliminated. Consequently, the noise level of the interferometric phase relies on the coherence of the image pairs. In this regard, the displacement phase is grounded on the repeat-pass interferometric phase after eliminating the topographic effect, the atmospheric effect, and baseline error. In other words, phase displacement is considered as the incoherent absolute of changes in the surface backscattering phase, volume backscattering phase, and double-bounce backscattering phase. The mathematical description of phase displacement is given by:

$$\varphi_{disp} = + \varphi_{surface} + \varphi_{volume} + \varphi_{double-bounce} \tag{9.9}$$

Equation 9.9 is the principle concept to determine topographic level changing from the interferometric phase value as a function of high coherent variation. In other words, the interferogram, φ, consists of the topographic, φ_{topo} and deformation, φ_{defo}

components and ignore the others. The topographic component can be calculated using the relationship between the known heights H, acquired from a known topographic map as given by:

$$\varphi_{topo} = \frac{4\pi B_{CT} \cos(\theta - \gamma_{CT})}{\lambda R_m \sin \theta} H \qquad (9.10)$$

Equation 9.10 demonstrates that the SAR wavelength is λ, $B_{CT}\cos(\theta-\gamma_{CT})$ is the perpendicular baseline of the two observations, θ is the incidence angle, and R_m is the slant range distance. In this regard, the deformation due to oil and gas explorations can be estimated by subtracting φ_{topo} from the interferogram. Therefore, the deformation phase value, φ_{defo}, corresponds to the shrink or extension in the line-of-sight distance between the satellite and ground targets when the deformation occurs between two SAR observations. As well as the change in the line-of-sight distance as ΔR is identified, the phase value can be given by:

$$\varphi_{defo} = \frac{4\pi}{\lambda} \Delta R \qquad (9.11)$$

Equation 9.11 reveals that the range of phase value is constrained between $-\pi$ and π, which does not allow to distinguish deformation greater than a quarter-wavelength. For instance, L-band wavelength is 24 cm. In the view of $\Delta R = 0, +/-12, +/-24...$ cm, φ_{defo} becomes 0 with indefinite $2m\pi$. Consequently, the deformation cannot be defined if the neighbouring pixels have more than a 6 cm line-of-sight difference. This indicates that long-wavelength has the advantage to measure a large deformation. Therefore, the absolute deformation can be estimated by unwrapping the phase as long as the deformation satisfies the sampling theorem.

9.8 Quantum of Differential-InSAR (QD-InSAR)

Let us assume that the dual antenna and a delay phase ϕ_{delay} is acquainted with one of the lengths of the interferometer. Consequently, the quantity ϕ_{delay} is computed by evaluating the strength of the microwave signal of the dual output beams. In this regard, the value ϕ_{delay} can be computed with a statistical error proportional to $\frac{1}{\sqrt{N}}$, as N is a non-entangled microwave signal. Precisely, the direct simulation of the mean traveling time is determined for N, a non-entangled signal. In this instance, the miscalculation is expressed as:

$$\partial R \approx O\left[\Delta\omega\sqrt{N}\right]^{-1} \qquad (9.12)$$

Equation 9.12 confirms that the inverse of the bandwidth $\Delta\omega$ affects the inaccuracy that arises on account of the oscillation of the range ∂R. Nonetheless, entangled signals can be utilized to overwhelm the conventional quantum constraint and achieve the Heisenberg limit. To handle the quantum mechanism of DInSAR, let us deliberate a dual-mode, path-entangled, signal-number state, a sort of Schrödinger

Cat state, which is thoroughly recognized as the NOON state and is mathematically given by:

$$|\psi_{NOON}\rangle = \left[\sqrt{2}\right]^{-1}\left(|NO\rangle + |ON\rangle\right) \tag{9.13}$$

Equation 9.13 designates that the number of backscattered signals generated by DInSAR are either completely in the first complex SAR image S_1 or entirely in the second complex SAR image S_2. On the other hand, it is entirely removed into the entangled-signal source, for generating superpositions of these dual states in the form:

$$|NOON\rangle = |S_1\rangle + |S_2\rangle = |N\rangle_{S_1}|O\rangle_{S_2} + |O\rangle_{S_1}|N\rangle_{S_2} \tag{9.14}$$

Consequently, each part of the entangled state tolerates diagonally a different dimension of the phase changing, which is accurately articulated as:

$$|\psi_{NOON}\rangle = \left(\sqrt{2}\right)^{-1}\left[|N\rangle_{S_1}|O\rangle_{S_2} + |O\rangle_{S_1}|N\rangle_{S_2}\right] \tag{9.15}$$

Nonetheless, the quantity states are in existing tremendously non-conventional states, to commence with. Their accomplishment in the phase-shifter is extremely unrelated. When a monochromatic beam of measure states approves across a phase shifter, the phase shift is proportional directly to N, the number of photons. Therefore, there is no n-dependence in the coherent state, where induce n is the steady signal intensity at DInSAR. In manifestations of a unitary development of the state, the growth of any interferometry state transient through a phase shifter ϕ is standardized by:

$$\hat{U}(\phi) \equiv e^{(i\phi\hat{n})} \tag{9.16}$$

where $\hat{n}$ is the average interferometry fringes, which are generated by a phase shift in the quantum DInSAR technique. In this perspective, the phase shift operator can be exposed to have the consequent dual disparate influences on coherent received signal in opposition to quantize the phase shift states as:

$$\hat{U}_\phi|O\rangle = e^{i\phi}|O\rangle, \tag{9.17}$$

$$\hat{U}_\phi|N\rangle = e^{i\phi}|N\rangle, \tag{9.18}$$

Equations 15.17 and 15.18 expose that the phase shift for the coherent state is of known quantity; nonetheless, there is an N dependence on the exponential for the quantum phase shift state. The quantum phase shift state develops in phase N-times, which are further rapidly than the coherent state. In this instance, the N00N state develops into:

$$|\psi_{NOON}^\phi\rangle = \frac{1}{\sqrt{2N!}}\left(|N\rangle_{S_1}|O\rangle_{S_2} + e^{iN\phi}|O\rangle_{S_1}|N\rangle_{S_2}\right) \tag{9.19}$$

Consequently, Equation 9.19 can be expressed based on annihilation operator, i.e., bosonic creation operator $a_\alpha^\dagger$ as:

$$|\psi_{NOON}^{(\phi)}\rangle = \frac{1}{\sqrt{2N!}}\left(\left(\hat{a}_{S_1}^\dagger\right)^N + e^{iN\phi}\left(\hat{a}_{S_2}^\dagger\right)^N|0\rangle_{S_1}|0\rangle_{S_2}\right) \tag{9.20}$$

where the bosonic creation operator $a_\alpha^\dagger$ is given by

$$a_\alpha^\dagger \left| n_1,.....,n_{\alpha-1},n_\alpha,n_{\alpha+1},.... \right\rangle = \sqrt{n_{\alpha+1}} \left| n_1,.....,n_{\alpha-1},n_\alpha,n_{\alpha+1},.... \right\rangle, \tag{9.21}$$

Let us suppose the topography subsidence owing to oil and gas explorations S_{OG} is delivered by:

$$\hat{S}_{OG} = \frac{1}{N!}\left(\left(\hat{a}_{S_1}^\dagger\right)^N |0\rangle\langle 0| \left(\hat{a}_{S_2}^\dagger\right)^N + \left(\hat{a}_{S_2}^\dagger\right)^N |0\rangle\langle 0| \left(\hat{a}_{S_2}^\dagger\right)^N \right) \tag{9.22}$$

The QDInSAR observation measurements are vital to approximate the phase shift ground on annihilation operators. In this concern, the fluctuation in land deformation or land subsidence $\Delta\hat{S}_{OG}$ can create imprecision in the calculation of the phase shift as obeys:

$$\partial\phi_{defo} = \frac{\Delta\hat{S}_{OG}}{\left| -N\sin N\varphi \right|} \tag{9.23}$$

Equation 9.23 exposes that the Heisenberg constraint as the QDInSAR phase is deliberated as enormously entangled states. In other words, Equation 9.23 can be conveyed as [25]:

$$\lim_{\alpha_1\to 1}\lim_{\alpha 2\to 1} \partial\phi_{defo} = \frac{\Delta\hat{S}_{OG}}{\left| -N\sin N\varphi \right|} \tag{9.24}$$

9.9 Quantum Hopfield Algorithm for DInSAR Phase Unwrapping

To create quantum neural networks, let us deliberate the task of expanding multi-qubit quantum systems. In this regard, the accurate correlation between neurons and qubits [26], unlocking access to quantum properties of entanglement and coherence of SAR data is instigated. In this view, let us instead encode the neural network into the energy of a quantum state. This is accomplished by accommodating a memory regulation between activation patterns of the neural network and pure states $|\phi\rangle$ of L-level of a quantum system. In this understanding, the determined state features of the current pattern can be expressed as $\phi \rightarrow |\phi\rangle|\phi|_2$ with $|\phi|_2 = \sqrt{\sum_{i=1}^{L}\phi_i^2}$ and $|\phi\rangle := |\phi|_2^{-1}\sum_{i=1}^{L}\phi_i|i\rangle$, which is encoded concerning the standard basis such that $\langle\phi|\phi\rangle = 1$. The L-level quantum system can be realized by a register of $N = [log_2\ L]$ qubits with the intention of the qubit directly above signifying such network scales logarithmically with the quantity of neurons [27].

The quantum Hebbian learning algorithm (qHob) is exploited for weighting matrix M of the Hopfield network. It relies on dual vital perceptions: (i) a direct alliance of the weighting matrix with a mixed state ρ, and (ii) one can efficiently perform quantum algorithms that harness the information contained in W. Consequently, ρ registers of N qubits are relative to:

$$\rho := \omega + \frac{\Pi_L}{L} = W^{-1}\sum_{m=1}^{W}\left|\phi^{(m)}\right\rangle\left\langle\phi^{(m)}\right| \tag{9.25}$$

where $\prod_d$ is the L-dimensional identity matrix and W is a training set of activation patterns of phase ATI ϕ, with $m \in \{1, 2, ..., W\}$. In this view, the Hopfield neural network can be taught training set using the Hebbian learning rule, which sets the weighting matrix elements ω_{ij} along with the number of occasions in the training set that the neurons i and j mix together. In this sense, Hebbian learning is employed to set the weighting matrix M from the L-length training phase unwrapping data $\{\phi^{(m)}\}_{m=1}^{W}$. In this regard, the qHop algorithm continues to compute $|v\rangle = \dfrac{1}{\Delta\phi_{ATI}}|\phi\rangle$ where the pure state $|\phi\rangle$ is primarily arranged, which comprises user-identified neuron thresholds of the DInSAR data inverse matrix ϕ_{defo}^{-1} of feature identifications and a partial memory pattern. In other words, the inverse matrix ϕ_{defo}^{-1} contains information on the training data and regularization γ. To this end, the output of the pure state $|v\rangle$ contains information on the reconstructed state of phase $|\phi_{defo}\rangle$ and Lagrange multiplier vector $\vec{L}_y$. The feature state matrix of DInSAR phase data can be expressed as a quantum state by:

$$\phi_{defo}|v|_2|v\rangle = |w|_2|w\rangle \qquad (9.26)$$

where

$$|v\rangle := \frac{1}{|v|_2}\left(\left|\phi_{defo}\right|_2|0\rangle \otimes \left||\phi_{defo}|\right\rangle + \left|\vec{L}_v\right|_2|1\rangle \otimes \left|\vec{L}_v\right\rangle\right) \qquad (9.26.1)$$

$$|w\rangle := \frac{1}{|w|_2}\left(\left|\Xi\right|_2|0\rangle \otimes |\Xi\rangle + \left|\varphi_{defo}^{(inc)}\right|_2|1\rangle \otimes \left|\varphi_{defo}^{(inc)}\right\rangle\right) \qquad (9.26.2)$$

where φ_{defo}^{inc} is the normalized quantum state corresponding to the incomplete activation pattern and $|\varphi_{defo}^{inc}|_2 = l$. The objective of the Equations 9.26.1 and 9.26.2 is to optimize the energy function E in Equation 9.26. Besides, Ξ presents each element of qHop which should be set so that its magnitude is of order at most 1 qubit. In other words, $\Xi =: \{\Xi_i\}_{i=1}^{d} \in \mathbb{R}^d$ is a user-specified neuronal threshold vector that determines the switching threshold for each neuron. Let us identify the set of M unitary operators $\{U_k\}_{k=1}^{M}$ acting on an $N + 1$ register of qubits consistent with

$$U_k := |0\rangle\langle 0| \otimes \prod + |1\rangle\langle 1| \otimes e^{-i\left|\varphi_{defo}^{(k)}\right\rangle\left\langle\varphi_{defo}^{(k)}\right|\Delta t} \qquad (9.27)$$

Equation 9.27 reveals that the unitaries implement various phase pattern detections in DInSAR data for a small difference time Δt under the circumstance of $|\phi_{defo}^{(k)}\rangle\langle\varphi_{defo}^{(k)}|$. Conversely, these unitaries can be simulated using the following mathematical equation:

$$U_s := e^{-i|1\rangle\langle 1| \otimes s\Delta t}$$
$$|0\rangle\langle 0| \otimes \prod + |1\rangle\langle 1| \otimes e^{-is\Delta t} \qquad (9.28)$$

where, $|1\rangle\langle 1| \otimes s\Delta t$ is 1-sparse and efficiently simulatable. Conversely, the trace tr_2 of U_s is over the second subsystem containing the state feature $|\phi_{defo}\rangle$ in the DInSAR data, which is expressed by:

$$tr_2\left\{U_s\left(|q\rangle\langle q|\otimes\left|\phi_{defo}^{(k)}\right\rangle\left\langle\phi_{defo}^{(k)}\right|\otimes\sigma\right)U_s^\dagger\right\}$$
$$= U_k\left(|q\rangle\langle q|\otimes\sigma\right)U_k^\dagger + \mathcal{O}\left(\Delta t^2\right).$$

(9.29)

In Equation 9.29, the subsystem of ancilla qubit $|q\rangle\langle q|$ and σ efficiently experiences time evolution $\mathcal{O}(\Delta t^2)$ with U_k. The unitary $e^{i\phi_{defo}t}$ is simulated to fix the error ϵ for arbitrary t. Indeed, ϵ can be estimated by:

$$\epsilon := \left\|\left(e^{-i\left|\phi_{defo}^{(1)}\right\rangle\left\langle\phi_{defo}^{(1)}\right|\frac{t}{nM}}..e^{-i\left|\phi_{defo}^{(M)}\right\rangle\left\langle\phi_{defo}^{(M)}\right|\frac{t}{nM}}\right)^n - e^{-i\phi_{defo}t}\right\| \in \mathcal{O}\left(\frac{t^2}{n}\right).$$

(9.30)

Equation 9.30 indicates that the repetition of $n \in \mathcal{O}\left(\dfrac{t^2}{\epsilon}\right)$ is required alongside M-sparse Hamiltonian simulations. The advantages of this approach are that the copies of the training states $|\phi_{defo}^{(m)}\rangle$ can be used as quantum software states and, in addition, does not require superpositions of the training states. Conversely, the DInSAR feature matrix ϕ_{defo} can be identified as:

$$\arg(S_1\,S_2^*) := \begin{pmatrix} 0 & P \\ P & 0 \end{pmatrix} + \begin{pmatrix} -\left(\gamma+L_v^{-1}\right)\Pi_{L_v} & 0 \\ 0 & 0 \end{pmatrix} + \begin{pmatrix} \rho & 0 \\ 0 & 0 \end{pmatrix}$$

(9.31)

In this regard, the simulation time is split into n small time steps $t = n\Delta t$. In this circumstance, the error ϵ can be extended into Taylor expansion as:

$$\epsilon_{\Delta t} := \left\|e^{i\phi_{ATT}\Delta t} - U_B\left(\Delta t\right)U_C\left(\Delta t\right)U_D\left(\Delta t\right)\right\| \in \mathcal{O}\left(\Delta t^2\right)$$

(9.32)

where $U_B(\Delta t)$, $U_C(\Delta t)$, and $U_D(\Delta t)$ are the operators of $\begin{pmatrix} 0 & P \\ P & 0 \end{pmatrix}$, $\begin{pmatrix} -\left(\gamma+L_v^{-1}\right)\Pi_{L_v} & 0 \\ 0 & 0 \end{pmatrix}$, and $\begin{pmatrix} \rho & 0 \\ 0 & 0 \end{pmatrix}$, respectively. In this understanding, at most $\mathcal{O}(\Delta t^2)$, the errors $e^{i\begin{pmatrix} 0 & P \\ P & 0 \end{pmatrix}\Delta t}$, $e^{i\begin{pmatrix} -\left(\gamma+L_v^{-1}\right)\Pi_{L_v} & 0 \\ 0 & 0 \end{pmatrix}\Delta t}$, and $e^{i\begin{pmatrix} \rho & 0 \\ 0 & 0 \end{pmatrix}\Delta t}$ are simulated, respectively, which allow simulating the unitary $e^{i\phi_{ATT}t}$ to an error of $\epsilon \in \mathcal{O}\left(n\Delta t^2\right)$. Then, the matrix of spectral energy of the qHop can be mathematically expressed as:

$$\phi_E = \sum_{j:\left|\lambda_j(\Delta\phi)\right|>\lambda} \lambda_j(\Delta\phi)\left|E_j(\Delta\phi)\right\rangle\left\langle E_j(\Delta\phi)\right| +$$
$$\sum_{j:\left|\lambda_j(\Delta\phi)\right|<\lambda} \lambda_j(\Delta\phi)\left|E_j(\Delta\phi)\right\rangle\left\langle E_j(\Delta\phi)\right|$$

(9.33)

Equation 9.33 demonstrates how to simulate the gradient energy of the various feature pattern variations in the DInSAR as a function of the size of the eigenvalues λ_j in comparison to a fixed user-defined number $\lambda > 0$. In this circumstance, qHop maintains the polylogarithmic efficiency in run time whenever λ is such that $\lambda^{-1} \in$

$O(poly(\log d))$. Therefore, the primary matrix inversion algorithm returns (up to normalization) as:

$$A_E^{-1}\left|E\right\rangle = \sum_{j:\left|\lambda_j(\Delta\phi)\right|\geq\lambda} \left\langle E_j\left(A_{SAR}\right)\middle|E\right\rangle\left(\lambda_j(\Delta\phi)\right)^{-1}\left|E_j(\Delta\phi)\right\rangle\left(\lambda_j(\Delta\phi)\right) \qquad (9.34)$$

The input state of the energy gradient $\left|E\right\rangle$ is first achieved and it contains the threshold data and incomplete activation pattern and considers it in the eigenbasis of the matrix of DInSAR image energies A_E. Therefore, obtaining information from each energy pattern requires $O(K)$ operations of qHop. The qHop algorithm is then initialized along with sparse Hamiltonian simulation [38] to perform quantum phase estimation, allowing $\sum_j\left\langle E_j(\Delta\phi)\middle|E\right\rangle\middle|\tilde{\lambda}_j(\Delta\phi)\right\rangle\otimes\left|E_j(\Delta\phi)\right\rangle$ to be obtained with $\tilde{\lambda}_j(\Delta\phi)$, an approximation of the eigenvalue $(\lambda_j(\Delta\phi))$ to precision ϵ.

9.10 Application of Quantum DInSAR Hopfield Algorithm in Land Deformation Owing to Oil and Gas Explorations

In this section, the Phased Array type L-band Synthetic Aperture Radar (PALSAR) is used. It is an active microwave sensor using L-band frequency to achieve cloud-free and day-and-night land observation. It provides higher performance than the JERS-1's synthetic aperture radar (SAR). However PALSAR data have advantageous observation of fine resolution mode. ScanSAR mode enables us to acquire a 250 to 350 km width of SAR images (depending on the number of scans) at the expense of spatial resolution. This swath is three to five times wider than conventional SAR images. The development of the PALSAR is a joint project between JAXA and the Japan Resources Observation System Organization (JAROS).

Three ALOS PALSAR satellite data were acquired on January 19th, 2007 (Figure 9.12a), October 22nd, 2007 (Figure 9.12b), and April 1st, 2008 (Figure 9.12b) and November 17th, 2008. The first two images were ascending, while the third image was descending with HH polarization. These data are single look complex formatted data, located along the East Coast of Dungun peninsular, Malaysia. This Dungun district involves several oil and gas companies; for instance, Kawasan Perindustrian Petroleum Petronas, which is located between latitude 4° 33' 54" N and 4° 35' 6" N and longitude 103° 26' 6" N and 103° 27' 54" E (Figure 9.13). Kawasan Perindustrian Petroleum Petronas company metal structures are dominated by a higher backscatter value of −10 dB than the surrounding environment. However, seawater is dominated by a lower backscatter value of −22 dB. The slick zones in the second image have the lowest backscatter values of −25 dB, which are found along the coastal zone of Kawasan Perindustrian Petroleum Petronas company. In fact, natural oil-seeps are escaping from the oil and gas company into coastal waters and causing the lower backscatter zone in SAR images.

It is interesting to find that the coherence image coincided with backscatter variation along the coastal zone. The highest coherence value of more than 0.8 and close to 1.0 dominated the features of the oil and gas company metal structures in addition to the top of mountains at the left of the coherence image. These interesting findings are proved by the coherence ratio of 2.5 as is allied with the highest

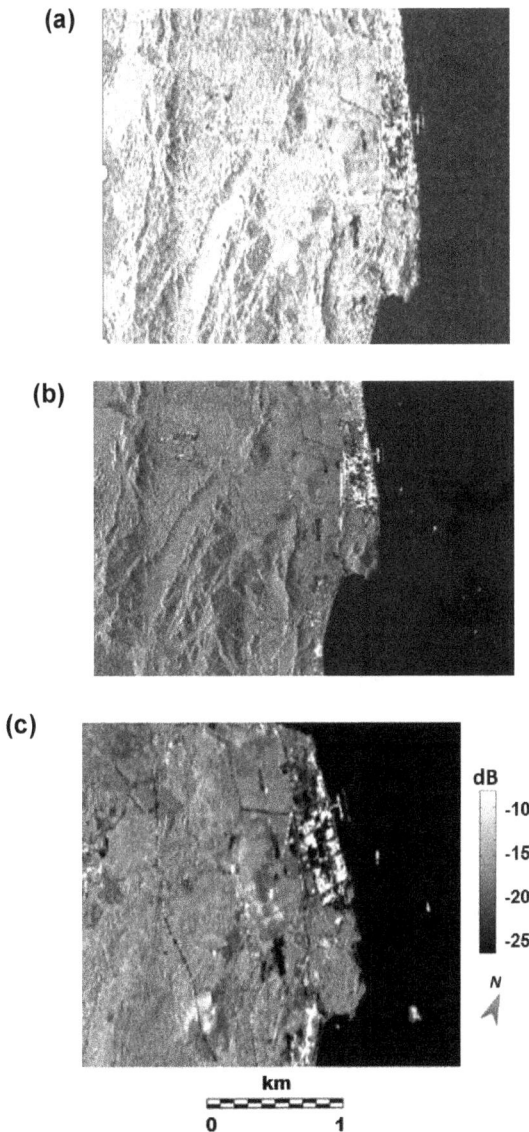

Figure 9.12. ALOS PALSAR satellite data were acquired on (a) January 19th, 2007, (b) October 22nd, 2007 and (c) April 1st, 2008.

coherence value of oil and gas company. The lowest coherence values of less than 0.4 and ratio coherence less than 1.5 are found in heavy vegetation covers and water body. In this regard, the water body has the lowest coherence value close to zero (Figure 9.14). Moreover, the ratio coherence image clearly indicates the total topographic decorrelation effects along the radar-facing slopes, which are dominant and are highlighted as brightest value of 3 of ratio coherence than the surrounding background.

Figure 9.13. The geographical location of oil and gas field company.

Coherence   Ratio Coherence

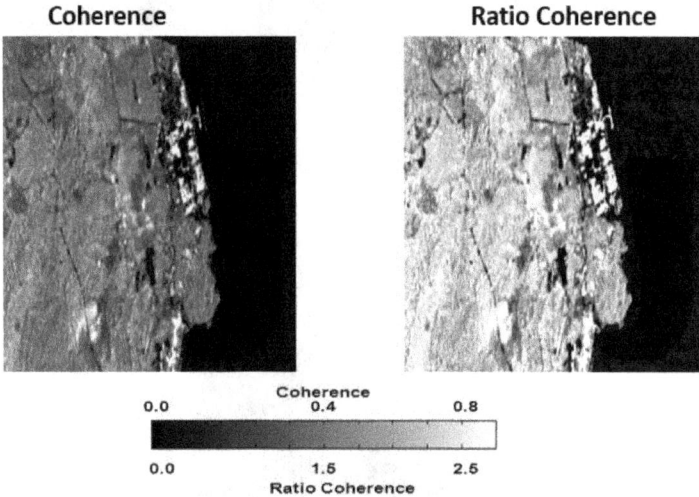

Figure 9.14. Coherence and ratio coherence images.

The interferogram phase derived by direct DInSAR phase measurement involves a high level of noise and the fuzzy pattern of the interferogram phase is ranged between $-0.7°$ to $+0.7°$ (Figure 9.15). Nevertheless, Figure 9.16 reveals that the interferogram phase is ranged between $-0.8°$ to $+0.8°$ derived from the qHop algorithm. A similar pattern is visible. This pattern signature represents deformation feature variations along with Kawasan Perindustrian Petroleum Petronas. Subsequently, qHop algorithm delivers an interferogram clear phase without noises.

Figure 9.17 demonstrates the rate of displacement or deformation along with Kawasan Perindustrian Petroleum Petronas. The deformation was modelled using qHop algorithm that ranges from -2 mm to $+2$ mm/year. The land subsidence occurs east of the Kawasan Perindustrian Petroleum Petronas owing to the existence of the major fault in this area (Figure 9.18), which is allied with the crosscut fault

-0.7°   +0.7°
Interferogram Phase

Figure 9.15. Interferogram phase retrieved by direct DInSAR phase estimation.

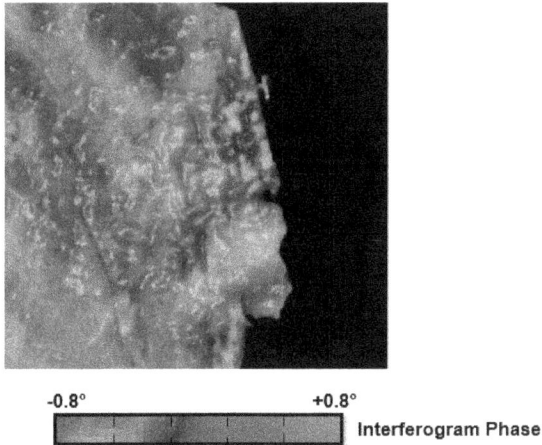

-0.8°   +0.8°
Interferogram Phase

Figure 9.16. Interferogram phase retrieved by qHop algorithm.

zone. In this view, the heavy density of northeast lineaments run towards Kawasan Perindustrian Petroleum Petronas.

Figure 9.19 reveals land displacement of +2 mm/year across Kawasan Perindustrian Petroleum Petronas due to heavy oil and gas activities along the East Coast of Malaysia, which stress the running lineaments and fault across oil and gas company. This scenario is proved by QDInSAR fringes that show land deformation extends in the west side of oil and gas company with 2 mm/year. In other words, the land deformation of –2 mm/year is allied with the lineaments and faults are demonstrated in Figure 9.20.

According to the above perspective, the qHop algorithm can be used to model land deformation owing to oil and gas exploration. Involving such quantum

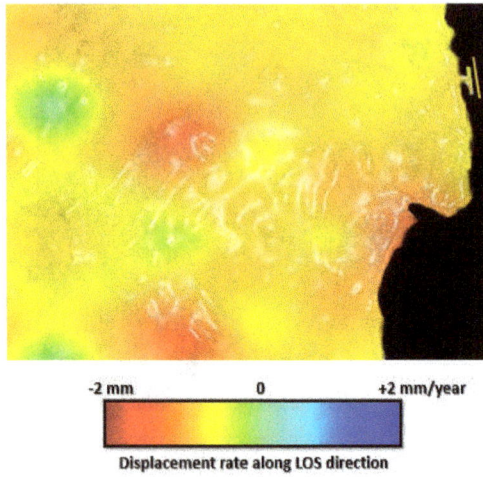

Figure 9.17. Deformation rate along with Kawasan Perindustrian Petroleum Petronas.

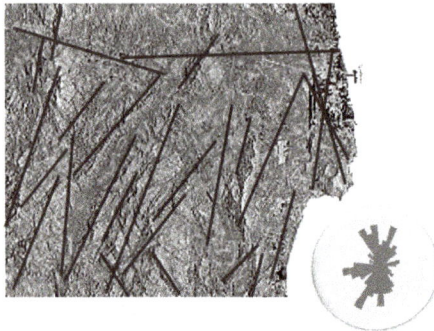

Figure 9.18. Faults and lineaments around Kawasan Perindustrian Petroleum Petronas.

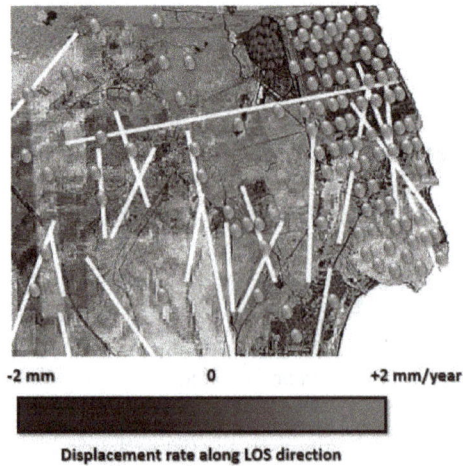

Figure 9.19. Land deformation rate crosses Kawasan Perindustrian Petroleum Petronas.

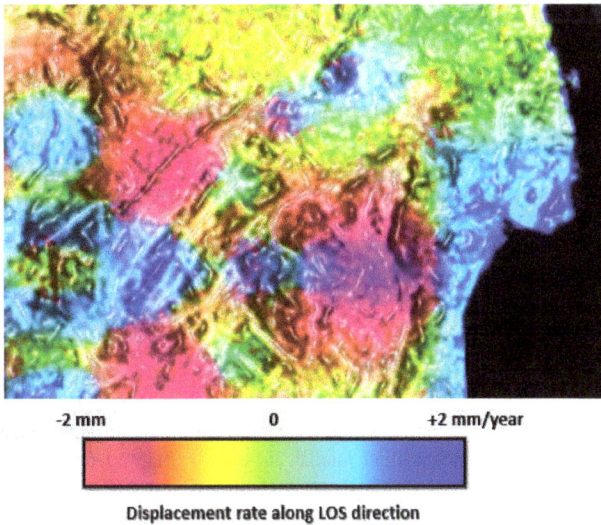

Figure 9.20. QDInSAR land deformation along with Kawasan Perindustrian Petroleum Petronas.

computing algorithm qHOP in phase unwrapping can determine accurate phase interferogram information that assists to reveal the land deformation across the oil and gas exploration company in the East Coast of Malaysia with the rate of land subsidence of –2 mm/year. Needless to say that integration qHOP algorithm with the QDInSAR technique can be exploited to monitor the ground movement and deformation across the oil and gas exploration company. This investigation confirms previous work that implemented the InSAR technique in monitoring oil and gas fields [25–28].

This chapter introduces a novel technique for interferometry synthetic aperture radar processing. The modification of Differential-InSAR (DInSAR) is performed using quantum mechanics. In this context, the quantum Hebbian learning algorithm (qHob) is exploited for weighting the matrix of the Hopfield network to reconstruct accurate phase unwrapping. The new technique, which is named quantum Differential-InSAR (QDInSAR), can detect the subsidence activity along with the oil and gas company.

References

[1] Zoback MD. Reservoir Geomechanics. Cambridge University Press; 2010.

[2] Zoccarato C. Data Assimilation in Geomechanics: Characterization of Hydrocarbon Reservoirs. Ph.D theses, Università degli Studi di Padova; 2016.

[3] Rosen PA, Hensley S, Joughin IR, Li FK, Madsen SN, Rodriguez E, Goldstein RM. Synthetic aperture radar interferometry. Proceedings of the IEEE. 2000 Mar; 88(3): 333–82.

[4] Zebker HA, Werner CL, Rosen PA, Hensley S. Accuracy of topographic maps derived from ERS-1 interferometric radar. IEEE Transactions on Geoscience and Remote Sensing. 1994 Jul; 32(4): 823–36.

[5] Rao KS, Al-Jassar HK. Error analysis in the digital elevation model of Kuwait desert derived from repeat pass synthetic aperture radar interferometry. Journal of Applied Remote Sensing. 2010 Sep; 4(1): 043546.

[6] Lu Z, Crane M, Kwoun OI, Wells C, Swarzenski C, Rykhus R. C-band radar observes water level change in swamp forests. EOS, Transactions American Geophysical Union. 2005 Apr 5; 86(14): 141–4.

[7] Ramsey III E. Radar Remote Sensing of Wetlands. Ann Arbor Press, Michigan, 1999.

[8] Marghany M. Synthetic Aperture Radar Imaging Mechanism for Oil Spills. Gulf Professional Publishing, 2019.

[9] Kwoun OI, Lu Z. Multi-temporal RADARSAT-1 and ERS backscattering signatures of coastal wetlands in southeastern Louisiana. Photogrammetric Engineering & Remote Sensing. 2009 May 1; 75(5): 607–17.

[10] Marghany M. Three dimensional coastline deformation from Insar Envisat Satellite data. pp. 599–610. In International Conference on Computational Science and Its Applications 2013 Jun 24. Springer, Berlin, Heidelberg.

[11] Marghany M. Hybrid genetic algorithm of interferometric synthetic aperture radar for three-dimensional coastal deformation. pp. 116–131. InSoMeT 2014 Aug 29.

[12] Marghany M. Three-dimensional visualisation of coastal geomorphology using fuzzy B-spline of dinsar technique. International Journal of Physical Sciences. 2011 Nov 23; 6(30): 6967–71.

[13] Marghany M. Simulation of three-dimensional of coastal erosion using differential interferometric synthetic aperture radar. Global NEST Journal. 2014 Jan 1; 16(1): 80–6.

[14] Marghany M. Three-dimensional coastal geomorphology deformation modelling using differential synthetic aperture interferometry. Zeitschrift fur Naturforschung A-Journal of Physical Sciences. 2012 Jun 1; 67(6): 419.

[15] Marghany M. DInSAR technique for three-dimensional coastal spit simulation from radarsat-1 fine mode data. Acta Geophysica. 2013 Apr 1; 61(2): 478–93.

[16] Marghany M. Four-dimensional earthquake deformation using ant colony based Pareto algorithm. Communications in Applied Sciences. 2019 Feb 20; 7(1).

[17] Van Genderen J, Marghany M. A three-dimensional sorting reliability algorithm for coastline deformation monitoring, using interferometric data. In IOP Conference Series: Earth and Environmental Science 2014 (Vol. 18, No. 1, p. 012116). IOP Publishing.

[18] Ferretti A, Monti-Guarnieri AV, Prati C, Rocca F, Massonnet D. INSAR Principles B. ESA Publications; 2007.

[19] Xu W, Cumming I. A region-growing algorithm for InSAR phase unwrapping. IEEE Transactions on Geoscience and Remote Sensing. 1999 Jan; 37(1): 124–34.

[20] Bechor NB, Zebker HA. Measuring two-dimensional movements using a single InSAR pair. Geophysical Research Letters. 2006 Aug; 33(16).

[21] Ferretti A, Prati C, Rocca F. Multibaseline InSAR DEM reconstruction: The wavelet approach. IEEE Transactions on Geoscience and Remote Sensing. 1999 Mar; 37(2): 705–15.

[22] Imel DD, Hensley S, Pollard B, Chapin E, Rodriguez E. AIRSAR Along-Track Interferometry. Inchez 4th European Conference on Synthetic Aperture Radar 2002 (Vol. 117).

[23] Suchandt S, Runge H. Along-track interferometry using TanDEM-X: first results from marine and land applications. pp. 392–395. In EUSAR 2012; 9th European Conference on Synthetic Aperture Radar 2012 Apr 23. VDE.

[24] Romeiser R, Johannessen J, Chapron B, Collard F, Kudryavtsev V, Runge H, Suchandt S. Direct surface current field imaging from space by along-track InSAR and conventional SAR. pp. 73–91. In Oceanography from Space 2010. Springer, Dordrecht.

[25] Tamburini A, Del Conte S, Ferretti A, Cespa S. Advanced satellite InSAR technology for fault analysis and tectonic setting assessment. Application to reservoir management and monitoring. pp. cp-395. In IPTC 2014: International Petroleum Technology Conference 2014 Jan 19. European Association of Geoscientists & Engineers.

[26] Ketelaar VG. Satellite Radar Interferometry: Subsidence Monitoring Techniques. Springer Science & Business Media; 2009 Apr 7.

[27] Klemm H, Quseimi I, Novali F, Ferretti A, Tamburini A. Monitoring horizontal and vertical surface deformation over a hydrocarbon reservoir by PSInSAR. First Break. 2010 May 1; 28(5).

[28] Tamburini A, Bianchi M, Giannico C, Novali F. Retrieving surface deformation by PSInSAR™ technology: A powerful tool in reservoir monitoring. International Journal of Greenhouse Gas Control. 2010 Dec 1; 4(6): 928–37.

CHAPTER 10

Quantum Machine Learning Algorithm for Iron, Gold, and Copper Detection in Optical Remote Sensing Data

10.1 How Copper and Gold Form in the Earth?

To comprehend how copper is created, it helps to watch how porphyry deposits—the major existing causes of copper ore—develop. In hydrothermal veins, the orebodies develop that are born in subversive magma chambers far beneath the deposit itself. In this view, the extremely high temperatures of volcanic magma generate hydrothermal veins, permitting some of the thermal energy to seepage near the higher layers of the Earth's crust.

This is why copper is frequently initiated in the sedimentary layer, wherever sand and mud are crushed and developed a deposit of sedimentary rock on the surface of the Earth. Therefore, copper ore can be developed and trapped in oxidized zones within these kinds of rocks. In this regard, copper is generally formed in basalt cavities that have been in contact with hydrothermal veins, and in the oxidized zones of mineral deposits. In this sense, the existence of volcanoes in a province is an excellent evidence of the existence of cooper since lava can create plenty of basalt nearby the sedimentary deposit layers on the Earth's crust [1, 2].

The vital query is now: where does all Earth's gold come from? In the creation of Earth, molten iron descended to its centre to form the core in which common massive valuable metals are located; for instance, gold and platinum. There are adequate precious metals in the core to conceal the entire surface deposits of Earth with four-metre thick layers. In this understanding, ancient Greenland's billion years old rocks demonstrate that gold and other precious elements have complied with the core of the Earth when gold formed. In other words, the excess of accessible gold on Earth is the fortunate by-product of meteorite bombardment. In this regard, the impacting meteorites were stirred into Earth's mantle by gigantic convection processes and then left behind the formation of the gold. Therefore, geological processes created the continents and converged the precious metals in ore deposits which are mined nowadays. Needle to say that gold has been added to the Earth by auspicious chance when the Earth was stricken by approximately 20 billion tonnes of asteroidal material [3].

In this understanding, while nuclear fusion within the Sun makes many elements, the Sun cannot synthesize gold. The considerable energy required to make gold only occurs when stars explode in a supernova or when neutron stars collide. Under these extreme conditions, heavy elements form via the rapid neutron-capture process or r-process.

The exclusion of gold to the core should keep the outer slice of Earth bereft of bling. Nevertheless, precious metals are tens to thousands of times additionally plentiful in Earth's silicate mantle than expected. In this view, this opportune over-abundance results from a cataclysmic meteorite rinse that struck Earth after the core came into existence A full load of meteorite gold was thus combined to the mantle alone and not dropped to the deep interior [3, 4].

10.2 How Copper and Gold are Mined?

In conjunction with numerous additional sorts of ore, copper is initiated. In this understanding, copper can occur near gold, silver, zinc, lead, and other sorts of metal deposits. Therefore, is copper abundant on earth? Nearly 80 percent of the copper that has been mined to date is still in use. Copper is a 100 percent recyclable metal. It is an abundant metal in Earth's crust, present at concentrations of 50 parts per million.

In anticipation of the progress of the recent copper mining procedure, it was naturally a derivative of mining for other sorts of metals. In this view, the deposits that include the major amounts of copper are termed porphyry deposits, and the procedures of mining generally comprise drilling an open mine into a deposit of sedimentary rock [2, 5].

Therefore, copper and gold have also been found near certain iron ores, a fact that these "big three metals" could be shaped at the same geological time, deep down in the crust of the Earth. In other words, iron illustrates the route to copper and gold. Consequently, the significant question is now: where does gold occur?

All of the gold found on Earth came from the debris of dead stars. As the Earth formed, heavy elements such as iron and gold sank toward the planet's core. If no other event had occurred, there would be no gold in the Earth's crust. But, around 4 billion years ago, Earth was bombarded by asteroid impacts. These impacts stirred the deeper layers of the planet and forced some gold into the mantle and crust.

Some gold may be found in rock ores. Therefore, the gold occurs with pyrite between the quartz pebbles (Figure 10.1). It makes occur as flakes, as the pure native element, and with silver in the natural alloy electrum. Erosion frees the gold from other minerals. Since gold is heavy, it sinks and accumulates in stream beds, alluvial deposits, and the ocean.

Earthquakes play an important role, as a shifting fault rapidly decompresses mineral-rich water. When the water vapourizes, veins of quartz and gold deposit onto rock surfaces. A similar process occurs within volcanoes. Consequently, gold cannot be formed through chemistry or alchemy reaction processes. Indeed,

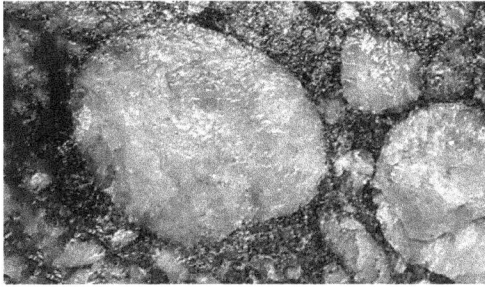

Figure 10.1. Quartex pebbles contain gold.

chemical reactions cannot modify the number of protons within an atom. The proton number or atomic number expresses an element's distinctiveness. In this understanding, attempts by alchemists to turn lead (or other elements) into gold were unsuccessful because no chemical reaction can change one element into another. Chemical reactions involve a transfer of electrons between elements, which may produce different ions of an element, but the number of protons in the nucleus of an atom is what defines its element. All atoms of gold contain 79 protons, so the atomic number of gold is 79. In this regard, making gold isn't as simple as directly adding or subtracting protons from other elements. The most common method of changing one element into another (transmutation) is to add neutrons to another element. Neutrons change the isotope of an element, potentially making the atoms unstable enough to break apart via radioactive decay.

10.3 What are the Characteristics of Copper and Gold?

Most copper ores are of hydrothermal origin and consist of various mixtures of copper sulfides such as chalcopyrite, $CuFeS_2$, bornite, Cu_5FeS_4, and chalcocite, Cu_2S. Native copper perhaps encompasses minor quantities of Ag, As, and Fe. The crystal structure of copper and gold is isometric with an all-face-centred lattice (F) and involves Cu atoms in cubic closest packing. Consequently, gold, silver, and copper are isostructural (Figure 10.2) [2, 5].

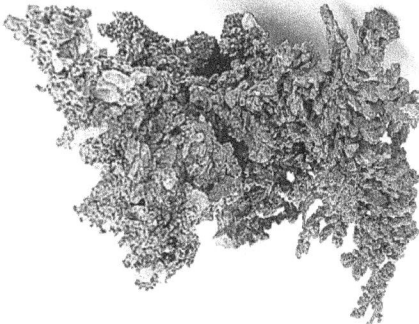

Figure 10.2. Isostructural of copper.

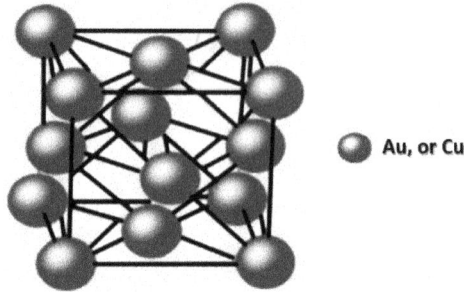

Figure 10.3. The crystal form of copper and gold.

Consequently, the crystal form of copper is isometric, $4/\bar{3}2/m$. Copper crystals are distorted cubes, octahedrons, and dodecahedrons in branching and arborescent groups (Figure 10.3), which is usually formed in irregular masses and scales.

The copper's hardens is 3½–3, while its specific gravity is approximately 8.9. Copper is also highly malleable, ductile and hackly fracture. Therefore, copper has a red colour on a fresh surface but is usually dark as a consequence of tarnish. It has a metallic luster and is opaque. Needle to say that native copper is recognized by its high specific gravity, malleability, and red colour on a fresh surface.

Most native gold, therefore, involves some Cu, Ag, or other metals. In this regard, small amounts of Cu and Fe may also be present in most of the native gold. In this understanding, in an alloy, the proportion of pure gold is expressed in carat (k). Subsequently, pure gold is 24 carat, where 18 k gold comprises 18 fragments of gold and 6 fragments of other metals.

It is isometric, $4/\bar{3}2/m$, which is rarely in virtuous crystals; for instance, well-developed octahedra in irregular plates, scales, or masses. Besides, actual gorgeous, museum-quality specimens involve arborescent crystal groups and irregular dendritic shapes (Figure 10.4) [3–5].

Gold's hardness (H) equals 2½–3, and the specific gravity of 19.3 when the gold is pure. However, its specific gravity becomes as lower as 15 in the presence of other metals in solid solution

Therefore, pure copper is soft, malleable, ductile with a hackly fracture. Besides, cooper and gold have a metallic luster and opaque as well as other minerals such as silver.

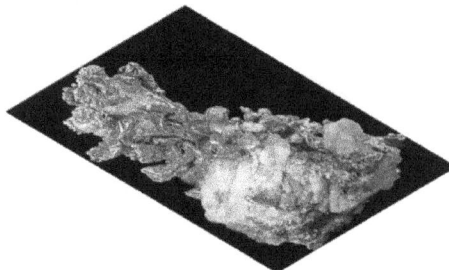

Figure 10.4. Crystal group form of gold.

Its colour ranges from various shades of yellow to very pale yellow as a function of increasing Ag content. Moreover, gold is recognized from yellow sulfides, for instance pyrite and chalcopyrite, by their high specific gravity and sectility. Sometimes, pyrite is denoted as fool's gold.

10.4 Remote Sensing for Copper and Gold Identifications

The keystone indicator in copper, iron, and gold exploration is the identification of hydrothermally altered rocks in remote sensing data. Porphyry copper deposits are copper ore bodies that are moulded from hydrothermal fluids created from a voluminous magma chamber more than a few kilometres beneath the deposit itself. Preexisting or allied with those fluids are upright dikes of porphyritic intrusive rocks which are formed earlier in the crystallisation sequence of the magma. In later phases, mingling meteoric fluids perhaps interrelate with the magmatic fluids. Consecutive wrappers of hydrothermal alteration naturally enfold a core of scattered ore minerals in regularly stockwork-creating hairline fractures and veins. On account of their great bulk, a little accumulation of copper of 0.15% on porphyry orebodies is considered as economic quantities of by-products; for instance, molybdenum, silver, and gold. In this understanding, spotting host rocks in remote sensing data have been widely and successfully implemented for the exploration of epithermal gold and porphyry copper deposits [7, 10].

The main question is: how remote sensing can recognize epithermal gold and porphyry copper deposits? The high spectral characteristics for remote sensing data such as ASTER shortwave infrared radiation subsystem are designed to gauge spectral signature reflection to identify Al–OH, Fe, Mg–OH, Si–O–H and CO absorption features. In this sense, distinguishing host rock and then alteration zone are the gateway in detecting a specific ore deposit kind. For instance, porphyry copper and epithermal gold mineralization deposits naturally arise in acidic igneous rocks to intermediate igneous rocks, for instance rhyolite, dacite, diorite and granodiorite, which are possibly sensed by their spectral absorption features. Electrons in mineral atoms vibrate owing to spectral absorption features that entangle with incidence photons of shortwave infrared radiation and form specific quantum spectral signatures, in which different sorts of rocks and minerals can be distinguished. Consequently, the spectral absorption characteristics of such rocks are the keystone to distinguish hydrothermal alteration as the gateway for ore deposit detections and mapping [8–11].

In hydrothermal and magmatic-hydrothermal gold deposits, potassic, sodic-calcic, sericitic, argillic, and propylitic alterations are the most prevalent hypogene alterations. In this view, iron oxide-hydroxide minerals are effective indicators for supergene alterations including hematite, goethite, and jarosite. Generally, oxidation of copper-iron and iron sulfide minerals, which are chiefly pyrite and chalcopyrite, form these minerals in supergene surroundings. In this understanding, gold deposits, pyrite and chalcopyrite are the most significant co-deposited and host minerals for gold particles in hydrothermal and magmatic-hydrothermal [7–10].

Consequently, integration between geochemical and remote sensing techniques are a valuable approach for gold exploration. At the regional scale, stream sediment geochemistry in the form of a conventional fine fraction, Bulk Leach Extractable

Gold (BLEG), cyanide leach or pan concentrate sampling are effective geochemical procedures for gold exploration. However, geochemical techniques, for instance soil, lag, and rock chip sampling, are the most accurate procedures in determining anomalies allied with outcropping or sub-cropping deposits at a local scale. The query is now: what is the keystone of the integration of geochemical and remote sensing techniques?

The cornerstone for geochemical exploration of gold is pathfinder elements since they have a significantly less erratic spreading, create greater halos, and arise in intensive concentrations. Besides, pathfinder elements perhaps are diverse in each precise gold territorial dominion. How pathfinder elements assist remote sensing technology in gold mineral exploration? VNIR-SWIR spectral regions of electromagnetic waves can recognize the chemical bonds and amalgamations of restricted ions owing to their electronic and vibrational processes. In other words, VNIR-SWIR spectroscopy is a quick, cost-efficient, and non-destructive technique for detecting many minerals and compounds. Nevertheless, the identification of pathfinder elements in visible-near infrared (VNIR) and shortwave infrared (SWIR) spectrometers is imperfect. This could be contributed to fact that they are not elemental detectors. Oxide-hydroxides, for instance, reveal diagnostic absorption features in the VNIR-SWIR region, although most sulfide minerals do not have such features in this spectral region [11].

Near 0.43 μm, 0.66 μm, 0.90 μm, 1.47 μm, and 2.27 μm, respectively, jarosite demonstrates absorption features along with a maximum reflectance near 0.70 μm. However, goethite reveals absorption features near 0.50 μm, 0.66 μm, and 0.90 μm, respectively, as well as its maximum reflectance in goethite rises close to 0.75 μm. Consequently, hematite has core stone absorption features close to 0.53 μm, 0.63 μm, and 0.88 μm, respectively.

Thus, alunite, kaolinite, montmorillonite, and muscovite reveal significant absorption features in the VNIR-SWIR region near 1.44 μm and 2.1–2.4 μm, respectively. Therefore, close to 2.17 μm, alunite reveals the strongest absorption and kaolinite illustrates a doublet diagnostic absorption feature at 2.165 and 2.2 μm. Moreover, near 2.2 μm, muscovite and montmorillonite show the diagnostic absorption feature.

Definitely, supergene alteration triggers spectrally imperceptible iron and iron-copper sulfide minerals that destruct to develop iron oxide-hydroxide minerals, which are recognized as gossans right atop of mineralized zones. Consequently, the VNIR-SWIR spectral region can sense gossans based on their specific spectral signatures. Needle to say that the hydrothermal alteration zones and iron gossans are common exploration indicators for recognizing areas of high-potential gold mineralization, as a result of their wide-ranging spreading and detectability. The abundant quantity of supergene iron minerals counting goethite, hematite, limonite, and jarosite are reliable exploration indicators of gold deposits [9, 12].

10.5   Conventional Image Processing Techniques for Gold, Iron, and Copper Explorations

Optical remote sensing data involve both multispectral and hyperspectral data handled by two main procedures: pre-processing, and post-image processing.

10.5.1 Preprocessing

The main aim of preprocessing is data preparing for actual image processing. This step involves radiometric calibration, atmospheric correction, geometric correction and noise removal. Radiometric correction, therefore, is a vital procedure in mineral exploration and lithological discrimination exploitation remote sensing techniques. Radiometric calibration is utilized to rectify image data to radiance, reflectance, or brightness temperatures and to diminish the errors in the digital numbers of images. The atmospheric correction technique is exploited to eliminate the impact of the atmosphere that contribute to the reflected signal restrained optical satellite sensors, for instance, ASTER. In the case of ASTER data, a cross-talk signal is implemented to overwhelm the signal scattering problem of the ASTER SWIR sensor. This phenomenon was revealed post-launching of ASTER aboard the Terra platform in December 1999.

The keystone tool in atmospheric correction is the Fast Line-of-sight Atmospheric Analysis of Spectral Hypercubes (FLAASH) technique, which delivers an accurate atmospheric correction of visible wavelengths through near-infrared and shortwave infrared regions, up to 3 μm. The FLAASH algorithm is the successful method for mineral explorations as claimed by Cooley et al. [13], Rajendran et al. [14], and Salem et al. [15]. In this regard, Tropical (T) atmospheric and rural aerosol models are the main core for FLAASH algorithm functionality. In this view, imaging spectrometer raw radiance data are turned into reflected spectral data using FLAASH algorithm. Consequently, image rectification is achieved. In this procedure, the satellite image is rectified to the same grid system of universe transverse Mercator (UTM), datum WGS 84, Zone 36 north, which is an initial step to ensure that all the geographical data and the satellite images are correctly overlaid.

Internal Average Reflectance (IAR) is reflectance calibration used for images normalization to a scene average spectrum. It is effective for reducing hyperspectral data to relative reflectance in an area where no ground measurements exist and little is known about the scene. It works best for arid areas with no vegetation. An average spectrum is usually calculated from the entire scene and then used as the reflectance spectrum before being divided into the spectrum at each pixel of the image.

Post geometric correction, the different spatial resolution of the one sensor must be merged with each other. For instance, Landsat-8 or ASTER data are required to implement spatial resolution merging as it uses different spatial resolution bands to merge the high-resolution bands with the coarser bands to acquire accurate spatial resolution for precise analysis. In the case of ASTER data, it is efficiently exploited to intensify the spatial resolution of the SWIR bands to 15 m in preference to 30 m and the TIR bands to 30 m rather than 90 m. Certainly, this could be beneficial in representing the lithological elements since the wide spectral range at the high spatial resolution of 15 and 30 m instead of 30 and 90 m, respectively, is qualified for more discrimination and characterization of the lithological components.

10.5.2 Post Image Processing

What are the image processing methods behind mineral explorations? Main spectral characteristics of alteration minerals and lithological components as indicators for gold exploration can be achieved by (i) False Colour Composite (FCC), (ii) Band Ratio (BR), (iii) Principal Component Analysis (PCA), (iv) Minimum Noise Fraction (MNF), and (v) Spectral unmixing in n-dimensional spectral feature space.

10.5.2.1 False Colour Composite

In these views, the FCC permits the allocation of three dissimilar sorts of spectral bands to the three-chief channels of red, green and blue (RGB) colours. The aim of colour composite image is just to acquire quick visual interpretation based on the variation of RGB through the requested data. This sort of image interpretation does not allow to acquire accurate information on gold, copper, and iron. As it still requires precise algorithms in addition to fieldwork validations to ensure accurate mineral explorations from remote sensing data. Although many researchers have used this technique based on the selected spectral characteristics to enhance the discrimination of the lithological, minerals and alteration zones, it is difficult to implement in heavy vegetation covers.

10.5.2.2 Band Ratio

The simplest post image processing technique for mineral exploration is the band ratios that permit the spreading and strength of exact absorption or emission features to be explored in a quasi-quantitative approach. The formula of hydrothermal alteration mapping in ASTER data can be expressed mathematically by band ratios as:

$$OHI = \left[\frac{Band7}{Band6}\right]\left[\frac{Band4}{Band6}\right] \tag{10.1}$$

$$KLI = \left[\frac{Band4}{Band5}\right]\left[\frac{Band8}{Band6}\right] \tag{10.2}$$

$$ALI = \left[\frac{Band7}{Band5}\right]\left[\frac{Band7}{Band8}\right] \tag{10.3}$$

$$CLI = \left[\frac{Band6}{Band8}\right]\left[\frac{Band9}{Band8}\right] \tag{10.4}$$

where OHI is the index for OH-bearing minerals, KLI is the kaolinite index, ALI is the alunite index, and CLI is the calcite index. Each index is thresholded and then merged spatially to map the alteration zones using the mineralogical indices for each alteration mineral [7]. Figure 10.5, for instance, reveals the Fe-oxides in the alteration zone as based on the band ratio of 4/8.

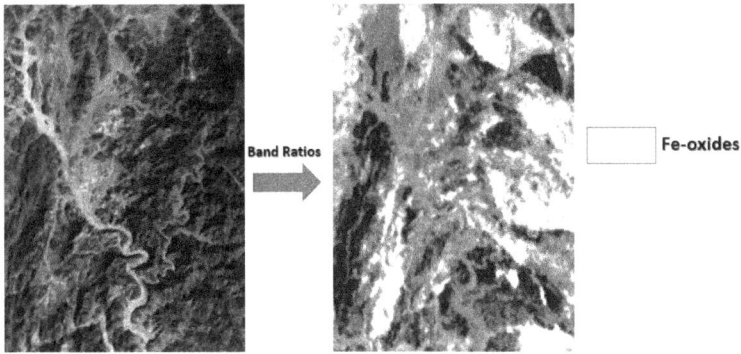

Figure 10.5. Example of band ratios for extracting Fe-oxides in the alteration zone.

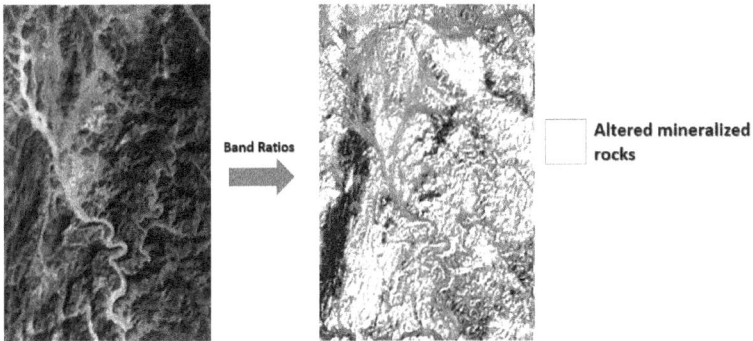

Figure 10.6. Band ratio of 4/2 for separation of altered mineralized rocks.

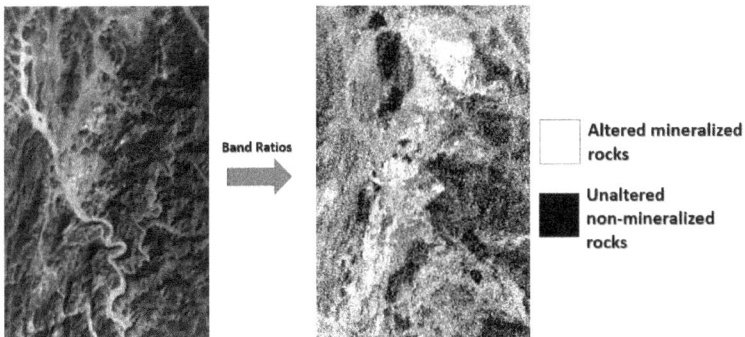

Figure 10.7. Band ratio of 8/9 for fine discrimination of rocks.

Therefore, Figure 10.6 demonstrates the separating of the altered mineralized rocks from all the other environment materials, which involve both the unaltered rocks and altered non-mineralized porphyry using a band ratio of 4/2. To acquire fine discrimination from entire-rock components in the ASTER image, the band ratio 8/9 is implemented (Figure 10.7).

10.5.2.3 Principal Component Analysis (PCA)

The PCA technique is a multivariate statistical widely exploited in the image processing industry currently, which converts several correlated spectral bands into a minor quantity of uncorrelated spectral bands known as principal components. Consequently, PCA aids to improve and separate specific sorts of spectral signatures from the surrounding environment components. PCA image with moderate to high eigenvector, therefore, is stacked for diagnostic reflective, and absorptive bands of mineral or mineral set with contradictory signs that improve mineral visibility. In this view, when the stacking is positive in the reflective band of a mineral, the image tone turns bright. However, if the stackings are negative, the image tone must be dark for the enhanced mineral element. Accordingly, in every PCA, eigenvector stacking would recognize the PCA image in which the spectral information of the mineral being scrutinized is stacked. Therefore, these statistics regularly signify, in quantitative terms, an identical minor fraction of the entire information involving the raw band data. However, it is predictable that the stacked information designates the spectral signature of the anticipated mineral.

PCA of individual spectral bands, nevertheless, cannot eliminate topographic and atmospheric impacts because it is merely a spin of spectral character space to the tips of maximum variance. In this view, the combination of PC images with other band ratios perhaps enhances mineral mapping [16]. PCA produces n independent PC images from m ($m \times n$) correlated bands of a multi-spectral image. In most image processing software packages, the PCA function produces exactly the same number of PCs as the image bands rather than fewer. The point here is that the high-rank PCs contain little information. Figure 10.8 reveals the PC images derived from PCA 4 to PCA8, which confirms the existence of Fe-oxides in alteration zones. In other words, PCA4 to PCA8 increases the response of the iron oxides in the altered mineralized rocks. In this understanding, PCA4 to PCA8 in ASTER image are therefore effective for indicating hydrothermal alteration minerals by its strong co-occurrence of clay minerals and iron oxide. The iron oxide-rich parts of the alteration are considered to be the main target for gold exploration.

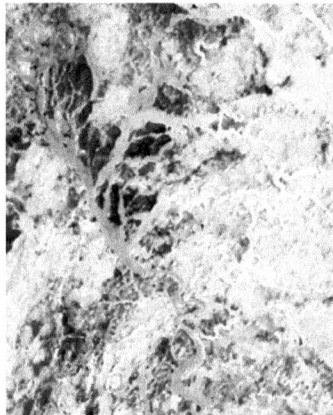

Figure 10.8. Example of PC image derived from PCA4 to PCA8.

10.5.2.4 Noise Fraction (MNF)

MNF plays a role in determining the inherent dimensionality of image data. Besides, MNF diminishes the computational necessities for succeeding image processing. Moreover, MNF transformation can recognize the sites of spectral signature variances. This process is of interest to exploration geologist because spectral anomalies are often indicative of alterations due to hydrothermal mineralization and lithological units.

10.5.2.5 Spectral Unmixing in n-dimensional Spectral Feature Space

The aims of MNF technique is decreasing profuse knowledge and data dimensionality using the Minimum Noise Fraction (MNF) transform. Subsequently, Pixel Purity Index-Mapping (PPI) is applied to resolve the purest pixels in the multispectral and hyperspectral data, which is illustrated as end members operation of the n-Dimensional-Visualizer tool (n-D-Vis).

Therefore, the extracted end members are then equated to recognized spectra, for instance, from USGS spectral libraries to further recognize and formulate for Spectral Angle Mapper (SAM) classification. However, this technique perhaps is not optimal for multispectral data. For instance, in the majority of ASTER data, the pixels are by some means detached by the system. In this understanding, the SAM creates a cluster image that contains all input pixels, however, the unclassified pixels are assigned as zero. In this view, the SAM class image assigned the whole thing detached or uncategorized since the "Maximum Angle" parameter for the SAM procedure is operated at the defaulting value of 0.10 radians. Nevertheless, the real angles that SAM considered, for instance, the ASTER data are approximately ranged from 0.3 and 0.4 rad. Accordingly, the majority of pixels are assigned to the "unclassified" class owing to the fact that none of the angles are smaller than the highest angle. On the other hand, for instance, the 0.10 rad maximum angle is acquired for hyperspectral data where perceptive variances in spectra may differentiate between diverse constituents. However, this is not an ideal opportunity for multispectral data, for example, the ASTER image. Increasing the maximum angle cannot be an ideal solution since the angle alterations are minor. In this understanding, implementing such "rule images" reveals the definite SAM values between each pixel and the specified target spectrum, delimited valued information. In this regard, SAM values can show sensible consequences. Therefore, the rule images are the images that comprise the definite spectral angles that are estimated in radians. They are commonly termed spectrally matched score images since they are created by spectrally matching the target spectra with identified spectra. Separately, pixel is counted between 0 and 1 consistent with its closeness to the maximum angle threshold in radians. In this understanding, if the "match" between the angle of target spectra with identified spectra in n-dimensional spectral feature space is accurate, then the rule image pixel value is 1. For all other circumstances, it diverges between 0 and 1.

In the next step, (1) n-dimensional spectral feature spaces are instigated to determine varieties of minerals. For instance, Figure 10.9 reveals the resulting alteration minerals' abundance image with a mineral fraction of the ASTER

Figure 10.9. Results of the partial spectral unmixing in n-dimensional spectral feature space.

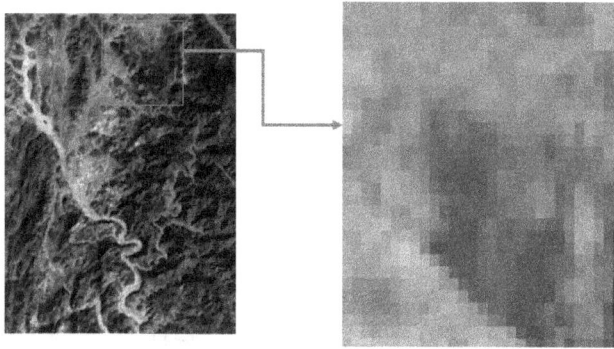

Figure 10.10. PCA image of the spectral unmixing scored images.

Alunite
Montmorillonite
Muscovite
Kaolinite

Figure 10.11. Combination of spectral unmixing in n-dimensional spectral feature space and the USGS spectral library.

reflectance image, exploiting the partial spectral unmixing in n-dimensional spectral feature space. Consequently, Figure 10.10 is the resulting PCA image of the spectral unmixing scored images. The resulting alteration minerals' abundance image with a mineral fraction of the ASTER reflectance image using the partial spectral unmixing in n-dimensional spectral feature space and the USGS spectral library is revealed in Figure 10.11.

However, in heavy vegetation cover zones, accurate identification of hydrothermally altered rocks is not possible, in which advanced image processing tool based on the quantum image processing algorithms is required.

10.6 Quantum Machine Learning

Quantum machine learning (QML) is constructed on two concepts: quantum information and hybrid quantum-classical models. Quantum information is any data furnish that takes place in a natural or artificial quantum system. This can be statistics generated with the aid of a quantum computer, as the samples gathered from the Sycamore processor for Google's demonstration of quantum supremacy. Quantum statistics reveals superposition and entanglement, leading to joint probability distributions that ought to require an exponential quantity of classical computational resources to represent or store in qubit. The quantum supremacy scan proved it is possible to be shaped from an extraordinarily problematic joint probability distribution of 2^{53} Hilbert space [17].

The quantum information generated with the resource of NISQ processors is noisy and generally is entangled simply rather than the classical computer computing procedures. Heuristic machine learning techniques can create models that maximize the extraction of beneficial classical files from noisy entangled data. The Tensor Flow Quantum (TFQ) library gives primitives to advance models that disentangle and generalize correlations in quantum data—opening up probabilities to enhance existing quantum algorithms or find out new quantum algorithms.

A quantum model can represent and generalize data with a quantum mechanical origin. Because near-term quantum processors are still fairly small and noisy, quantum models cannot generalize quantum data using quantum processors alone. NISQ processors must work in concert with classical co-processors to become effective. Since TensorFlow already supports heterogeneous computing across CPUs, GPUs, and TPUs, it is used as the base platform to experiment with hybrid quantum-classical algorithms [17–19].

A quantum neural network (QNN) is used to describe a parameterized quantum computational model that is best executed on a quantum computer. This term is often interchangeable with the parameterized quantum circuit (PQC).

10.7 Classifier Architecture

Quantum encoding and processing of information is a prevailing alternative to conventional machine learning quantum classifiers. In particular, it permits us to encode data in quantum catalogues that are concise, compared to the quantity of features, analytically engaging quantum entanglement as a computational source and retaining quantum quantity for category extrapolation. Therefore, circuit centric quantum classifier is a comparatively straightforward quantum solution that syndicates data encoding with a quickly entangling/disentangling quantum circuit grasped by quantity to deduce category labels of data illustrations. In this view, the aim is to certify classical portrayal and loading of subject circuits, along with hybrid quantum/classical working out of the circuit constraints even for tremendously huge feature dimensions [18–20].

Let us assume that $y_1, y_2,...., y_d$ are the class labels of the alteration hydrothermal zone elements as discussed in Section 10.5.2.2. In this perspective, the "training data set" is a gathering of samples $D = (x, y)$, with known pre-assigned labels. In this understanding, x presents the data sample of alteration zone elements such as OHI, KLI, ALI, and CLI. Therefore, y is known as a "training label". Consequently, quantum classification involves three stages: (i) data encoding, (ii) preparation of a classifier state, and (iii) measurement owing to the probabilistic nature of the measurement. Subsequently, these three stages must be replicated several times. Both the encoding and the computing of the classifier state are completed employing *quantum circuits*. While the encoding circuit is regularly data-driven and parameter-free, the classifier circuit comprises an adequate convention of learnable parameters.

In the recommended solution, the classifier circuit is constituted of binary-qubit controlled rotations and single-qubit rotations. In this circumstance, the learnable parameters are considered as the rotation angles. Both rotation and controlled rotation gates are identified as universal for quantum computation, which entails that any unitary weight matrix can be disintegrated into an elongated sufficient circuit entailing such gates.

In the proposed version, simply one circuit grasped on its frequency guesstimate is braced. Accordingly, the solution is a quantum analogue of a support vector machine with a low-degree polynomial kernel. In this view, a classical support vector machine (SVM) solution can be compared to a simple quantum classifier design. Consequently, the inference for a data sample x in the case of SVM is completed exploiting an optimal kernel form $\sum_j \alpha_j k(x_j, x)$, where k is a certain kernel function. On the contrary, the quantum classifier exploits the predictor of wanted minerals in optical remote sensing sensors as given by:

$$p(y|x, U(\theta)) = \langle U(\theta)x|M|U(\theta)x \rangle \tag{10.5}$$

Equation 10.5 reveals that when a forthright amplitude encoding is exploited, $p(y|x, U(\theta))$ is a quadratic form in the amplitudes of x (Figure 10.12). However, the coefficients of this form are no longer learned independently; they are instead aggregated from the matrix elements of the circuit $U(\theta)$ (Figure 10.13), which naturally has pointedly less learnable constraints θ than the dimension of the vector x. Consequently, the polynomial degree of $p(y|x, U(\theta))$ in the unique features can be enlarged 2^l by implementing a quantum product encoding on l duplicates of x [17–21].

This architecture, therefore, investigates comparatively narrow circuits, which hence must be speedily entangled to portray all the associations between the data features at all series. For instance, Figure 10.14 reveals the useful rapidly entangling circuit component of geometry consists of only 3n+1 gates, the unitary weight matrix that it calculates confirms noteworthy cross-talk between 2^n gold indices features. Figure 10.14 reveals the computed gold indices using speedily entangled circuits with 3n+1 gates. Therefore, the unitary weight matrix confirm accurate cross-talk between 2^n gold feature indices.

Therefore, the quantum circuit of gold mining perhaps involves 6 single-qubit gates ($G_1,...,G_5$, G_{16}) and 10 two-qubits gates ($G_6,...,G_{15}$). Assuming that

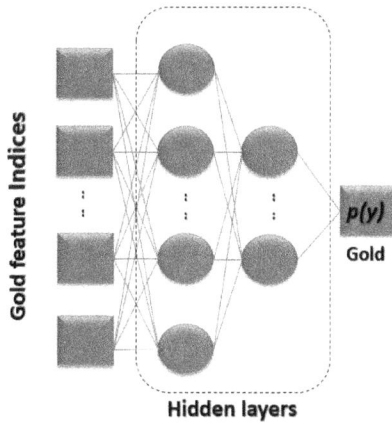

Figure 10.12. Quantum neural network architecture for gold mining.

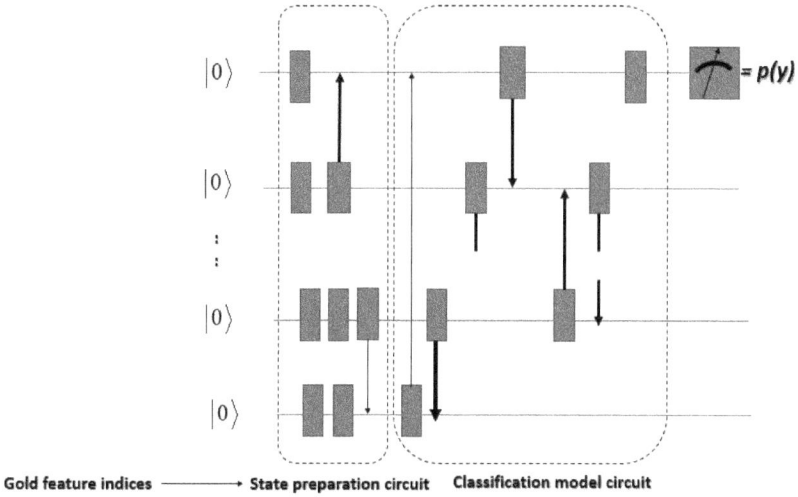

Figure 10.13. Quantum circuit encoding for gold mining prediction.

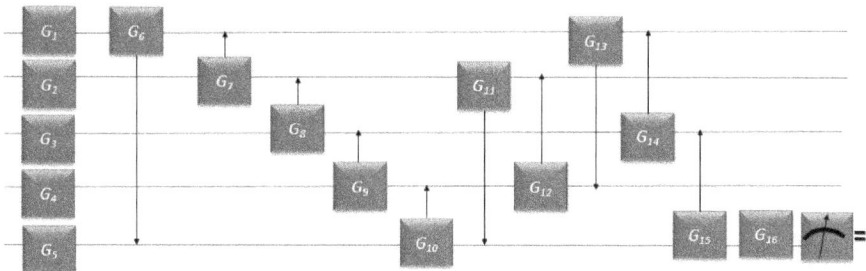

Figure 10.14. Quantum circuit for gold mining.

every one of the gates is well-defined with one learnable parameter, for instance, 16 learnable parameters, while the dimension of the 5-qubit Hilbert space is 32 to determine the accurate gold feature indices. In this regard, such circuit geometry can be straightforwardly sweeping to any *n*-qubit catalogue, when *n* is odd, generating circuits with 3n+1 parameters for 2n-dimensional gold mining feature indices variation in optical remote sensing data [18–20].

10.8 Classifier Training as a Supervised Learning Task

Quantum training of a classifier model comprises acquiring optimal values of its operational gold mining parameters, such that they exploit the average likelihood of deducing the precise training labels are diagonally matched up the exact alteration zone features in remote sensing data. In this view, four-level classifications can be determined, for instance, the case of *d* = 4 and only four classes with labels OHI (y_1), KLI (y_2), ALI (y_3), and CLI (y_4). In this understanding, a fundamental approach for simplifying this technique to a random number of sets is to swap qubits with qudits, which are quantum units with *d* source statuses, and the two-approaches of measurement with *d*-approach quantity. In this understanding, the qubits with qudits spin up and down through the image to determine that pixels belong to specific mineral spectral signatures (Figure 10.15).

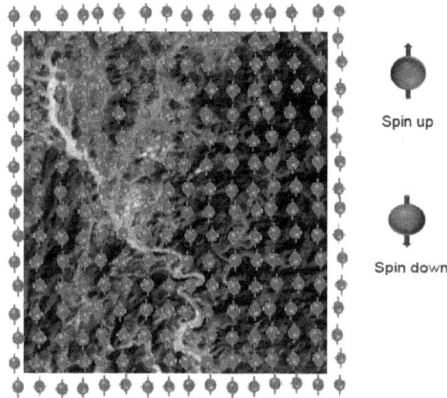

Figure 10.15. Qubits and qudits spin up and spin down to determine mineral spectral signatures.

Consequently, the training goal is considered as the likelihood that delivers a learnable quantum circuit $U(\theta)$, where θ is a vector of parameters [18, 20]. In this understanding, θ signifies the ending measurement by M. The mathematical expression of the average likelihood of the accurate label extrapolation is given by:

$$\Im(\theta) = \frac{1}{|D|}\left(\sum_{(x,y_1)\in D} P\left(M = y_1 \middle| U(\theta)x\right) + \sum_{(x,y_2)\in D} P\left(M = y_2 \middle| U(\theta)x\right) \right) \tag{10.6}$$

Equation 10.6 says that in the quantum state $|0\rangle$ and $|1\rangle$, $P(M = y|0)$, $M = y|0\rangle)$ is the probability measuring the possibilities of gold mining occurrences as based on the event of the feature indices such as OHI (y_1), KLI (y_2), ALI (y_3), and CLI (y_4). In this perspective, it suffices to say that the likelihood function $\mathfrak{I}(\theta)$ is smooth in (θ) and its derivative in any (θ_j) can be calculated by basically the identical quantum procedure as exploited for calculating the likelihood function itself [20–23].

Conversely, in the circumstance of convergence to the true spectral signature quantum state upon the convergence of the training, the value of loss function can deliver the free energy of the real spectral signature state, which is proportional to the log of the spectral signature partition function as follows:

$$\mathfrak{I}(\theta) = -\log \mathbb{Z}_S. \tag{10.7}$$

Here, $\mathbb{Z}_S$ is known as the spectral partition function that is considered to phrase the quantum simulation task as a quantum-probabilistic variational learning task (Figure 10.16). In this circumstance, a quantum-probabilistic ansatz for the true spectral signature states $\bar{\rho}_{\theta\phi}$ with parameters (θ, ϕ). Therefore, the given spectral signature of a variational cluster for the quantum spectral states (Figure 10.17) is achieved by minimizing the free energy as loss function,

$$\mathfrak{I}(\theta, \phi) = \mathbb{Z}_S tr(\bar{\rho}_{\theta\phi} \hat{H}) - M(\bar{\rho}_{\theta\phi}), \tag{10.8}$$

Thus, a recovery of the loss function for the ground state Variational Quantum Eigensolver (VQE), and thus the ground state variational principle is recovered as follows:

$$\mathfrak{I}(\theta, \phi) \xrightarrow{\; \mathbb{Z} \to \infty \;} \left\langle \hat{H} \right\rangle_{\theta\phi}, \tag{10.9}$$

Equation 10.9 says that in restricted $\left\langle \hat{H} \right\rangle_{\theta\phi}$ of zero spectral signature, there is no need for hidden spaceparameters as there is no information, and the hidden state is

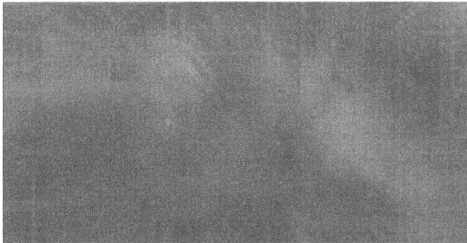

Figure 10.16. Spectral partition function.

Figure 10.17. Minimizing the loss function of the quantum spectral state.

unitarily equivalent to any first course pure state. In this regard, system dimension can be simulated on the classical computer. In fact, the wavefunction is required to be simulated in the Hilbert space directly [18, 21, 24]. Consequently, Quantum Hamiltonian-Based Models $\left\langle \hat{H} \right\rangle_{\theta\phi}$ operate to an abundant wide-ranging class of quantum spectral states (Figure 10.18). In other words, $\left\langle \hat{H} \right\rangle_{\theta\phi}$ acts as a kernel window detector to determine the true spectral partition for gold feature indices, as will be described and discussed in the following sections.

Figure 10.18. Quantum Hamiltonian for searching spectral signature classes.

10.9 Training Score and Classifier Bias

Both training score and classifier bias rely on the accurate gradient of the partition function of the true spectral of gold feature indices. In this understanding, Hamiltonian-Based Models $\left\langle \hat{H} \right\rangle_{\theta\phi}$ are considered for an analytical expression for the partition function. Consequently, the spectral partition function $\mathbb{Z}_S$ can be transferred into the following mathematical form:

$$\log \mathbb{Z}_\theta = \sum_j \log \left(\sum_{kj} k_j \left(\theta_j \right) \right) \tag{10.10}$$

Equation 10.10 says that $k_j(\theta_j)$ is the key index of the jth parameterized modular Hamiltonian $\hat{K}_j(\theta_j)$. Thus, the knowledge of true spectra of gold feature indices can be expressed mathematically in the form of parameterized modular Hamiltonians:

$$\partial_\theta \log \mathbb{Z}_\theta = \left[\mathbb{Z}_m \left(\theta_m \right) \right]^{-1} tr \left(\hat{M}_\theta e^{-\theta_m \hat{M}_\theta} \right) \tag{10.11}$$

Equation 10.11 can be used directly to appraise, analytically, information of the spectrum $\hat{M}_\theta$ of gold feature indices in multi-spectral and hyperspectral remote sensing data. Thus, by merely training on the Hamiltonian-gradient $\left\langle \partial \hat{H} \right\rangle_\theta$, the gradient variation between different gold spectral feature indices is being constructed. In other words, Equation 10.11 construct the distinguished boundary between different spectral signature clusters as a function $\left\langle \partial \hat{H} \right\rangle_\theta$. In this view, Figure 10.19 demonstrates the

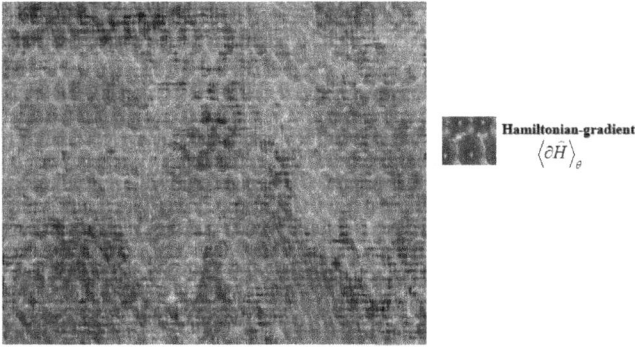

Figure 10.19. Sharpest boundary gradient between the variety of classes in Hamiltonian-gradient mapping.

gradient boundary variation owing to Hamiltonian-gradient between different spectral feature classes.

The different classes, therefore, can be identified as well as boundary gradients between different spectral classes, are determined.

Simulation of the spectral feature using Hamiltonian-gradient can reduce the bias [24–27].

To this end, let us assume that b is the single real value of gold mining feature indices, which is known as classifier bias, to do the inference for final values of the parameters in (θ, ϕ). In this regard, we can establish a frame rule for the label interference of the gold mining feature indices as a function $P(M = y|0), M = y|1))$. To this end, a frame rule can be designed as follows:

Let sample x is assigned to label y_2 and satisfies that:

$$P(M_{\theta\phi} = y_2|U(\theta, \phi)x + b) > 0.5 \tag{10.12}$$

In the circumstance of Equation 10.12, if this condition is not achieved then sample x is otherwise assigned label y_1. Consequently, the accurate gold mining mapping must be satisfied so that b value is constrained between significant interval of -0.5 and 0.5. Figure 10.20 reveals the probability variation for quantum spectral state construction with bias. It is worth noticing that a high probability of 0.8 leads to a low bias of -0.5. The most tremendous role for bias in determining a misclassification is if the label inferred for x as per Equation 10.7 is dramatically different from y. In other words, the training case $(x, y) \in D$ is counted as a misclassification given the bias b. In this circumstance, the inclusive number of misclassifications in the training score of the classifier is specified with the bias b. In this regard, the optimal classifier bias b diminishes the training score. Consequently, it is simple to find that certain precomputed probability approximations can be expressed by:

$$P(M_{\theta\phi} = y_2|U(\theta, \phi)x|(x,*)) \in D \tag{10.13}$$

In the circumstance of Equation 10.13, the optimal classifier bias can be initiated by binary examination in the interval $(-0.5, +0.5)$ by forming at most $\log_2(|D|)$ steps. Figure 10.21 illustrates the perfect quantum spectral state classes with the probability of 1 and zero bias. In this view, every class of alunite, montmorillonite, muscovite,

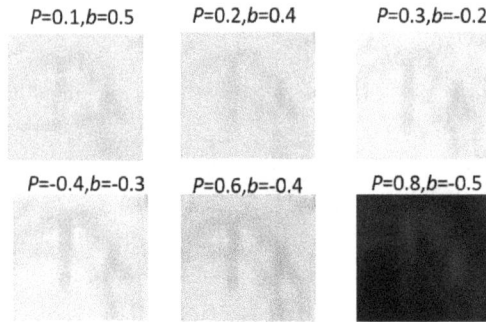

Figure 10.20. Probability and bias of spectral state constructions.

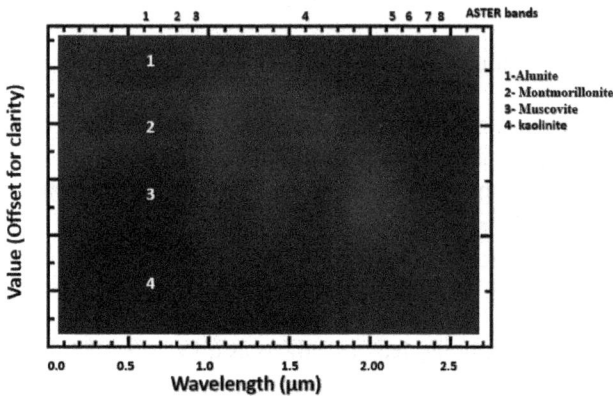

Figure 10.21. Hamiltonian-gradient for well-identification of gold feature indices.

and kaolinite is well-identified and distinguished from each other through quantum spectral states. Needless to say that involving the Hamiltonian-gradient can reduce the bias and well-identified each gold feature indices to its real spectrum signature with a high probability of 1 [17, 20, 23].

10.10 Gold Mining Simulation Using Quantum Machine Learning

Figure 10.22 reveals the quantum image of the ASTER data. In this view, the quantum ASTER data involves 16×16 spins ½ (qubits) that equals $2^{(16 \times 16)}$, which is approximately 10^{77} worth of data. Consequently, Figure 10.23 illustrates the quantum image feature occurrences that are simulated using quantum machine learning classes. It is worth noticing that quantum machine learning delivers the sharpest features within 100 training iteration using the Hamiltonian gradient algorithm. It is worth noticing that both spectral signature indices and image feature boundaries can be reconstructed exploiting the Hamiltonian-gradient algorithm.

Figure 10.24 demonstrates clear categorization of gold feature indices in the ASTER image compared to the conventional method such as PCA and spectral unmixing protocols. In this view, every class of alunite, montmorillonite, muscovite, and kaolinite is well-identified and distinguished from each other.

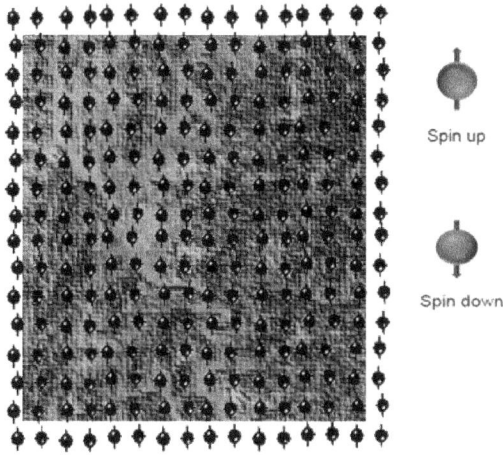

Figure 10.22. Quantum ASTER image reconstruction.

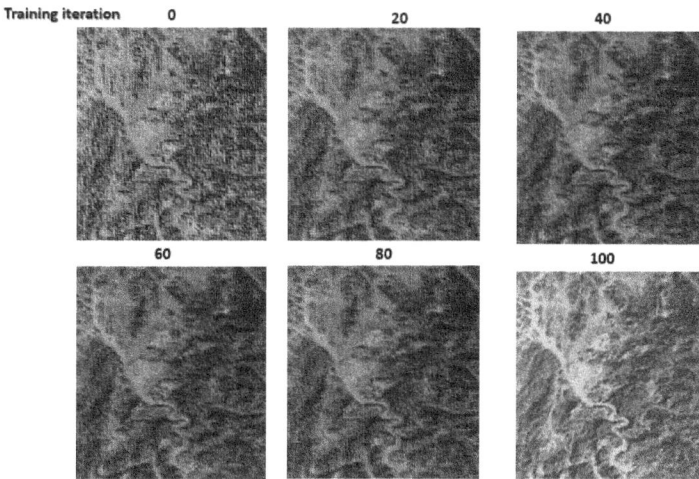

Figure 10.23. ASTER image feature reconstruction from 0 to 100 training iterations.

Variational Quantum Spectral Classifier Training (VQSCT) for the d-spectral signature state is exploited to distinguish between the spectral signature variation of the different gold feature indices. In this concern, VQSCT is mathematically expressed by:

$$\widehat{H}_{HEIS} = \sum_{\langle ij \rangle_h} J_h \widehat{S}_i . \widehat{S}_j + \sum_{\langle ij \rangle_v} J_v \widehat{S}_i . \widehat{S}_j \tag{10.14}$$

where $h(v)$ denotes horizontal (vertical) bonds, and $\langle \cdot \rangle$ signifies nearest-neighbour pairs. The Variational Quantum spectral classifier is simulated using an imagined one-dimensional array of qubits, which is parametrized with only single-qubit rotations and two-qubit rotations between adjacent qubits. In this understanding, VQSCT perhaps is used to formulate an approximate spectral signature state in

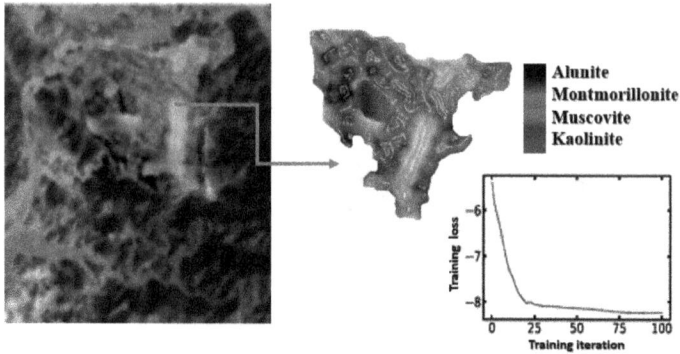

Figure 10.24. Quantum machine learning classification of gold feature indices in ASTER image.

ASTER image. Additionally, modular Hamiltonian learning could facilitate the simulation of out-of-equilibrium time-evolution, going beyond what is feasible with classical methods such as PCA [24–27]. In particular, Modular Hamiltonian Learning gives one access to the eigenvalues of the density matrix and the unitary, which diagonalizes the modular Hamiltonian. Besides, the modular Hamiltonian itself provides invaluable information related to topological properties, spectral signature variation, and nonequilibrium dynamics [23].

10.11 Quantum Artificial Neural Network (QANN) for Gold Exploration

The gold exploration relies on the occurrences of alteration zone elements such as OHI, KLI, ALI, and CLI. The vital question is: can QANN detect gold mining automatically in optical remote sensing data? The keystone answer to this question relies upon the implementation of the quantum artificial neural network (QANN). In this view, the quantum artificial neural network (QANN) formulates the routine of virtual quantum powers for gold mining as based on alteration zone elements. Therefore, the topology structure of a quantum feed-forward artificial neural network does not diverge from a conventional one. On the contrary, QANN involves linear superposition, which makes QANN different from the conventional one. In this regard, the linear superposition or coherence, which results in quantum parallelism, is defined by the quantum state as:

$$|\psi\psi\rangle = \sum_i cc_{ii} |\psi\psi_{ii}\rangle \tag{10.15}$$

Equation 10.15 says that in the Hilbert space, with complex coefficients c_i and onset of states ϕ_i in space, the state superposition is achieved. In this regard, the annihilation of state superposition is formed when the quantum state interacts with its environment, which is the fundamental principles of quantum computer. The coefficients c_i indicates the indication of the system being allied with the state ϕ_i when the quantum computing occurs. Consequently, the sum of the coefficients would sum to unity, which is formed based on a quantum state of the system. In this sense, summing up, the linear superposition involves entirely conceivable formations of a quantum system as soon as the measurement is done, which directs to a collapse or decoherence. In particular, the superposition tolerates quantum parallelism. Consequently, let us assume that the training set for a QANN has hundreds of input data sets separately containing numerous attributes. On this occasion, the input training data would initially cultivate an initial cultivate and then U_f in a conventional computer would have to be smeared to each single input data set repeatedly. In this view, 200 Hadamard transformations (gates) H would be exploited to each Qbit prior to the application of U_f, as:

$$U_f(H^{\otimes n} \otimes 1_m)(|0\rangle_n |0\rangle_m) = \sum_{0 \leq x < 2n} U_f(|x\rangle_n |0\rangle_m) \tag{10.16}$$

Equation 10.16, however, states that quantum parallelism permits the routine of an exponentially high quantity of U_f in unitary time. Furthermore, the simulation of quantum bits for processing the information associated along with the notion of quantum parallelism, for example, can be designated by a wave function ψ that occurs in Hilbert space. Consequently, entanglement is a vital procedure in QANN, which occurs between dual Qbits. This procedure is known as the Bell-states [19, 25, 27] and also acknowledged as maximally entangled dual-Qbit states. The quantum entanglement states are mathematically expressed by:

$$|\phi^+\rangle = \frac{1}{\sqrt{2}}\left(|0\rangle_A \otimes |0\rangle_B + |1\rangle_A \otimes |1\rangle_B\right) \tag{10.17}$$

$$|\phi^-\rangle = \frac{1}{\sqrt{2}}\left(|0\rangle_A \otimes |0\rangle_B - |1\rangle_A \otimes |1\rangle_B\right) \tag{10.18}$$

$$|\phi^+\rangle = \frac{1}{\sqrt{2}}\left(|0\rangle_A \otimes |1\rangle_B + |0\rangle_A \otimes |1\rangle_B\right) \tag{10.19}$$

$$|\phi^-\rangle = \frac{1}{\sqrt{2}}\left(|0\rangle_A \otimes |1\rangle_B - |0\rangle_A \otimes |1\rangle_B\right) \tag{10.20}$$

Equations 10.17 to 10.20 reveals two individuals A and B, individually possessing one of the dual entangled Qbits, designated in the Bell-states by the subscripts A and B. For instance, if A adopts to determine the associated Qbits, then the consequence cannot be foretold as:

$$\left|\frac{1}{\sqrt{2}}\right|^2 = 0.5 \tag{10.21}$$

Therefore, Equation 10.21 says that the probabilities of the measurements would be either $|0\rangle$ or $|1\rangle$. Nevertheless, as stated by quantum entanglement, B would now measure the similarity owing to the ending state $|0\rangle$ that can simply be the consequence of A's Qbit ensuing in the state $|00\rangle$. Consequently, the input layers and output registers would be in the state $|0\rangle$; subsequently on both applications of XX of the input and output registers use HH.

In this circumstance, the input for U_f develops into:

$$(H \otimes H)(X \otimes X)(|0\rangle \otimes |0\rangle) = \left(\frac{1}{\sqrt{2}}|0\rangle - \frac{1}{\sqrt{2}}|1\rangle\right)\left(\frac{1}{\sqrt{2}}|0\rangle - \frac{1}{\sqrt{2}}|1\rangle\right)$$

$$= 0.5\left(|0\rangle|0\rangle - |1\rangle|0\rangle - |0\rangle|1\rangle + |1\rangle|1\rangle\right)$$

(10.22)

Therefore, Equation 10.22 leads to the quantity of U_f in unitary time as follows:

$$0.5U_f\left(U_f(|0\rangle|0\rangle) - U_f(|1\rangle|0\rangle) - U_f(|0\rangle|1\rangle) + U_f(|1\rangle|1\rangle)\right)$$

(10.23)

Summing up Equations 10.22 and 10.23, we can obtain:

$$(H \otimes 1)U_f(H \otimes H)(X \otimes X)(|0\rangle \otimes |1\rangle) =$$

$$\begin{cases} |0\rangle\frac{1}{\sqrt{2}}|f(0)\rangle - |\widehat{f}(0)\rangle, & f(0) \neq f(1) \\ |1\rangle\frac{1}{\sqrt{2}}|f(0)\rangle - |\widehat{f}(0)\rangle, & f(0) = f(1) \end{cases}$$

(10.24)

Equation 10.24 reveals that the input register is either $|0\rangle$ or $|1\rangle$. In this circumstance, two of four possible interpretations of the function f are abolished, impartial by one operation in a quantum computer-specific facility. In stipulations of the QANN, the equivalent would be achieved in further complex approaches, as not simply one Qbit requests to be considered. Consequently, the interior controls of an ANN require more unitary transformation.

Let us assume that a further unitary transformation W_f, since the U_f transformation does not gauge further Qbits, which are necessary in running QANN computing. In this view, U_f transformation is just deliberated the Qbits separately for either the input or output registers, while W_f considers all Qbits. Consequently, gold exploration in remote sensing data would be considered in the form of quantum multi-layer perceptron, which is mathematically expressed as:

$$|f(x_t)\rangle = \varsigma_2\left(w_0 + \sum_{j=1}^{1}|w_j\rangle\varsigma_1\left(w_{0j} + \sum_{i=1}^{p}|w_{ij}x_{t-i}\rangle\right)\right) + \varepsilon_t$$

(10.25)

Equation 10.25 demonstrates that the input layer has p inputs $x_{t-1},....., x_{t-p}$, the hidden layer has one hidden node and the output, and there is a single output for the output layer x_t. In this view, layers are entirely allied by weights $|w\rangle_{ij}$, which are created in linear superposition and represents i^{th} input for the jth node in the hidden layer. In this regard, $|w\rangle_j$ is the weight assigned to the jth node in the hidden layer for the output. Consequently, w_0 and w_{0j} are the biases, while ς_1 and ς_2 are activation functions. Consistent with this perspective, linear superposition is the main formulation of dynamic action potentials of the neurons, which directs to multiple patterns of both weights and the neurons. Therefore, a quantum neural network involves a layered structure with parameter-dependent unitary operations. In this view, Figure 10.25 demonstrates that the lines relate to qubits, in which the lowest line presents the readout qubit, while the highest line presents the existence sustenance qubits [26].

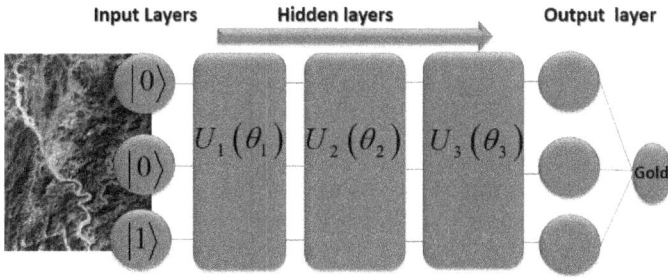

Figure 10.25. Structure of QANN.

Figure 10.26 reveals the automatic detection of the gold mining potential zones in ASTER data. Thus, gold occurrences were coded as $|1\rangle$ and the remaining locations as $|0\rangle$. These values constituted the desired output. Input variables were linearly rescaled to the range $|0\rangle|1\rangle$. QANN is also applied to a different zone with a different optical sensor to ensure its accurate performance. In this regard, Figure 10.27 illustrates composite VNIR bands of Hyperion data along the Bau gold mining district in Sarawak, Malaysia. It is noticeable that the gold potential zones have the highest probability value close to 1 along with the entire Hyperion image. In fact, this Bau gold mining is dominated by the highest occurrences of iron oxide minerals and hydroxyl-bearing (clay) alteration mineral assemblages.

Consequently, iron oxides are one of the important mineral sets that are allied with hydrothermally altered rocks over porphyry copper-gold ore bodies. In this view, QANN is also able to map the iron oxides potential zone in the Hyperion image (Figure 10.28). It is worth noticing that the iron oxides are well-identified using QANN in Hyperion image with the highest probability of 1, which allied with porphyry copper-gold ore bodies. This occurs extremely with advanced argillic alteration, decalcification, brecciation, and silicification of the sedimentary host rocks (Figure 10.29).

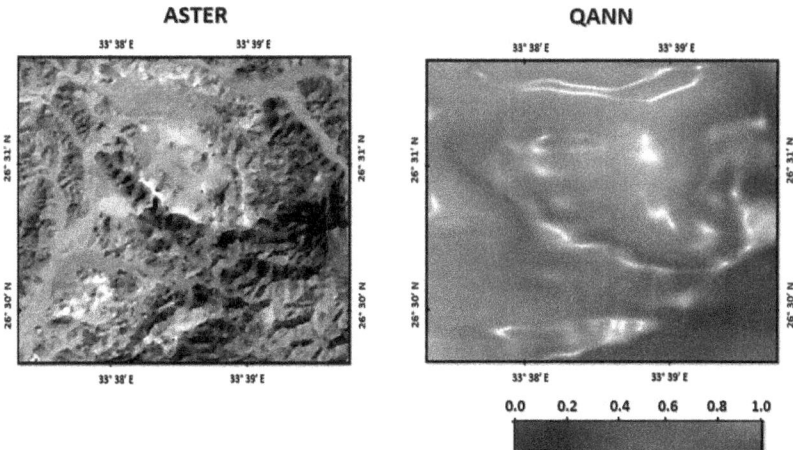

Figure 10.26. QANN gold mining potential zone in ASTER image.

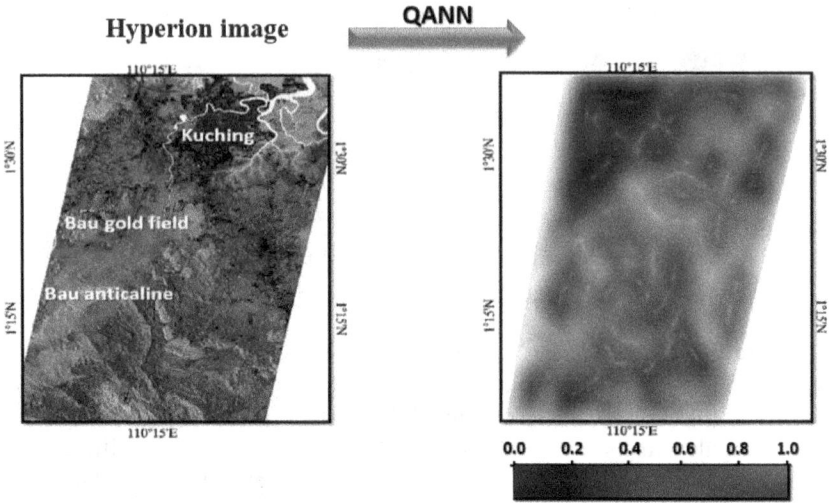

Figure 10.27. QANN gold mining potential zone in VNIR bands of Hyperion data.

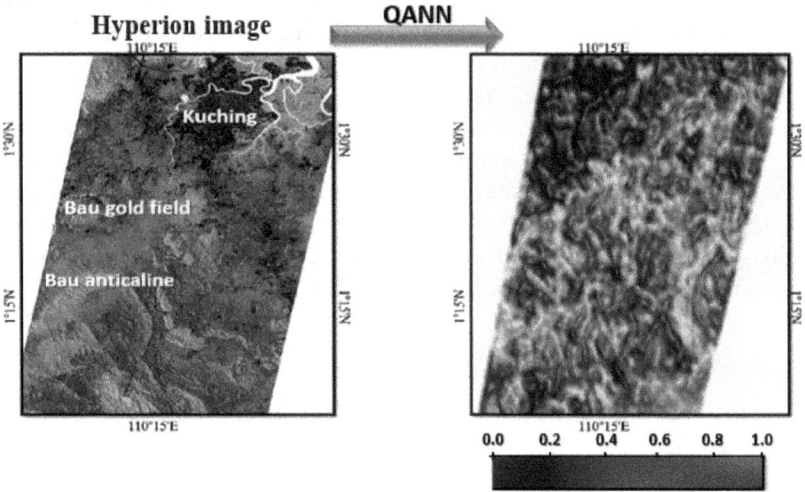

Figure 10.28. QANN for iron oxides probability occurrences in Hyperion image.

Cross-validation specified that the QANN yielding the lowest sweeping statement error is one with three hidden layer units (Figure 10.25). While adding more hidden units diminished the training error, it created an insignificant variance to the precision calculation mistake. QANN eliminates and optimizes bias in data classification. The average accuracy assessment root means square error RMSE is ± 0.13 and the average RMSE for the independent test set is ±0.19 (r^2 = 0.90). In this view, the testing precision attained is 92%. In the final training step, the RMSE for the training set is ±0.23 and its r^2 value is 0.92, while the average RMSE for the accuracy assessment set (in this case the test set) is ±0.113 and its r^2 value is 0.93. Training and accuracy assessment set accuracies are 87.5% and 92.32%, respectively.

Figure 10.29. Advanced argillic alteration with ore mineralization of gold-bearing arsenopyrite.

Table 10.1. QANN accuracy performance.

Performance	RMSE±	r²	Accuracy (%)
Average accuracy	0.13	0.89	91.50
Independent test set	0.19	0.90	91.82
Training set	0.23	0.92	91.53
Test set	0.113	0.93	92.32

These consequences recommend that the QANN based machine learning could be exploited to determine the presence/absence occurrence patterns of gold mining potential zones in remote sensing data.

10.12 QANN for Copper Mining Potential Zone

The utilization of ASTER data in copper mining potential zones is well-addressed. However, most of these studies rely on the feature indices of the copper occurrences such as muscovite/kaolinite. Consequently, QANN delivers accurate copper mining potential zones with a probability close to 1 as illustrated in Figure 10.30. In this regard, QANN automatically identifies the copper zone potential zone from the

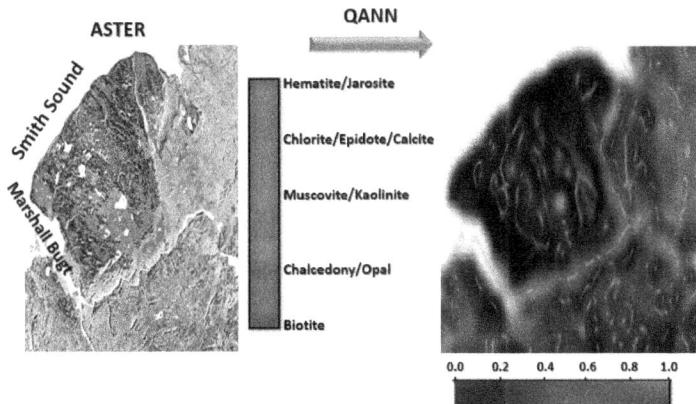

Figure 10.30. QANN for automatic detection of the copper potential zone in ASTER data.

surrounding indices occurrence features of hematite/jarosite, chlorite/epidote, chalcedony/opal, and muscovite/kaolinite as revealed along the Franklinian basin (Figure 10.30).

Consequently, implementation of the QANN for automatic copper potential zone detection in WV-3 data reveals a great promise. In this view, a high concentration of hematite, jarosite, and ferric silicates are allied with the highest probability of copper occurrences within 0.9 (Figure 10.31). Alternatively, carbonates (calcite and dolomite) also exist in alliance with ferric silicate, particularly in the essential and northwestern parts of the Franklinian basin. Consequently, the majority of the ferrous silicates are distinguished in the drainage systems and geological structures.

According to the above perspective, the multispectral WorldView-3 (WV-3) sensor comprises of the highest spatial, spectral resolutions and radiation in the VNIR of eight bands with 1.2 m spatial resolution. Furthermore, SWIR of eight bands with 3.7 m spatial resolution is currently considered as the fragments amongst the multispectral satellite sensors [28]. The advancement of WV-3 is demonstrated with a swath width of 13.2 km [29–31]. Both VNIR and SWIR bands of WV-3 have the great promises in mineral explorations. In this understanding, the VNIR and SWIR bands of WV-3 are successfully exploited for mineral exploration and mapping of hydrothermal alteration zones and lithologies [28–31].

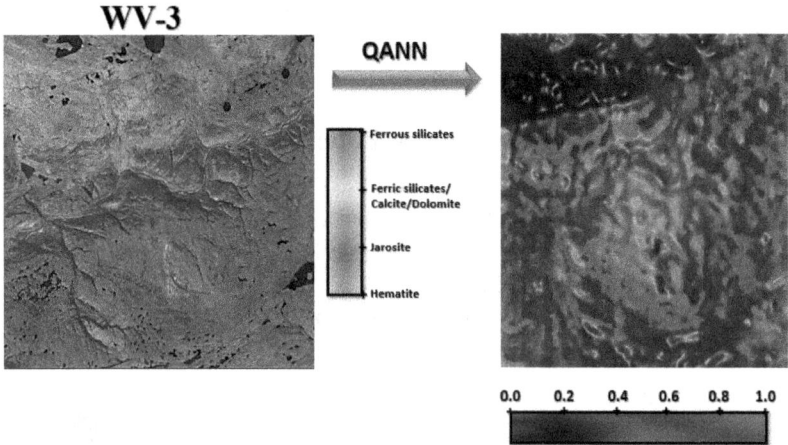

Figure 10.31. QANN for automatic detection of the copper potential zone in WV-3 data.

10.13 Why Quantum Machine Learning can be Used for Mineral Exploration?

This chapter introduced a quantum machine learning based on the neural network, QANN, that can signify categorized data, conventional or quantum, and be accomplished by supervised learning. In this regard, the quantum machine learning based on QANN consists of a sequence of parameter-dependent unitary transformations, which performs on an input quantum state. For binary classification, a single Pauli operator is measured on a designated readout qubit. The measured output, consequently, is the quantum neural network's predictor of the binary label of the input state for mineral exploration. Quantum machine learning based on the QANN eliminate the undesirable configuration using low estimated probability and the time required for quantum search, which does not expose the anticipated configuration [17, 20, 26].

In this circumstance, further registration is compulsory for gauging the performance of the quantum artificial neural network, for which permanent entanglement with the computation register is required. Subsequently, having directed the unitary transformations netting away from the entanglement between the input, computation, and output registers, entanglement of the calculation and performance registers would merely continue. When exploiting a quantum exploration on the performance register, which operates the phases of the conceivable performance values until the probability of gauging the desired output is close unity, the sequence measurements in QANN also permits the computing register collapse, which then illustrates the desired function f; thus, the quantum artificial neural network one is searching for accurate mineral exploration potential zones.

In this view, quantum machine learning based on QANN offers some very interesting effects in terms of processing information, like entanglement, which optimizes the accurate mineral exploration potential zones. However, QANN entanglement does not have a counterpart in conventional computation. In terms of quantum artificial neural networks, the struggle is to gauge the system that is to abolish the superposition, precisely when the system's state accomplishes the necessitated learning criteria; for instance, numerous of the training classes such as gold, iron and copper would be included. Likewise, when taking the internal computations, dynamic thresholds, and the weights into consideration, the conceivable conformations of the QANN within its superposition develop exponentially in the dimensions of the shared input vector as a function of threshold values. In this circumstance, the determination of accurate mineral potential zones is optimized owning to QANN entanglement and superposition. In other words, QANN optimization can allow for accurate automatic detection of the mineral exploration potential zones since it enhances the learning search procedures through quantum machine learning algorithm.

Consequently, the accurate performance of QANN is achieved owing to the following algorithm design phases as is well demonstrated in Figure 10.32.

Stage 1: Input image is embedded into a quantum circuit.
Stage 2: A quantum computation, allied with a unitary matrix (U_f), is performed on the system.   **Stage 3:** Generate the unitary transferring: let quantum operations have a unitary matrix (U_f) description applied to gate, operations and circuit that represents specific mineral elements.
Stage 4: The quantum system is finally measured, obtaining a list of conventional probability values [0–1].
Stage 5: Similarity of a conventional convolution input layer, every prospect probability value mapped to a different channel of a single output pixel.
Stage 6: Iterating the same procedure over dissimilar selected kernel windows, one can scan the complete input image.   **Stage 7:** Creating an output object, which can be organized as a multi-channel image.
Stage 8: End.   **Stage 9:** Run.

Figure 10.32. Algorithm design stages for QANN.

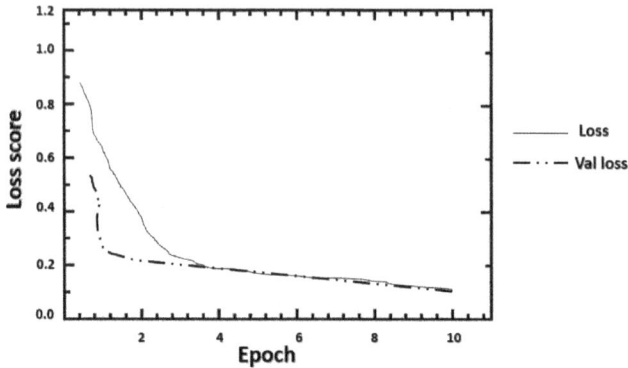

Figure 10.33. Modification in loss per epoch for training and validation.

Therefore, the modification in epoch lowers the loss score and recovers the complete hinge precision. In this understanding, the alteration in the precision score is perceived noteworthy boost after the third epoch and steadily improves, reducing the *loss* to 0.12 and the percentage change of loss score is shown in the analysis in Figure 10.33. The complete *accuracy* of the instigated QANN is 98.23%, whereas *recall* is 98.87%, respectively.

Subsequently, the designated metric is "Hinge Loss" for the QANN mineral potential zone explorations as the problem formulation is a binary classification problem. The loss signifies how much of the prediction fluctuates from actual data and, meanwhile, justification loss is not cast-off to renew weights. On the whole, hinge loss becomes the right measure of any neural network model. A hinge loss is estimated by comparing the prediction mineral exploration zones with the specific object for the likelihood and further subtract the value from 1. Subsequently, computing the maximum value between 0 and the result of the earlier computation offers an accurate view of any mineral potential zone exploration in multispectral and hyperspectral data. Overall, this proposed QANN algorithm acts as a supervised quantum machine learning algorithm, which emerges promising for operation on genuine quantum devices soon [20–26].

This chapter has demonstrated advanced quantum computing based on the quantum machine learning algorithm. However, QANN reveals the ability in determining the specific mineral potential zone in various optical remote sensing data across different example zones in the world than the quantum machine learning algorithm. The next chapter will demonstrate the most advanced quantum image processing technique in the automatic detection of specific mineralization using radar satellite data.

References

[1] Shuo G, Yuanyi Z, Huanchun Q, Dexin W, Hong X, Chao L, Yan LI, Xiaoyun ZH, Zengke WA. Geological characteristics and ore-forming time of the dexing porphyry copper ore mine in Jiangxi Province. Acta Geologica Sinica-English Edition. 2012 Jun; 86(3): 691–9.
[2] Hitzman M, Kirkham R, Broughton D, Thorson J, Selley D. The sediment-hosted stratiform copper ore system. Economic Geology. 2005; 100.

[3] McHugh JB. Concentration of gold in natural waters. Journal of Geochemical Exploration. 1988 Jan 1; 30(1-3): 85–94.

[4] Sherr R, Bainbridge KT, Anderson HH. Transmutation of mercury by fast neutrons. Physical Review. 1941 Oct 1; 60(7): 473.

[5] Willbold M, Elliott T, Moorbath S. The tungsten isotopic composition of the Earth's mantle before the terminal bombardment. Nature. 2011 Sep; 477(7363): 195–8.

[6] Klein C, Philpotts AR. Earth Materials: Introduction to Mineralogy and Petrology. Cambridge University Press; 2013.

[7] Gabr S, Ghulam A, Kusky T. Detecting areas of high-potential gold mineralization using ASTER data. Ore Geology Reviews. 2010 Oct 1; 38(1-2): 59–69.

[8] Gupta RP. Remote Sensing Geology. Springer; 2017 Nov 24.

[9] Saeedi M, Tangestani MH, Gourabjeri A. Prospecting for gold mineralization using geochemical, mineralogical, and WorldView-2 data: Siyah Jangal area case study, Northern Taftan Volcano, SE Iran. Natural Resources Research. 2021 Feb; 30(1): 129–52.

[10] Sabbaghi H, Moradzadeh A. ASTER spectral analysis for host rock associated with porphyry copper-molybdenum mineralization. Journal of the Geological Society of India. 2018 May; 91(5): 627–38.

[11] Clark RN. Spectroscopy of rocks and minerals, and principles of spectroscopy. Manual of Remote Sensing. 1999 Jun 25; 3: 3–58.

[12] Clark RN, King TV, Klejwa M, Swayze GA, Vergo N. High spectral resolution reflectance spectroscopy of minerals. Journal of Geophysical Research: Solid Earth. 1990 Aug 10; 95(B8): 12653–80.

[13] Cooley T, Anderson GP, Felde GW, Hoke ML, Ratkowski AJ, Chetwynd JH, Gardner JA, Adler-Golden SM, Matthew MW, Berk A, Bernstein LS. FLAASH, a MODTRAN4-based atmospheric correction algorithm, its application and validation. pp. 1414–1418. In IEEE International Geoscience and Remote Sensing Symposium 2002 Jun 24 (Vol. 3). IEEE.

[14] Rajendran S, Nasir S, Kusky TM, Ghulam A, Gabr S, El-Ghali MA. Detection of hydrothermal mineralized zones associated with listwaenites in Central Oman using ASTER data. Ore Geology Reviews. 2013 Sep 1; 53: 470–88.

[15] Sabins FF. Remote sensing for mineral exploration. Ore Geology Reviews. 1999 Sep 1; 14(3-4): 157–83.

[16] Zhang X, Pazner M, Duke N. Lithologic and mineral information extraction for gold exploration using ASTER data in the south Chocolate Mountains (California). ISPRS Journal of Photogrammetry and Remote Sensing. 2007 Sep 1; 62(4): 271–82.

[17] Schuld M, Bocharov A, Svore KM, Wiebe N. Circuit-centric quantum classifiers. Physical Review A. 2020 Mar 6; 101(3): 032308.

[18] Havlíček V, Córcoles AD, Temme K, Harrow AW, Kandala A, Chow JM, Gambetta JM. Supervised learning with quantum-enhanced feature spaces. Nature. 2019 Mar; 567(7747): 209–12.

[19] Pérez-Salinas A, Cervera-Lierta A, Gil-Fuster E, Latorre JI. Data re-uploading for a universal quantum classifier. Quantum. 2020 Feb 6; 4: 226.

[20] Du Y, Hsieh MH, Liu T, Tao D, Liu N. Quantum noise protects quantum classifiers against adversaries. arXiv preprint arXiv:2003.09416. 2020 Mar 20.

[21] LaRose R, Coyle B. Robust data encodings for quantum classifiers. Physical Review A. 2020 Sep 29; 102(3): 032420.

[22] Chen H, Wossnig L, Severini S, Neven H, Mohseni M. Universal discriminative quantum neural networks. Quantum Machine Intelligence. 2021 Jun; 3(1): 1–1.

[23] Verdon G, Marks J, Nanda S, Leichenauer S, Hidary J. Quantum hamiltonian-based models and the variational quantum thermalizer algorithm. arXiv preprint arXiv:1910.02071. 2019 Oct 4.

[24] Neukart F, Moraru SA. On quantum computers and artificial neural networks. Journal of Signal Processing Research. 2013 Mar 1; 2: 1.

[25] Mermin ND. Quantum Computer Science: An Introduction. Cambridge University Press; 2007 Aug 30.

[26] Neukart F, Grigorescu CM, Moraru SA. High order computational intelligence in data mining a generic approach to systemic intelligent data mining. pp. 1–9. In 2011 6th Conference on Speech Technology and Human-Computer Dialogue (SpeD) 2011 May 18. IEEE.

[27] Mermin ND. Quantum Computer Science: An Introduction. Cambridge University Press; 2007 Aug 30.

[28] Salehi T, Tangestani MH. Large-scale mapping of iron oxide and hydroxide minerals of Zefreh porphyry copper deposit, using Worldview-3 VNIR data in the Northeastern Isfahan, Iran. International Journal of Applied Earth Observation and Geoinformation. 2018 Dec 1; 73: 156–69.

[29] Kruse FA, Perry SL. Mineral mapping using simulated Worldview-3 short-wave-infrared imagery. Remote Sensing. 2013 Jun; 5(6): 2688–703.

[30] Kruse FA, Baugh WM, Perry SL. Validation of digital globe WorldView-3 Earth imaging satellite shortwave infrared bands for mineral mapping. Journal of Applied Remote Sensing. 2015 May; 9(1): 096044.

[31] Asadzadeh S, Filho CRS. Investigating the capability of WorldView-3 superspectral data for direct hydrocarbon detection. Remote Sensing Environment. 2016; 173: 162–173.

CHAPTER 11

Four-Dimensional Hologram Interferometry for Automatic Detection of Copper Mineralization Using Terrasar-X Satellite Data

11.1 What is the Real Age of Copper?

What is the age of copper? Pre-dynastic Egyptians recognized copper incredibly meticulously. In hieroglyphic writing, while the symbol exploits copper to symbolize eternal existence, the ankh, was correspondingly exploited to denote copper. Later, the Greek philosophers espoused the symbol, faintly revised, as ♀. The relationship between eternal resilience and the lifetime cost-efficiency of copper and its composites is confidently not spontaneous!

Consequently, the Egyptians attained most of their copper from the Red Sea Hills; however, copper predates ancient Egypt by numerous millennia, and it is nowadays recognized that an older civilization grounded on the Euphrates similarly newfangled copper and exploited well-developed melting practices. Copper is the origin of artifacts made up from melted metal, and archaeological site at Catal Huyuk nearby Konya in the Southern Anatolia has yielded slags resulting from the melting of copper that has been dated since 7,000 BC. Moreover, other civilizations in the Near- and the Middle East, Hindustan and China correspondingly established the consumption of copper as vital metal [1].

11.2 Occurrences of Copper

Previous chapters demonstrated that the zone's porphyries are mainly exaggerated by potassic alteration. Composed completely of alkali feldspar polymorphs and albite proxy of plagioclase, the alteration is subjugated by an additional prevalent biotite alternate of hornblende with extremely conserved igneous texture. In this view, granular quartz apart from quartz-K feldspar veinlets similarly occur along with the alteration.

A noteworthy percentage of the mine's copper is comprised of veins (Figure 11.1), veinlet packing faults and fault-interrelated with shatter zones. In this sense, orebodies of the mine are structured by a fault system, besides bearing a regularly composite kinematic primitive counting slanting and inverse interchanges. Therefore, the mine's copper is similarly categorized by the superposition principle of pyritic foremost-period veins with prevalent quartz-sericite. Moreover, the foremost-phase veins, encompassing quartz, pyrite, chalcopyrite and bornite, are spread throughout the dextral shear of the fault system. Sinistral strike-slip fault is formed due to shear stress across the faults. In this view, enargite veins can be developed across the other side of fault. Consequently, copper minerals and ores originate in both igneous and sedimentary rocks. This directs to the vital question: are copper and gold found together? Copper originates together with numerous other sorts of ore. It can originate near gold, silver, zinc, lead, and other kinds of metal deposits. When it is mixed in with other ores, copper is not regularly initiated in excessive copiousness [2].

Figure 11.1. The copper mine comprises of veins.

In this regard, what are the types of copper deposits? There are two distinct kinds of copper deposits: sulfide ore and oxide ore. Currently, the most common source of copper ore is the sulfide ore mineral chalcopyrite, which accounts for about 50% of copper production. Sulfide ores are processed via froth flotation to obtain copper concentrate. Therefore, copper is a copper red to reddish-brown, soft, ductile and malleable metal. Its colour as copper red is best seen by reflected light.

However, the laborious task of mining copper ore by hand made it difficult to obtain large quantities for production. During the Industrial Revolution, coal- and steam-powered machinery paved the way for a huge increase in copper production with mines smelting between 200 and 300 tonnes of copper ore per week.

11.3 Conventional Methods for Copper Extraction

The question is now: how do we mine copper? The ore is detached from the pounded sedimentary rocks in either open-pit or underground mines. Underground—sinking

Figure 11.2. Open-pit method.

a vertical shaft into the earth to reach the copper ore and driving horizontal tunnels into the ore for copper mining. Open-pit—90% of ore is mined using the open-pit technique (Figure 11.2).

Basically, can be mechanically creased. They can be also grounded in order to realize the autonomy of the copper mineral particles. Flotation by the injection of air and violent agitation is carried out with the crushed ore retained in suspension in water, to which surface-active agents have been adjoined. The procedure yields quintessence of copper mineral particles comprising roughly 30% copper, which is successively nourished to a *smeltery*, a boiler in which some iron and sulfur are detached, then to a *converter* or *converting* heater, where most of the residual iron and other scums are detached. Relying on the sort of smelting and converting heater exploited, as much as 99+% of the sulfur can be retrieved. It is exploited to form sulfuric acid, which is sold or exploited to leach copper from suitable ores directly. To this end, skirting is the perfect smelting-converting cycle for copper mining. The consequence is a contaminated (98+%) mixture of metal recognized as blister copper.

In this view, the blister is then fire-refined further to regulate its sulfur and oxygen contents, yielding metal that is adequately pure for numerous other-than-electrical consumptions. Nevertheless, since the fire-refined metal may encompass commercially practical absorptions of costly metals (generally silver and gold), most of it is cast into thick sheets identified as anodes that are directed to the huge electrolytic cells, where ultimate filtering acquires copper deposits. A dc electric current passes through the cells, dissolves the anodes and deposits the copper on cathodes. The final product of the refining process is electrolytic tough pitch (ETP) copper, normally containing between 99.94 and 99.96% Cu. Cathodes are re-melted under controlled conditions and cast into forms suitable for further extraction of copper mineral particles.

Modern procedures of mining permit economic leaching and electro-winning of copper from low-grade ores and mining techniques are unceasingly being polished and settled to attain the most effective elimination of copper from a broad diversity of ores from causes around the globe. The methods for mining copper from oxidized ores are fairly dissimilar from those retained for the sulfide ores. The oxidized ores, involving the silicates, carbonates and sulfates, are preserved by numerous approaches, which is approximately including all the procedure of leaching of the crushed ore with sulfuric acid to yield mixture solutions of copper sulfate. Nowadays, more than 13% of entirely "newfound" copper is twisted from leachates using both procedures of purified and conventional electrowinning for assembling of the copper mineral particles. However, sulfide ores are not proficiently tackled by sulfuric acid, but they can be leached if primarily oxidized by long exposure to the atmosphere and by exchange with rough stirring of thiobacillus thiooxidans and thiobacillus ferrooxidans bacteria.

11.4   What is the Major Challenge with Optical Remote Sensing and Microwave Radar Data?

A major challenge for using optical remote sensing for the mapping of surface mineralogy is the presence of lichen (particularly in Arctic regions) and vegetation. When the vegetation only partially covers the soil or rock surface, the measured reflectance is a mixture of the reflectance of the vegetation and that of the soil/rock. For example, through spectral mixture analysis, it may be possible to distinguish the contributions of the different land cover classes and extract information from the soil/rock component. In this regard, airborne hyperspectral data and spectral mixture analysis are used to quantify the degree of mine pollution (acid rock drainage) around a closed copper and zinc mine in a highly vegetated area to monitor the rehabilitation process. Moreover, airborne hyperspectral data and spectral mixture analysis are used to map rock types associated with nickel and copper mineralisation where lichen cover is abundant [3].

Geological mining detections and identifications in SAR images are required standard procedures and accurate algorithms. The main disadvantage of SAR images is speckling. The speckles do not allow any accurate retrieving of information from SAR data. According to Marghany [4], the high speckle noise in SAR images has posed great difficulties in inverting SAR images for geological features detection and mapping. A speckle pattern, consequently, is a random intensity pattern fashioned by the mutual interference of a set of wavefronts having different phases. Under this circumstance, they add together to give a resultant wave whose amplitude, and therefore intensity, varies randomly. In this context, Lopes et al. [5] stated that if each wave is modelled by a vector, then it can be seen if several vectors with random angles are added together. The length of the resulting vector, therefore, can be anything from zero to the sum of the individual vector lengths—a 2-dimensional random walk, sometimes known as a drunkard's walk. Further, when a surface is illuminated by a microwave, spectra according to diffraction theory, each point on an illuminated surface acts as a source of secondary spherical waves. The microwave spectra at any point in the scattered microwave field are made up of waves that have been scattered from each point on the illuminated surface. If the surface is rough enough to create path-length differences exceeding one wavelength, giving rise to phase changes greater than 2π, the amplitude, and hence the intensity, of the resultant backscatter microwave varies randomly [5–8].

Most of the target detection problems in SAR data are characterized by multiple objectives, which often oppose and contend with one another. Such complications arise in SAR applications, where more objective functions have to be minimized or maximized concurrently. Due to the multi-criteria nature of SAR speckles, the optimality of a solution has to be redefined, giving rise to the concept of Pareto optimality [4]. Evolutionary algorithms are appropriate to multi-objective problems because of their proficiency in synchronous exploration for multiple Pareto optimal solutions. Also, evolutionary algorithms do not rely on the precise features of the objective function to accurately perform, for instance, continuity, convexity and concavity. Therefore, Particle Swarm Optimization (PSO), Shuffled Complex Evolution Metropolis (SCEM), and Complex Evolution Metropolis (CEM) are major evolutionary algorithms. In this regard, Particle Swarm Optimization has been implemented in many different research fields because of its accurate performance in solving numerous single and multi-objective optimization problems such as despeckles [9].

11.5 Underground Mines and Open Pits Identification and Monitoring by InSAR

In Chapter 9, we learned the ability of InSAR to distinguish millimetric distortion throughout huge zones deprived of field access requisites. In this view, the elongated temporal sequences of SAR data, SAR remote sensing, and specifically InSAR have countless potential in the mining sector [10].

Over underground mines, the foremost matters are associated with ground subsidence owing to the extraction operations, and distressing the stability of constructions, and infrastructure. In this context, InSAR has archetypally been exploited to spot the potential subsidence, and map the level of the subsidence area. Moreover, the utilisation of the SAR time series can deliver the variation of the deformation [11]. Consequently, over open pits, stability matters occur owing to steep slopes regularly necessitated by the mining techniques and/or the geometry of the orebody (Section 11.2). In this regard, satellite InSAR can deliver evidence about the spreading, fullness, and growth of ground distortion over open pits [12]. Thrilling topography features, however, can cause geometrical deformation and diminish InSAR consequences at approximate sites. Besides, the decrease of coherence, perhaps, arises because of noteworthy ground fluctuations, and distortion can surpass the operative wavelength between two SAR flight acquisitions and cause phase ambiguity. In this regard, Paradella et al. [12] delivered a stimulating amalgamation of procedures to overwhelm this matter, involving InSAR, coherence, and SAR backscatter evaluation. On the other hand, the main issue in dealing with InSAR family techniques requires a careful solution for phase ambiguity in phase unwrapping techniques. Consequently, the next section will address the InSAR processing challenges.

11.6 InSAR Processing Challenges

As introduced in Chapter 9, the phase information of a single image is not directly usable. Yet, phase alterations between two (or more) dissimilar acquisitions are consequential. This is the foremost issue of SAR interferometry (InSAR). By calculating phase alterations between two acquisitions from two dissimilar sites, it is conceivable to formulate a Digital Elevation Model (DEM), while by calculating phase fluctuations between two acquisitions at two changing times, it is likely to perceive the ground deformation. In this understanding, the investigation of phase variations between two acquisitions at two dissimilar times can deliver information about ground distortion at a millimetre scale. As well as the excellent precision, the foremost return of the technique is the quantity competence over huge and rigid-to-access zones, throughout day and night and by any sort of weather conditions. However, InSAR is exaggerated by continuous restrictions are dominated InSAR products.

The first challenge is associated with the shifts that have a constituent in the line-of-sight (LOS) direction, which is contingent on the flying orientation of the satellite and the incidence angle of the radar beam. Figure 11.3 demonstrates that an area on the ground is imaged at two different time, t_0 and t_1. In this regard, the steeper the incidence angle, the accurate is the sensitivity to vertical dislocations. Therefore, looking toward the west, for instance, a descending orbit provides primarily nondistorted coverage in the west-facing slope, and an ascending orbit covers primarily east-facing slopes (Figure 11.4). In this view, sensitivity is identically low in instances where the real surface dislocation vector is perpendicular to the LOS. Owing to the north-south orbit direction, the sensitivity to a surface shift in this plane is close to zero.

The second challenge contains phase ambiguity because of the cyclic shape of the wave. The phase difference between the two acquisitions is higher than half of an effective wavelength, i.e., 0.5λ. In this understanding, a sensor with a wavelength of 5.55 cm with a revisit time of 24 days such as RADARSAT-2, a moving area with a velocity higher than 2.8 cm in 24 days will cause a phase ambiguity and InSAR cannot deliver applicable information about ground or object distortion.

Signal delay is the third challenge in the InSAR process owing to the atmospheric impact. In the other words, surface deformation occurring between the

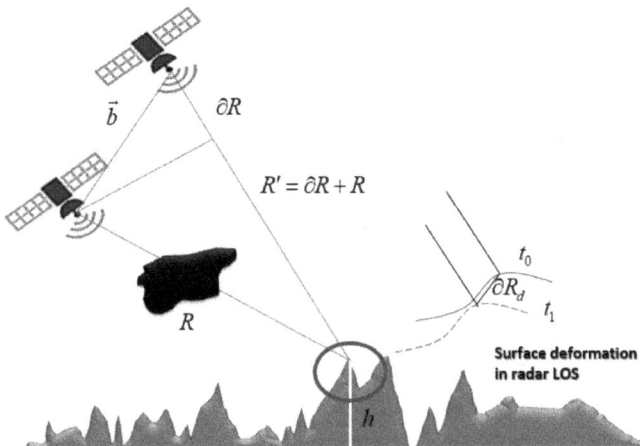

Figure 11.3. Surface detection displacement with InSAR.

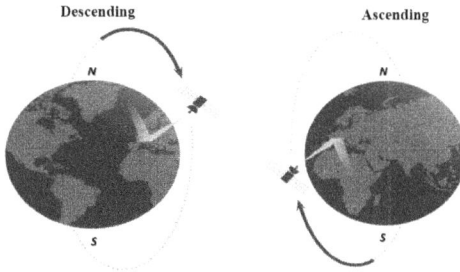

Figure 11.4. Descending and ascending orbits.

acquisition times will lead to an interferometric phase term ΔR_d. The cloud at the acquisition t_0 illustrates that atmospheric effects can affect the radar propagation producing an additional path delay (Figure 11.3). In this circumstance, the phase differences for measuring topographic components can be achieved. Besides, the large baseline is also caused by decorrelation, which distorts surface displacement information. This causes the fourth challenge that is known as phase decorrelation. In this regard, it occurs owing to fluctuations in the location of individual scatters within the resolution cell. Therefore, decorrelation is large because of the impacts of SAR imaging geometric,and is known as spatial decorrelation. In this understanding, the spatial decorrelation is interrelated to the baseline between the sensor at the dissimilar acquisitions. Moreover, temporal decorrelation is because of fluctuations in geometrical or electrical characteristics of the surface between the different acquisitions. Consequently, these fluctuations can be caused by shifting parts of vegetation, noteworthy fluctuations of water content or soil characteristics. Consistent with this perspective, a superior quantity of the interferometric phase is the complex correlation coefficient, or complex coherence $|\gamma|$. The values of $|\gamma|$ are between 0 and 1, where a coherence value of 1 corresponds to the perfect phase correlation between the two measurements. The coherence is frequently high on rocks or infrastructure and low in vegetated or wet zones. In this view, the calculation of the spatial spreading and temporal development of the coherence has a value to recognize and enumerate zones with large deformation or changes of surface properties such as open-pit mining.

11.7 Why Do We Still Need to Identify Well-known Open-Pit Mining?

Open-pit mines are, therefore, more susceptible to illegal activities, requiring effective monitoring. Conventional approaches such as field research and monitoring using unmanned aerial vehicles (UAVs) are time-consuming, labour-intensive, and difficult to apply at a large scale. In comparison, surface change detection using optical satellite images can facilitate monitoring on a large scale, yet optical images are susceptible to adverse weather conditions as well as sun illumination. Additionally, most of these methods are two-dimensional, making it impossible to detect the vertical changes caused by open-pit mining. Worse still, to avoid surveillance, many illegal mining activities are undertaken at night, making them even more difficult to detect remotely [13].

In this understanding, the association between SAR data and conventional *in-situ* measurements can deliver harmonizing information to endorse the evolution of mining areas, increase safety and avoid significant damages to surrounding infrastructure and environment. In this understanding, automatic detection of open-pit mining is important not only for nonidentified ones but also for historical monitoring and detecting the land subsidence owing to mining processes.

11.8 What are the Advantages of TanDEM Data?

The Tan Digital Elevation Models (DEM)-X functions are based totally on (i) cross-track two synthetic aperture radar (SAR) interferometry, (ii) along-track SAR interferometry, and (iii) new SAR techniques. The three radar strategies evolve from the machine specification described with the aid of the TerraSAR-X satellite and the interferometric configuration itself. Due to its manifold machine configurations, TanDEM-X is a bendy and multimode mission, which grants a broad variety of application possibilities. Dual cross-track SAR interferometry is an identified method to decide the terrain topography. The usage of this technique is mounted for the calculation of phase variances calculated with two SAR antennas separated by a terrific baseline. This permits approximating the radar elevation angle to the phase centre of every photo decision cell, where the precise elevations are derived from the interferometric section alternate [14, 20]. Dual along-track SAR interferometry is used to compute the velocities of shifting objects, which are a feature of a segment modification measurement, whereby the two SAR antennas achieve complex SAR images of the identical location with a quick time lag. Therefore, new SAR strategies will establish the possibility of superior SAR systems that have not yet or only incompletely been mounted on the ground or with aeroplanes.

The TanDEM-X operational consequence entails the synchronized operation of dual satellites hovering in a contiguous configuration (Figure 11.5). The amendment compressions for the development are as follows: (i) the orbits ascending nodes, (ii) the angle between the perigees, (iii) the orbital eccentricities, and (iv) the phasing between the satellites [14–16]. The main aim of the TanDEM-X mission is to create a unique third-dimensional (3-D) image of the Earth, which is regular in superiority and excellence in precision. At present, the DEMs are offered free, which are of the stumpy resolution, erratic, or defective. Likewise, DEMs are regularly set up on numerous databases and surface survey techniques [14–18]. In these regards, accurate DEM can be delivered by TanDEM-X satellite data. In other words, TanDEM-X satellite data deliver a homogeneous elevation model as an indispensable basis for many commercial applications and scientific questions.

DEMs provide a necessary footing for all topics in geological science; as a result, the demand for particular and straightforward DEMs is of accurate prominence (Figure 11.6). DEMs, for instance, are a requirement for the improvement of geological maps. Up-dated accurate high-resolution DEM is required in both forecasting of the earthquake occurrence and the volcanic eruption.

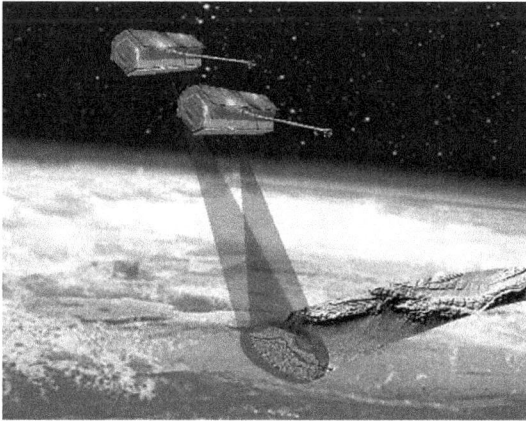

Figure 11.5. Formation of TanDEM-X satellite mission.

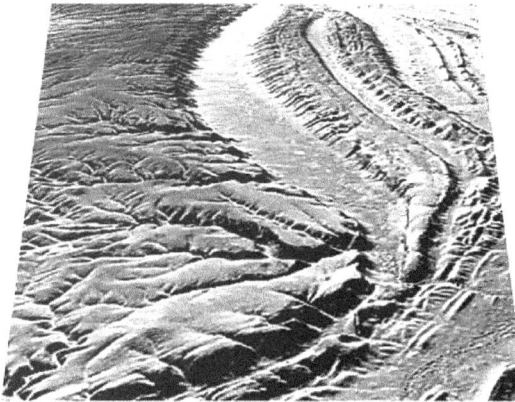

Figure 11.6. Accurate DEM construction using TanDEM-X mission.

Moreover, straightforward and unique DEMs are required for the recognition of perilous developed zones being affected by way of failures [14, 17]. The international coverage of topographic data at an ample splendid 3-D resolution is at present not reachable and would be delivered using the TanDEM-X mission [18–20]. The precise DEMs are necessary for geological, mining detections despite the negative aspects of SAR records due to speckles and object geometry distortions.

11.9 What is Meant by Four-Dimensional and Why?

The conventional well-recognized dimension is running from one dimension (1-D) to three-dimension (3-D). In this view, these dimensions are well known as a spatial dimension that involves x for 1-D, x,y for 2-D and x,y,z for 3-D. Needless to say, 2-D is encoded in the 1-D object and the 3-D object is encoded in the 2-D object (Figure 11.7). In this understanding, 4-D would be encoded into 3-D. A significant question can be raised: how is 4-D encoded and reconstructed in 3-D? If the

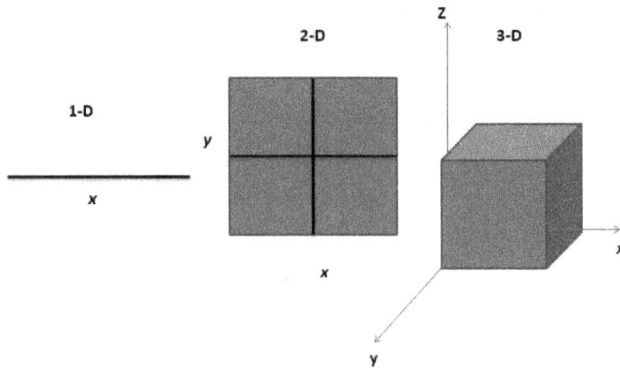

Figure 11.7. Development from 1-D to 3-D.

conventional dimensions are defined by space, and any dynamic fluctuations are restricted to space as a function of the time change. Then, time is considered as another dimension besides 3-D. In this circumstance, the time dimension begins to be exclusive of the three dimensions of space. In previous chapters, the time fluctuations as the inverse of the frequency lead to the Doppler frequency shift as a function of a delay time of the backscattered signal. This delay time instigates relativity of velocity bunching due to length contraction, then it can be useful to investigate the possibility of 4-D imaging in SAR.

Consequently, there are solely three spatial dimensions determined in the physical space. The fourth dimension originates in time, which is genuine "variation" to act with the three spatial dimensions. The distinct speculation of relativity verifies a relationship of time with the three spatial dimensions. Utilization of the 4-D in geological mining studies must be considered with the comprehending of the full scenario of the InSAR techniques. The critical question is about: can 4-D deliver perfect copper mining fluctuations and how is 4-D reconstructed in 3-D? This chapter delivers a transitory circumstantial technique to rationalize the physics of the commenced speculation on the 4-D copper mineralization reconstruction based on the SAR interferometry [14]. The 4-D speculation, perhaps, can be a novel model of copper mine deformation and retrieving in SAR images.

11.10 Does N-dimensional Exist?

It is a critical question about: do other dimensions, which are called 5th, 6th, 7th, 8th, 9th, 10th, and 11th exist? In this view, these mentioned dimensions have assumed to occur in the space-time dimensions and can be explained by the hyper dimensional manifolds of the string theory. It develops as constituted of entirely these dimensions, and at present is well-known as 11-D. The 'Calabi-Yau' manifolds are the mathematical hyperdimensional structural entities composed of the extra dimensions that exist within all things according to the super-string theory or theory of everything. In other words, in algebraic geometry, a Calabi–Yau manifold, additionally recognized as a Calabi–Yau space, is a unique sort of manifold which has properties, such as Ricci flatness, yielding functions in theoretical physics. Particularly in superstring theory, the greater dimensions of space-time are once in a while conjectured to take the structure of a 6-dimensional Calabi–Yau manifold, which lead to the concept of replicate symmetry (Figure 11.8). They must also exist in space and time. The eleven dimensions have to occur composed, the four of which are at ease to recognize 6-dimensional Calabi–Yau manifold, and the residual seven is extremely complicated. From the point of view of the Hilbert space, and the Hilbert-Einstein collaborations, an infinite number of vectors can exist around the Hilbert space. In this circumstance, the perception is extended to dimensions; hypothetically, an infinite number of dimensions can be generated. It leads to mathematical abstractions that do not have mathematical schemes to explain or describe.

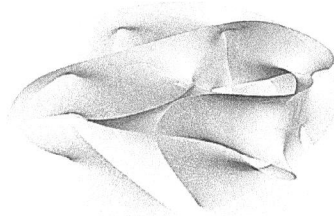

Figure 11.8. 6-dimensional Calabi–Yau manifold.

Consequently, hypothetically and abstractly, inter-dimensional existences are perhaps possible, and they possibly will or perhaps not have complexity intermingling with the recent shape of the universe. The growth in sublimation technology of nanotechnology, and Femto-technology that is involved in quantum computing, perhaps can overcome the absence of the specific mathematical algorithms to reconstruct inter-dimensional of any object on the Earth's surface.

11.11 What is Hologram Interferometry?

The important interrogative is: what is meant by hologram and can it be used in remote sensing processing as a novel tool? The applicable explanation of a hologram is grounded in Greek. In this tendency, the Greek word 'holos' indicates a total, area 'gramma' designates a message. It is outstanding of HOL-o-gram, which is a 3-D image view, formed by photographic propulsion. Different from 3-D or virtual practicality on an n-dimensional machine assemblage, a hologram is an imitation three-dimensional and unfastened-status image that doesn't emulate the spatial powerfulness or necessitates an inimitable observing tool.

In the view of physics, the coherent electromagnetic wave is the cornerstone of the hologram performance, as well as laser, illuminates the object, and its image is chronicled. Certainly, if the laser is visible in the film, it is reflected from the object and to a direct beam of the laser. In this regard, the interference of the laser beam of coherent electromagnetic beams on the film illuminates the object and generates a 3-D image [21, 23].

Holography is an exercise of the interference physical characteristics of the beam. By superimposing two sets of coloured and coherent wavefronts on a photographic plate, a microscopic interference pattern is produced. The developed plate, called a hologram, contains a permanent record of the interference pattern, i.e., it stores both amplitude and phase information. When the hologram is situated in an experience of the same seamless, coloured signalling, the diffracted photons produce a set of wavefront on photo termed as fringes. Consequently, when viewed, the diffracted wavefronts offer a remarkably realistic three-dimensional render of the object.

11.12 Marghany's 4-D Hologram Interferometry Theory for Copper Mineralization

The foremost interrogation is how to reconstruct 4-D from 3-D? This chapter postulates that 4-D can be implied from 3-D phase unwrapping of SAR hologram interferometry. SAR interferometry is a powerful approach to measure shifts in the stability of electromagnetic wavelength spectra. One significant limitation of common interferometric methods is that they require specular reflectors. This limitation can be removed by utilizing holography, allowing very small motions of arbitrary, diffusely reflecting, objects to be detected.

The keystone novelty of this chapter is to derive a new formula for 4-D hologram interferometry phase unwrapping using Hamming Graph. The main objective is to reconstruct the fourth dimension of copper mineralization using TanDEM-X satellite data by optimization of 4-D hologram interferometry.

Incidentally, computer-generated holograms do not compel definite targets to produce the hologram which provides information of the light scattered or diffracted off the object. In this view, mathematical routes can be involved to perform these decoherence hologram procedures. Under these circumstances, the SAR backscattering properties of the object are no longer a problem for the ideal object wave which can be computed mathematically. Consequently, a computer formed holographic image can be imitated by being numerically established by the physical phenomena of SAR interferometry.

Coherent signal sources such as radar signal are suitable in forming the holographic image. Similar to traditional interferometric synthetic aperture radar (InSAR), two radar complex signal is then exploited to construct holographic interferometry. In this view, interference pattern consequences are encoded in either a 2-D or 3-D medium that forms a hologram. In 3-D, space involves three dimensions, which are x,y,z. Further, time (t) is also a dimension. Therefore, space and time are not concepts that can be considered independently of one another when we are looking for change detection from satellite data [14–20].

In this understanding, the mathematical description of the two complex SAR hologram interferometry image $H(x, y, z, t)$ is formulated as:

$$H\left(x,y,z,t\right)=\varepsilon_0 2c\left|S^2\right|=\varepsilon_0 c\left|\left(S_1+S_2\right)^2\right|=\varepsilon_0 c\left|S_1^2+S_2^2+2S_1\bullet S_2\right|$$

$$=\varepsilon_0 c\left(S_1^2+S_2^2+2S_1 S_2\cos\vartheta_{12}\right)$$

$$=\varepsilon_0 c\left[\begin{array}{l}S_{01}^2\cos^2(kz-\omega_1 t)+\dfrac{S_{02}^2}{r^2}\cos^2(kr-\omega_2 t)+\\[2mm]\dfrac{2S_{01}S_{02}}{r}\cos(kz-\omega_1 t)\cos(kr-\omega_2 t)\cos(\vartheta_{12})\end{array}\right] \qquad (11.1)$$

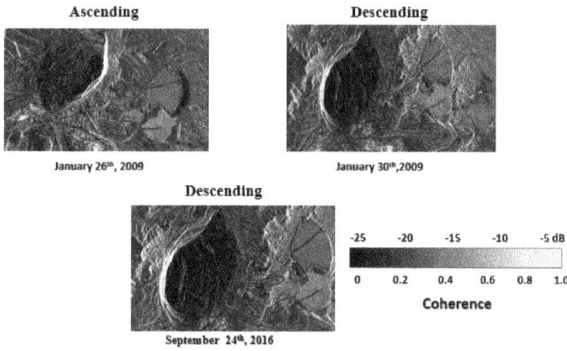

Figure 11.9. Three TerraSAR-X spotlight mode for copper mine of Chuquicamata, Chile.

Equation 11.1 says that the two complex radar images S_1 and S_2 are implemented to construct a 3-D hologram radar image. This hologram radar image involves the two signal wavenumber k, the baseline between both acquired SAR images r over the time of acquisition t and the radial frequency ω. In this view, designated microwave band spectra is present in the dual TerraSAR Spotlight mode data (Figure 11.9). These data were acquired on January 26th 2009, January 30th 2009, and September 24th, 2016, respectively. The first image is ascending, while the second and third images are descending along with the copper mine of Chuquicamata, Chile. These data have 1 m resolution with HH polarization. Their swath width is 10 km with an incident angle, which ranges from 20° to 55°.

The TerraSAR-X data is distinctive in that it has a characteristic speckled effect similar to other SAR sensors which looks like 'salt and pepper noise' (Figure 11.9) in the centre of the Atacama desert near South America's west coast where lies the world's largest open-cast copper mine. The mine was founded by Guggenheim Brothers at the beginning of the 20th century.

The images are the resulting grainy appearance that can be seen in Figure 11.9. This speckle noise is caused by random interferences when the signals arrive in the roughness of the studied areas and objects and that results in high spatial-frequency wavefront deformations. The speckle pattern superimposed on the images makes it very difficult to be used in many applications. Therefore, morphological features such as lineaments exist in TerraSAR-X data, which cannot be extracted and interpreted because of the speckle impacts (Figure 11.9). The highest coherence of 1 corresponds to the highest backscatter of −5 dB, which is noticed more within descending data along with the edge of open-pit mining. The low coherence is found inside the open-pit mining due to the deepest length of open-pit mining of approximately 800 m.

Besides, Equation 11.1 contains a part that just relies on both radar complex signal, plus an interferometry term involving the cosine of the angle ϑ_{12} between the two vector amplitudes (which may be a function of position x,y,z) [21, 24]. Generation of the realistic 3-D object in hologram radar interferometry is made up by wrapped phase $\phi_i(x, y, z)$, which is given by:

$$\phi_i(x, y, z) = \arctan\left(\frac{H_0(x, y, z) - H_2(x, y, z)}{H_1(x, y, z) - H_3(x, y, z)}\right), \quad i = 1, 2, 3 \tag{11.2}$$

Estimation of phase maps of different hologram radar images then can be achieved as:

$$\phi_{12}(x, y, z) = \arctan\left(\frac{\sin(\phi_2(x, y, z)) \times \cos(\phi_1(x, y, z)) - \cos(\phi_2(x, y, z)) \times \sin(\phi_1(x, y, z))}{\cos(\phi_2(x, y, z)) \times \cos(\phi_1(x, y, z)) + \sin(\phi_2(x, y, z)) \times \sin(\phi_1(x, y, z))}\right) \quad (11.3)$$

$$\phi_{23}(x, y, z) = \arctan\left(\frac{\sin(\phi_3(x, y, z)) \times \cos(\phi_2(x, y, z)) - \cos(\phi_3(x, y, z)) \times \sin(\phi_2(x, y, z))}{\cos(\phi_3(x, y, z)) \times \cos(\phi_2(x, y, z)) + \sin(\phi_3(x, y, z)) \times \sin(\phi_2(x, y, z))}\right) \quad (11.4)$$

$$\phi_{123}(x, y, z) = \arctan\left(\frac{\sin(\phi_{12}(x, y, z)) \times \cos(\phi_{23}(x, y, z)) - \cos(\phi_{12}(x, y, z)) \times \sin(\phi_{23}(x, y, z))}{\cos(\phi_{12}(x, y, z)) \times \cos(\phi_{23}(x, y, z)) + \sin(\phi_{12}(x, y, z)) \times \sin(\phi_{23}(x, y, z))}\right) \quad (11.5)$$

Equations 11.3 to 11.4 demonstrate that $\phi_{123}(x, y, z)$ diverges from 0 to 2π without 2π discontinuities, and then the image of the wrapped phase can be an expression. Consequently, reconstruction of the 3-D surface directly from $\phi_{123}(x, y)$ cannot deliver excellent consequences owing to $\phi_{123}(x, y, z)$ that does not reveal 2π discontinuities since SAR long-wavelength regularly rises in low quantity precision because of the nature of its coherence signal. In other words, the phase $\phi_1(x, y)$ has the wavelength ratio of three complex SAR images, which are $\dfrac{S_{12}(\lambda)}{S_1(\lambda)}$ discontinuities with equal spacing $S_{12}(\lambda)$. In this circumstance, inside every $S_{12}(\lambda)$, there are discontinuities of $\dfrac{S_{123}(\lambda)}{S_{12}(\lambda)}$ on $\phi_{12}(x, y, z)$. In other words, the conjugate of three complex SAR images $\dfrac{S_{123}(\lambda)}{S_{12}(\lambda)} \otimes \dfrac{S_{12}(\lambda)}{S_1(\lambda)}$ forms discontinuities $\phi_1(x, y, z)$. In general, the hologram interferometry image can be expressed as:

$$H_k(x, y, z) = \alpha(x, y, z) + \beta(x, y, z)\sin(\phi(x, y, z) + 0.5(\pi k)) + G(x, y, z). \quad (11.6)$$

Equation 11.6 demonstrates the occurrence of white Gaussian noise $G(x, y, z)$ since Equations 11.3 to 11.5 contain degradation in the computing of the absolute phase [25].

Succeeding Marghany [21] and Shaked and Rosen [22], the multiple viewpoint projection (MVP) holograms are generated by initial attaining of multiple projections of a 3D scene from various prospective viewpoints, and then digitally processing the acquired projections to yield the digital hologram of the scene. In this sense, the superposition of these TanDEM-X pulse electrons at any setting in space can be computed to attain the interference pattern from Fourier, Fresnel, image or other kinds of holograms.

Computer-generated holograms do not compel definite targets to produce the hologram, which provides information of the light scattered or diffracted off the copper mining location. In this regard, mathematical procedures can be involved to present these incoherent hologram procedures. Under these circumstances, the microwave pulse transmission and reflection properties of the copper mining in TanDEM-X satellite data are no longer a problem for the ideal object wave, which can be computed mathematically. Consequently, a computer-created holographic image can be simulated by being numerically based on the physical phenomena of SAR pulse diffraction and interference.

To understand the performance of a real gradient measurement interferometer, one needs to include decoherence. The object being measured will not have a completely uniform phase gradient. This is a potential advantage for the quantum Fourier transform interferometer, as it intrinsically uses multiple channels to analyze the gradient. A multiple channel interferometer can be averaged throughout any introduced phase noise, thus producing a robust measurement even in the presence of external noise.

An imperative inquiry is whether the scaling of directing superlative hologram interferometry can be decoded to a real-world occasion with phase noise and loss, directing to inadequate associations. Though, in the circumstance of a huge QFT network, the computation contains weighing a large number of SAR satellite data to overcome phase noise. In this understanding, abundant QFT interferometers up to M = 200 channels can be considered in the case of copper open-pit mining. The impacts of noise can be estimated using coherent state expansions. In fact, this approach demonstrates that the QFT interferometry is curiously robust against phase noise, with only an actual regular deprivation of peak visibility with phase-noise, not a disastrous collapse when deficiencies occur. The quantum Fourier transform interferometry (QuFTI) that measures the gradients of a phase-shift (Figure 11.10) is given by:

$$QFTI = \prod_{j=1}^{M-1} \frac{2j(M-j)\cos(M\phi) + M^2 - 2jM + 2j^2}{M^2} \tag{11.7}$$

Consequently the decoherence effects due to lowest dielectrical materials such as hydrocarbon micro-seepages are encompassed by using an autonomous phase noise term ξ at every site with a normal variation, having zero mean and variance σ^2. This is imperative since countless interferometry applications are delicate contrary to decoherence. In this regard, every mode has a noise term estimated by $\varphi_j = j\varphi + \xi_j$, $j = 1, 2, \ldots, M$.

Figure 11.11 displays the 2-D hologram resulting from QFT. It is interesting to attain that the computer-generated hologram rooted in the QFT algorithm can track copper mining variations or deformations. Nevertheless, this approach is not talented enough to reform the recreation of the information, which subsists in the TerraSAR data. This is because the amplitude and phase information about the footage pulse hologram interference vanishes.

Figure 11.10. QFTI generated phase-shift gradient.

Open -pit mining

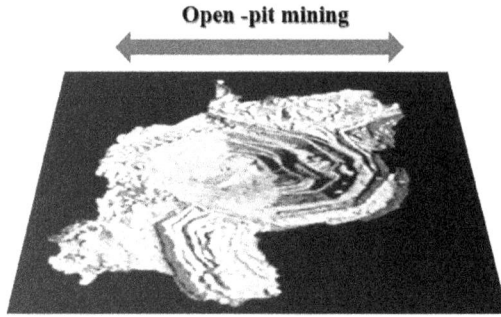

Figure 11.11. Computer-generated hologram.

11.13 Marghany' 4-D Phase Unwrapping Algorithm

The absolute phase cannot be used to determine the 3-D object reconstruction in the complex TerraSAR-X data but it can be determined using the phase unwrapping technique. To this end, the 4-D phase unwrapping can be expressed as:

$$\Psi_3(x,y,z,t) = \phi(x,y,z,t) + 2\pi \left(\begin{array}{l} \mathrm{int}\left(\dfrac{\phi_{12}(x,y,z,t)}{2\pi}\right) \otimes \left(S_1(\lambda) \times S_2(\lambda)^{-1}\right) \otimes \left(S_2(\lambda) \times S_3(\lambda)^{-1}\right) + \\ \mathrm{int}\left(\dfrac{\phi_{23}(x,y,z,t)}{2\pi}\right) \otimes \left(S_2(\lambda) \times S_3(\lambda)^{-1}\right) + G(\sigma) \end{array} \right) \quad (11.8)$$

The novelty of Equation 11.8 is based on the exchange of the absolute phase estimations of the three complex TerraSAR-X images with different acquisition time. In other words, the novel equation of Marghany's 4-D phase unwrapping is also considered the white Gaussian noise as a function of standard deviation σ. The aim of Equation 11.8 is to reconstruct phase unwrapping in 4-D and also to remove a discontinuity.

In this circumstance, let the Laplacian of the real phase ϕ_w be articulated in expressions of the wrapped phase, which is calculated using:

$$\nabla^2 \phi = \cos \Psi_w \nabla^2 (\sin \Psi_w) - \sin \Psi_w \nabla^2 (\cos \Psi_w) \quad (11.9)$$

where ∇^2 is the Laplace operator and by using an inverse of ∇^{-2}, the real phase ϕ' can be approximated in 4-D by assembling quantum Fourier-domain forward and ∇^{-2} as follows [26],

$$\phi'(i,j,k,t)_Q = QFT^{-1}\left[\frac{QFT[\cos\phi_w (QFT^{-1}[| p^2 + q^2 + r^2 + s^2 \rangle QFT(\sin\phi_w)])}{p^2 + q^2 + r^2 + s^2} \right]$$

$$-QFT^{-1}\left[\frac{QFT[\sin\phi_w (QFT^{-1}[(p^2 + q^2 + r^2 + s^2)_Q QFT(\cos\phi_w)])}{|p^2 + q^2 + r^2 + s^2 \rangle} \right] \quad (11.10)$$

where QFT^{-1} and QFT are quantum Fourier inverse and forward cosine or sine transforms, respectively, in 4-D of time-domain coordinates (i, j, k, t) and the Quantum Fourier-domain coordinates $|p^2 + q^2 + r^2 + s^2\rangle$. Equation 14.10 involves three types of operations: quantum forward and inverse cosine transforms, trigonometric operations, and the masking expression $|p^2 + q^2 + r^2 + s^2\rangle$.

The quantum Fourier transform is defined as:

$$QFT = \sum_j \alpha_k |j\rangle \to \sum_k \tilde{\alpha}_k |k\rangle, \text{where} \quad \tilde{\alpha}_k \equiv N^{-1} \sum_{j=1}^{N-1} e^{\frac{2\pi ijk}{N}} \alpha_j \tag{11.11}$$

Equation 11.11 reveals that in quantum computing, loading the coefficient (α_j) for the computation bases is not very simple. Also, there is no effective way to retrieve the coefficient (α_k) of the computation bases after the transformation. The measurement returns the computation basis, not the coefficient! To understand the inverse quantum Fourier transform (IQFT), let us consider a matrix M_Q, which represents QFT, then $M_Q M^\dagger = M^\dagger M_Q = I$, where $M^\dagger$ is the Hermitian adjoint of M_Q and I is the identity matrix. It follows that $M_Q - 1 = M^\dagger - 1$. Since there is an efficient quantum circuit implementing the QFT, the circuit can run in reverse to perform the Inverse Quantum Fourier Transform (IQFT). Thus, both transforms can be efficiently performed on a quantum computer. Then, IQFT can be estimated using:

$$QFT^{-1} (|\phi\rangle) = \frac{1}{2^N} \sum_{J=0}^{2^n-1} \sum_{k=0}^{2^n-1} e^{\frac{-2\pi ijk}{2^n}} |j\rangle \tag{11.12}$$

The discrete Fourier transform on 2^n amplitudes can be implemented as a quantum circuit consisting of only $O(n^2)$ Hadamard gates and controlled phase shift gates, where n is the number of qubits, in contrast with the classical discrete Fourier transform, which takes $O(n^2 n)$ gates, wherein the classical case n is the number of bits. Moreover, the precise QFT algorithms are just required $O(n\log n)$ gates to attain an effective calculation.

Following Marghany [21], the relative InSAR phase difference can be associated with a physical displacement through the sensitivity vector found in the hologram interferometry in two satellite data which can be expressed in 4-D as:

$$\begin{pmatrix} \Delta\Phi_1 \\ \Delta\Phi_2 \\ \Delta\Phi_3 \\ \Delta\Phi_4 \end{pmatrix} = \frac{2\pi}{\lambda} \begin{pmatrix} \vec{d}_{1i} & \vec{d}_{1j} & \vec{d}_{1k} & \vec{d}_{1p} \\ \vec{d}_{2i} & \vec{d}_{2j} & \vec{d}_{2k} & \vec{d}_{2t} \\ \vec{d}_{3i} & \vec{d}_{3j} & \vec{d}_{3k} & \vec{d}_{3t} \\ \vec{d}_{4i} & \vec{d}_{4j} & \vec{d}_{4k} & \vec{d}_{4t} \end{pmatrix} \begin{pmatrix} p \\ q \\ r \\ s \end{pmatrix} (t_n) \tag{11.13}$$

where d is the displacement along with orthogonal components of p, q, r, and s, in i, j, k, and t, respectively. Consistent with Marghany [21], the phase unwrapping can mathematically be extended into fourth-dimensional as:

$$\sum_{i,j,k,t} W_{i,j,k,t}^x \left| \left| \Delta\phi_{i,j,k,t}^x \right\rangle - \left| \Delta\psi_{i,j,k,t}^x \right\rangle \right|^L + \sum_{i,j,k,t} W_{i,j,k,t}^y \left| \left| \Delta\phi_{i,j,k,t}^y \right\rangle - \left| \Delta\psi_{i,j,k,t}^y \right\rangle \right|^L$$
$$+ \sum_{i,j,k,t} W_{i,j,k,t}^z \left| \left| \Delta\phi_{i,j,k,t}^z \right\rangle - \left| \Delta\psi_{i,j,k,t}^z \right\rangle \right|^L + \sum_{i,j,k,p} W_{i,j,k,p}^z \left| \left| \Delta\phi_{i,j,k,t}^t \right\rangle - \left| \Delta\psi_{i,j,k,t}^t \right\rangle \right|^L \tag{11.14}$$

where $|\Delta\phi\rangle$ and $|\Delta\psi\rangle$ are the quantum unwrapped and wrapped phase differences in x, y, z, t respectively, and W represents the user-defined weights. The simulation of

4-D open-pit mining is computed based on the standard deviation of its geometry shape displacement over the three SAR image acquisitions contains the 4-D phase unwrapping Ψ as $g_{123}(\sigma) \subset \Psi(x, y, z, t)$. In this understanding, the 4-D open-pit mining variations into hologram interferometry can be expressed as:

$$
\begin{pmatrix} \Delta\Phi_1 \\ \Delta\Phi_2 \\ \Delta\Phi_3 \\ \Delta\Phi_4 \end{pmatrix} = \begin{pmatrix} \vec{g}_{1i} & \vec{g}_{1j} & \vec{g}_{1_k} & \vec{g}_{1t} \\ \vec{g}_{2i} & \vec{g}_{2j} & \vec{g}_{2_k} & \vec{g}_{2t} \\ \vec{g}_{3i} & \vec{g}_{3j} & \vec{g}_{3_k} & \vec{g}_{3t} \\ \vec{g}_{4i} & \vec{g}_{4j} & \vec{g}_{4_k} & \vec{g}_{4t} \end{pmatrix} \otimes \frac{2\pi}{\lambda} \begin{pmatrix} \vec{\Psi}_{1i} & \vec{\Psi}_{1j} & \vec{\Psi}_{1k} & \vec{\Psi}_{1t} \\ \vec{\Psi}_{2i} & \vec{\Psi}_{2j} & \vec{\Psi}_{2k} & \vec{\Psi}_{2t} \\ \vec{\Psi}_{3i} & \vec{\Psi}_{3j} & \vec{\Psi}_{3k} & \vec{\Psi}_{3t} \\ \vec{\Psi}_{4i} & \vec{\Psi}_{4j} & \vec{\Psi}_{4k} & \vec{\Psi}_{4t} \end{pmatrix} \otimes \begin{pmatrix} W^x_{i,j,k,t} \\ W^y_{i,j,k,t} \\ W^z_{i,j,k,t} \\ W^t_{i,j,k,t} \end{pmatrix} \quad (11.15)
$$

Here, W represents the user-defined weights, and i, j, k, and t are cartesian coordinates of 4-D vector representation in hyperspace hologram interferometry. In this regard, this new formula is Marghany's 4-D hologram SAR interferometry.

11.14 Particle Swarm Optimization Algorithm

Kennedy and Eberhart in 1995 insinuated Particle Swarm Optimization (PSO) as a population-based optimization method. In this investigation, PSO is utilized to optimize the phase unwrapping issue and also to optimize the spatial variation of open-pit mining geometry modification. In other words, every particle in PSO is considered as the solution to the phase unwrapping and open-pit mining geometry modification (Figure 11.12).

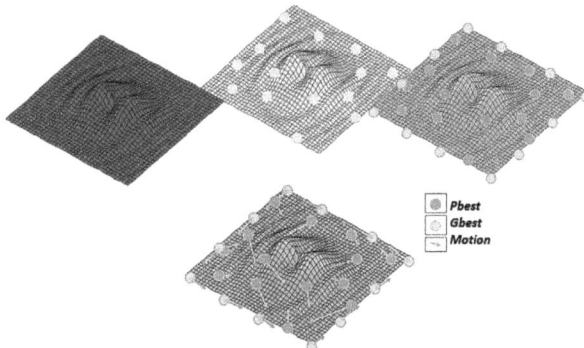

Figure 11.12. Optimization of PSO concept.

11.14.1 Optimization of 4-D Phase Unwrapping

Consequently, particle swarm can deliver accurate results due to its existing velocity in approaching optimal solution, i.e., its neighbours. Then the question is now: why do select PSO rather than other evolutionary computation techniques such as genetic algorithm, and multiobjective algorithms, for instance genetic algorithm for searching accurate copper open-pit mining in SAR data. PSO offers a quick ideal solution to any problem rather than other evolutionary computation techniques. PSO also delivers steady convergence properties, and it is straightforwardly executed.

A set of random particles is used at the PSO algorithm for automatic detection of copper open-pit mining in the SAR data. Consequently, these particles search the optimal 4-D phase unwrapping in order to construct 4-D of copper open-pit mining deformation.

In this regard, particles in a swarm come close to the optimal result through their contemporary speed, their earlier practice and the talent of their neighbours (Figure 11.12). In every initiation, every particle in a swarm is rationalized by dual superlative rates. The first one, therefore, is the finest result (best fitness) it has accomplished hitherto, which is termed as *Pbest*. Moreover, the alternative best rate that is traced by the particle swarm optimizer is the best fitness, attained hitherto by a little particle in the population set. In this view, the best fitness is a comprehensive best and termed as *gbest*. Consistent with each particle's flying experience and neighbour's flying experience, it moves its site in the search space along the 4-D phase unwrapping image and apprises its rapidity. In this understanding, the determination of the *Pbest* and gbest is achieved by:

$$V_{i,j,k,t}^{K+1} = \omega \times V_{i,j,k,t}^{K+1} + c_1 \times n \times (Pbest_{i,j,k,t}^{K} - \Psi_{i,j,k,t}^{K}) + c_2 \times r_2 \times (gbest_{i,j,k,t}^{K} - \Psi_{i,j,k,y}^{K}) \quad (11.16)$$

Equation 11.16 reveals that the PSO particles move through 4-D in the phase unwrapping space image i,j,k,t within velocity V through iteration K, which is regulated by the inertia weight factor ω. In this sense, c_1 and c_2 are the acceleration coefficients, r_1 and r_2 are positive random numbers between 0 and 1 whereas $Pbest_{i,j,k,t}^{k}$ is the best site of particle i,j,k,t in the phase unwrapping space image at iteration K, and $gbest_{i,j,k,t}^{K}$ is the best site of the set at iteration k. Consequently, the inertia weight ω is given by:

$$\omega = \omega_{max} - \frac{\omega_{max} - \omega_{min}}{K_{max}} \times K \quad (11.17)$$

where ω_{min} and ω_{max} present the minimum and maximum value of inertia weight factor, respectively. k_{max} matches the maximum iteration number and k is the existing iteration number.

Following Marghany [21, 23], the initial swarm particles are initialized to contain 3000 facts of particles to be implemented in the 4-D quality map $Q_{i,j,k,t}$ of phase unwrapping as:

$$Q_{Vol,t} = \frac{1}{Vol \otimes t} \otimes \left[\begin{array}{l} \sqrt{\sum (\Delta\Psi_{i,j,k,V}^{x} - \overline{\Delta\Psi_{i,j,k,V}^{x}})^2} + \sqrt{\sum (\Delta\Psi_{i,j,k,V}^{y} - \overline{\Delta\Psi_{i,j,k,V}^{y}})^2} + \\ \sqrt{\sum (\Delta\Psi_{i,j,k,V}^{z} - \overline{\Delta\Psi_{i,j,k,V}^{z}})^2} + \\ \sqrt{\sum (\Delta\Psi_{i,j,k,V}^{t} - \overline{\Delta\Psi_{i,j,k,V}^{t}})^2} \end{array} \right]. \quad (11.18)$$

where *Vol* is the 3-D volume of the hologram image to be encoded in 4-D, the time t is considered as 4th dimension, where the spatial variation of open-pit mining geometry deformation grows extremely within the time.

11.14.2 Optimization of Open-pit Mining Geometry Deformation

Determination of the open-pit mining geometry deformation in the 4-D model is required to initialize the variation of the deformation location in spatial(space) and time dimensions. To this end, let us assume v_s is the deformation rate in a particular position $v_{i,j,k,t}$, which is mathematically expressed as:

$$v_{i,j,k,t} = v_{i,j,k,t}\left(\min\right) + \left(v_{i,j,k,t}\left(\max\right) - v_{i+n,j+n,k+n,t+n}\left(\min\right)\right) \otimes rand(O,D) \qquad (11.19)$$

$$v_{s_{i,j,k,t}} = v_{s_{i,j,k,t}}\left(\min\right) + \left(v_{s_{i+n,j+n,k+n,t+n}}\left(\max\right) - v_{s_{i,j,k,t}}\left(\min\right)\right) \otimes rand(O,D) \qquad (11.20)$$

Both deformation growth and the actual position of open-pit mining are constrained to a minimum and maximum boundary in the circumstances of random mining deformation variation of locations and dimension, respectively. Consequently, the open-pit mining geometry deformation's new location $v_{i+1,j+1,k+1,t+1}$ is determined by:

$$v_{i+1,j+1,k+1,t+1} = rand(O, 1) - D^n \qquad (11.21)$$

Equation 11.21 reveals the random deformation of open-pit mining over any dimension number n. In this regard, $\Psi \subset v_{i,j,k,t} \subset v_{s_{i,j,k,t}}$, and $\Psi \cup v_{i+1,j+1,k+1,t+1}$. In this understanding, the 4-D phase unwrapping deformation image must be accomplished by random variations of open-pit mining geometry modification from one place to another in any space-dimension in the open-pit and its surrounding zones. This section is developed by the author in this recent study based on the work of Hudaib and Hwaitat [27].

Figure 11.13 demonstrates the results of the PSO algorithm for 4-D phase unwrapping optimization. The clear appearance of the copper mine in spotlight

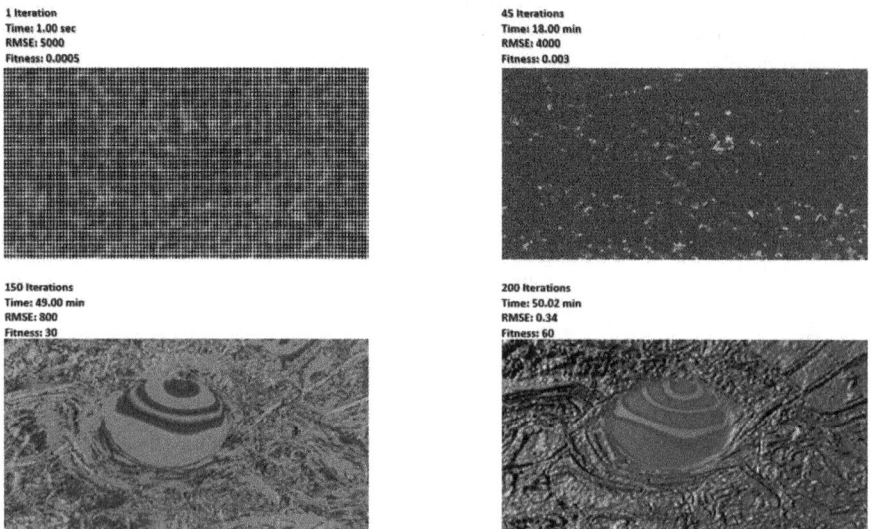

Figure 11.13. Particle Swarm Optimization for copper mine from spotlight mode with different iterations.

mode data is obvious. The PSO circumvents a decreasing resolution by making a weighted combination of running average with the neighbour surrounding pixels. This reduced the noise in the features' edge areas without losing edge sharpness. PSO within approximately 3 hours within 200 iterations can reconstruct the copper mine hole with some surrounding features with an RMSE of 0.34 with a fitness of 60.

11.15 Hamming Graph for 4-D Formation from Quantum Hologram Interferometry

Let us deduce that the vertices of the hypercube are $V \in H_C$, where H_C is the hypercube. In this regard, the hypercube graph is G_{H_C}, which is finite. The simplification expression of the graph is $G_{H_C} = (V, E_V)$, which is rooted in the adjacency matrix M_A as

$$M_A G[u, v] = [[(u, v) \in E_V]] \tag{11.22}$$

In fact, V is a set of vertices and a set of edges E_V between these vertices, where each edge is an unordered pair of vertices. In this circumstance, G_{H_C} required to be vertex-transitive, that is, for any $a, b \in V$, there is an automorphism $\pi \in M_A ut(G_{H_C})$ with $\pi(a) = b$ $\pi(a) = b$. The cartesian product $(G_{H_C} \oplus H_m)$ of graphs hypercube graph (G_{H_C}) and Hamming graph (H_m) (Figure 11.14), can be presented in a graph whose adjacency matrix is:

$$M \otimes M_A G_{H_C} + M_A H_m \otimes M \tag{11.23}$$

Let G_{V_n} designate the wide-ranging graph on n vertices. Then, the binary n-dimensional hypercube H_{C_n} may be defined recursively as

$$H_{C_n} = H_{C_n} - 1 \oplus G_{V_n} \quad \text{For } n \geq 2, \text{ and } H_{C1} = G_{V2}. \tag{11.24}$$

A quantum walk on $G_{H_C} \oplus H_m$ starting at the vertex (g,h) satisfies

$$G_{H_C} \oplus H_m \left| \phi_{G_{H_C} \oplus H_m}(t) \right\rangle = \left| \phi_{G_{H_C}}(t) \right\rangle \otimes \left| \phi_{H_m}(t) \right\rangle \tag{11.25}$$

where $\phi_{G_{H_C}}$ and ϕ_{H_m} are quantum phase unwrapping walks on G_{H_C} and H_m starting on vertices g and h, respectively. The adjacency matrix $G_{H_C} \oplus H_m$ is given by

$$M_A \otimes H_m + G_{H_C} \otimes M_A \tag{11.26}$$

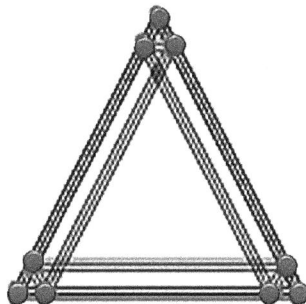

Figure 11.14. Hamming graph with H_m (3,3).

Since $M_A \otimes H_m$ and $G_{HC} \otimes M_A$ transform, then quantum phase unwrapping walks is formulated as:

$$\left| \phi_{G_{HC} \oplus H_m}(t) \right\rangle = \left| \phi_{G_{HC}}(t) \right\rangle \otimes \left| \phi_{H_m}(t) \right\rangle \tag{11.27}$$

In quantum mechanics, the natural approach to syndicate dual systems is across the tensor product $\otimes$. In this understanding, the Cartesian graph artifact attends a comparable function for quantum walks of the phase unwrapping. Equation 11.27 is the key ingredient to construct hypercube hologram interferometry through the Hamming graphs. In general, the Hamming graphs deliver tight characterization of quantum uniform mixing [21].

Therefore, the propagator can consistently be assumed as recitation of a quantum random walk on a systematic graph involving n-dimensional hypercube having a self-loop edge devoted to its separately vertices. The final version of the quantum walk for optimal search $2r$ can, therefore, be expressed by the alternating sequence of the unitary operators $U_m''^{(+)}$ as:

$$\left| \phi^{(e)}{}_{G_{HC} \oplus H_m}(2r) \right\rangle = (U_m^{(+)} U_m''^{(+)})^r \left| \phi^e{}_{G_{HC}}(t) \right\rangle \otimes \left| \phi_{H_m}(t) \right\rangle \tag{11.28}$$

where

$$U_m''^{(+)} = S^{(+)} C''^{(+)} \tag{11.28.1}$$

Here $C'''^{(+)}$ is the coin operator, which acts on the total Hilbert space $H^C \otimes H^V$ and S is a propagator of a quantum walk across the hypercube vertices V as:

$$S^{(+)} = \sum_{d,\vec{x}} \left| d, \vec{x} \oplus \vec{e}_d \right\rangle \left| d, \vec{x} \right\rangle \tag{11.28.2}$$

Here d is the direction of propagation and $\vec{e}$ is the edge of the vertices and $\oplus$ denotes the bitwise addition modulo 2 operator. Moreover, $\vec{x}$ is the Hamming weight of an integer that equals 1's in its binary string.

The coin operator $C'''^{(+)}$ is then determined from:

$$C'''^{(+)} = C_0 \otimes M_A + (C_1 - C_0) \otimes \sum_{j=1}^{m} \left| \vec{x}_{tg}^{(j)} \right\rangle \left\langle \vec{x}_{tg}^{(j)} \right| \tag{11.29}$$

C_0 presents the n-dimensional Grover operator, which is also known as the Grover diffusion operator" and C_1 is chosen to be -1. However, due to the symmetry of the hypercube graph (Figure 11.15), the vertices can always be relabeled in such a way that the marked vertex becomes $\vec{x}_{tg}^{(j)} = 0$. To formalize the task of finding multiple marked vertices, let us denote the number of elements marked by the oracle by m, and their labels by $\vec{x}_{tg}^{(j)}$ and $j = 1, \ldots, m$ [13].

Figure 11.15 explains how the 4-D is encoded into a 3-D hypercube. The effect of operator X is switching between phase unwrapping and vertices. In other words, the processors in the cubes of dimension 1, 2, and 3 are categorized with integers, which are represented as binary numbers. Therefore, those dual processors are neighbours in dimension $\vec{x}_{tg}^{(j)}$ in a hypercube of dimension, the encoded phase unwrapping that can be routed between any pair of processors in at most X hops.

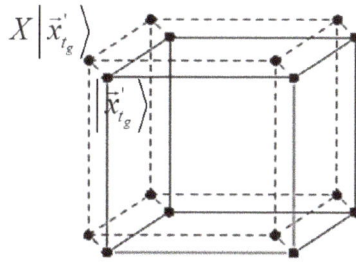

Figure 11.15. 3D encoded into a 4-D hypercube.

11.16 4D Hologram Interferometry of Open-Pit Mining

The keystone for achieving the hologram interferometry is rooted in the phase unwrapping technique. By using the hamming weight graph as discussed in the previous section, the 4-D phase unwrapping is performed. It is interesting to find that the proposed approach has created well-defined fringe patterns as compared to 2-D and 3-D phase unwrapping (Figure 11.16). Indeed, 4-D phase unwrapping using Hamming weight and a quantum walk algorithm is more vibrant as compared to 2-D and 3-D phase unwrapping. Particularly, 4-D phase unwrapping delivers well-identification of the complete cycle of hologram interferometry fringe patterns. Moreover, the proposed algorithm has minimized the error in the interferogram cycle due to the low coherence through copper mineralization depth due to incident angle impact. This could be an improvement of such previous works of Marghany [21, 12], Abdul-Rahman et al. [24], Marghany [28], Marghany [29], and Herráez [30].

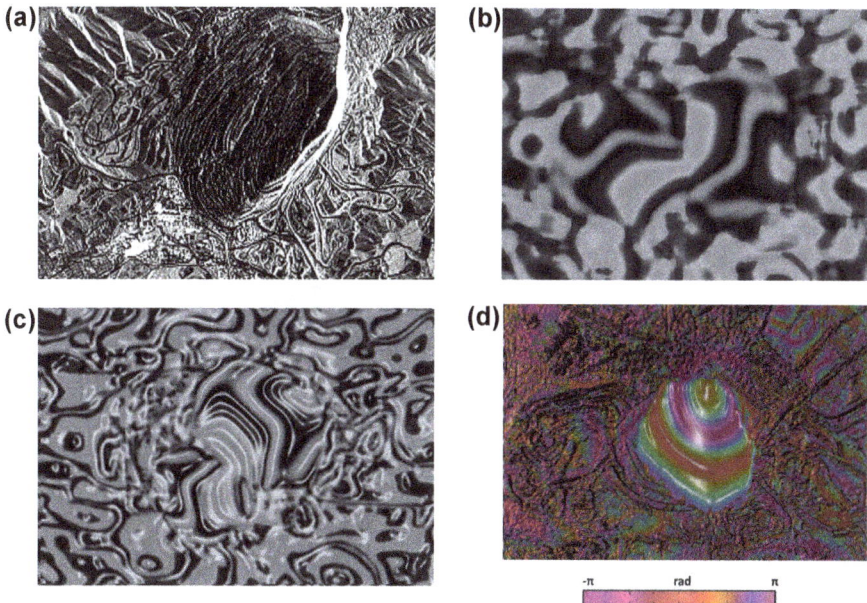

Figure 11.16. Phase unwrapping (a) original data; (b) 2-D; (c) 3-D; and (d) 4-D.

Figure 11.16d reveals the 4-D hologram interferometry phase unwrapping in which the full cycle of the 4-D phase unwrapping is perfectly completed and indicates a great deformation across the open-pit mining. However, 4-D phase unwrapping, yet, cannot distinguish any geological features but is able to determine the deformation owing to PSO optimization that is assisted by hamming weight graph. Consequently, Figure 11.17a demonstrated the 3-D of the copper mineralization of Chuquicamata, Atacama Desert, Chile that is delivered by TanDEM-X SAR digital elevation model. In this view, copper open-pit mining is well recognized but the geomorphological deformation is not well identified. On the contrary, Figure 11.17b reveals the 4-D visualization of copper open-pit mining with a complete volume geometry description. The 4-D visualization demonstrates ±40 m/year deformation that occurs along with the open-pit mining and downward owing to intensive mining activity. The dominant feature of the copper mine of Chuquicamata is an oval structure, which is the largest-ever depression in the Earth's surface ever produced by human effort.

The automatic distinction of pen-pit from the surrounding environment could ally with the PSO optimization algorithm. In this circumstance, PSO is allowed to state dual state open-pit geometry, its deformation, and surrounding environment state in a 4-D image. The fourth dimension, therefore, demonstrates the volumetric deformation of copper open-pit mining. The open-pit has the deepest depression of approximately 804.672 m, which is dominated by –60 m/year deformation downward and +40 m/year deformation upward across the copper open-pit mines.

These operational results confirm the study of Herráez et al. [30]. As a result, the hologram interferometry is deliberated as a deterministic set of rules, which is

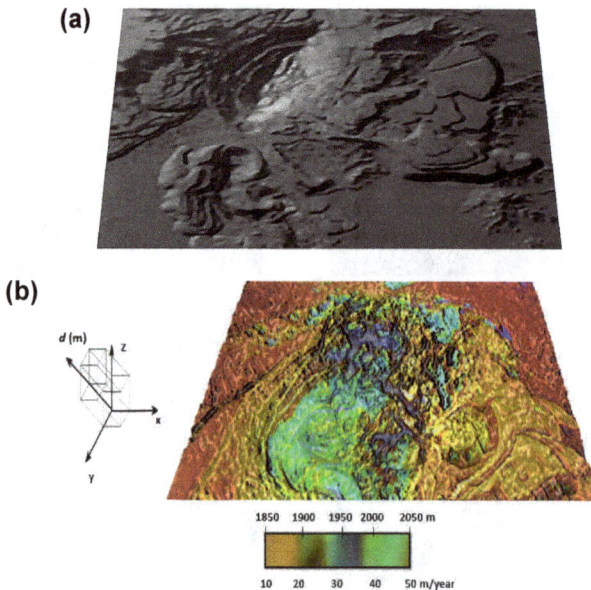

Figure 11.17. Comparison between (a) 3-D TanDEM-X and (b) 4-D visualization, created by hologram interferometry.

designated here to adjust a triangulation locally between binary diverse objects. It could precisely be articulated as an entire three-dimensional image (Figure 11.17a) proved on a high resolution. The hologram archives every slight detail open-pit mining position in space beside the texture and surface features of the copper mine of Chuquicamata.

All the hologram characteristics of the copper mine of Chuquicamata are preserved, and to an extent, enhanced. This corresponds to the feature of deterministic strategies of finding only sub-optimal solutions usually using Hamming graph. This confirms that quantum hologram interferometry is a robust process for evaluating displacements of the order of a photon. Therefore, involving a quantum hologram can assist in recovering the matter of specular reflection and decoherence along with the copper mine of Chuquicamata, besides permitting the tiny modification in random diffuse replication, which allows such highest coherence of copper mine of Chuquicamata to be identified precisely [21–30].

The visualization of the copper mine of Chuquicamata is extremely sharp by the reconstruction of the 4-D phase unwrapping by Hamming graph algorithm as rooted in quantum processing. These results deliver an accurate geometry morphology of the copper mine of Chuquicamata, which is indicated by the interference deformation and digital elevation model of open-pit mining. Needless to say that 4-D reconstruction of the open-pit using the quantum Hamming graph algorithm rooted in PSO delivers a perfect description of copper mineralization deformation geometry than 3-D of TanDEM-X SAR data.

11.17 Can Relativity Theory Explain 4-D Quantum Geometry Reconstruction?

Figure 11.18 spectacles the 4-D copper mineralization image from 3-D TanDEM-X data. In this view, 4-D delivers perfect morphological feature detections from diverse view angles [Figure 11.18]. This incorporates a subterranean copper mine within 800 m depth among contiguous mountains. Moreover, cavernous portrayals of the superiority of the substructures are well-identified in the 4-D image. The geomorphology of copper mineralization is to be more obvious with the rotation angle of 180° to 360°. Therefore, the geomorphology features of copper mineralization can be deeply recognized with altered view angles from 0° to 360°. The bright colour

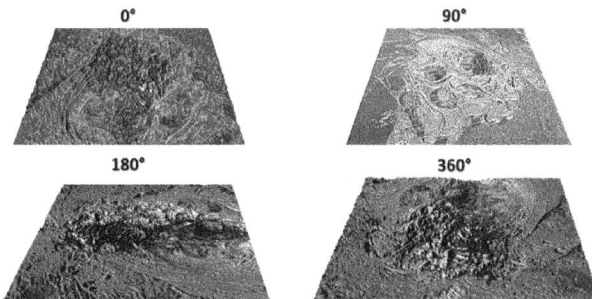

Figure 11.18. Different view angels of 4-D hologram interferometry.

along the object edges corresponds to the fourth dimension, which represents the time variation of data acquisitions.

From point of view of the hypercube, the geometry structure of the open-pit mining is clearly distinguished. This is well recognized in Figure 11.19. Four-dimensional open-pit mining is well-identified form different viewing angles as every side of the hypercube discloses a different geological geometry view of copper mineralization. More specifically, as geological geometry features of open-pit mining are extended up and down, right and left, the fourth dimension, which corresponds to tesseracts, extends. Lastly, as a result, they develop a cubic mass of identical trivial tesseracts, and when the tesseract is prearranged in the space world, it performs on the surfaces enclosing the upward and the right and left dimensions. From the point of view of duality, the local copper mineralization geometry and surrounding geological environmental pattern turn into extending on one side of the hypercube and topological objects on the other side. In this view, the geometry of copper mineralization features is considered as the unchaining of dual connection.

Moreover, the performance of a quantum random walk is rooted in the propagator across the n-dimensional hologram interferometry of hypercube vertices. The base of the 4-D construction of copper mineralization geometry is achieved by a quantum random walk search self-loop, which determines the edges of hypercube vertices. The final version of the quantum walk for optimal search is based on the sequence of the unitary operators, which assist to reconstruct a 4-D open-pit mining image. Therefore, for a quantum walk search with open-pit mining, the querying operator inverts the phase of the bases at each step where open-pit mining and surrounding geological environmental features such as faults and lineaments are located [24]. This phase inversion creates the probability amplitudes of the copper mineralization and its surrounding environmental geological feature and infrastructure vertices fluctuate more violently across the hypercube to form 4-D of open-pit mining.

In 4-D, the copper mineralization geometry deformation reveals vortex converts extended features, for instance, lines or surfaces as they are remarked in Figure 11.20. With the understanding that vortex, copper mineralization geometry deformation and, for instance, surrounding geological features such as lineaments are distinguished from each other, it indicates a strong combination constraint of the vortex with the copper mineralization geometry. In other words, the duality generates excellent dual gauge fields arbitrary exchanges between individual pixel scatter from different objects

Figure 11.19. Hypercube of copper mineralization geometry features in TerraSAR-X data.

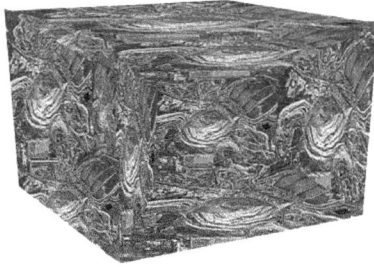

Figure 11.20. 4-D hypercube for copper mineralization geometry vortex pattern.

surrounding the open-pit mining. In this view, the dual gauge shown in Figures 11.17 to 11.20 is because of the hologram modelling. In this respect, holographic duality is known as a gauge. Lastly, the boundary system is referred to as a "hologram" of the bulk system. This confirms the work of Marghany [21].

The diverse observing angles revolve around a 3-D image, distance into the thickness and its thickness into complexity. By a natural spin, some of the three spatial dimensions can be transacted. At the moment, if rapidity or period is the fourth dimension, at that split second it is plausible to generate "spins" which restore the space into time or rapidity and vice versa [31–33]. In this context, the denotation of the period or velocity as the existence of the fourth dimension is that period and space can swirl into each other in specific mathematical rules. In other words, four-dimensional spins are exactly the modifications of space and period or rapidity, which are addressed by special relativity. In this view, space and time can be committed in a supreme regulation of the special relativity theory [21, 32].

Figure 11.21 reveals other clues of the features of four-dimensional space that can be protracted by pondering the obstruction of the assortment of duality. In this deference, the 4-D creates the spikiest structures, for instance, the sharp boundries of copper mineralization geometry (Figure 11.21). It is obviously implicit that the description of 4-D space is a variety of spaces, which embodies four coordinates. On the contrary, 3-D presents only the positions of three coordinates. The duality, in this manner, is also proven between a hologram and a hypercube. Consequently, this duality can be observed more clearly with the rotation of 4-D hypercube within 40° (Figure 11.22). The clear interaction between geometrical pattern of open-pit with surrounding infrastructures and also within its depth is addressed probably.

Figure 11.21. 4-D hypercube of the sharp boundaries of copper mineralization geometry.

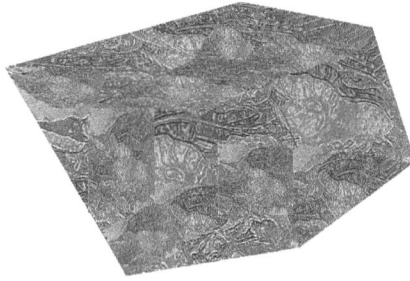

Figure 11.22. Oblique rotation of 4-D hypercube.

Lastly, in Einsteinian theories of relativity, space-time is 4-D since a fourth dimension that matches with a period is devoted to the 3-D space. In this sense, the space-time is 4-D, not space alone. The conception of space-time is established to be 4-D as a function of the Lorentz Transformations. In this circumstance, the physical morphological features of the copper mineralization geometry are relocated from an inertial reference frame (x,y,z), for instance, to another (x',y',z',d_t) [21, 31–34]. Figure 11.23 confirms the 4-D over time with spatial deformation through the copper mineralization as it is demonstrated that the deformation of –60 m occurs within its depth of 800 m. It can be concluded that the 4-D hologram quantum interferometry could be a new approach for automatic discrimination of open-pit mining in SAR images due to the involvement of PSO, Hamming graph, quantum walk search, and quantum hypercube algorithms. It is promising to transfer or convert 2-D and 3-D images into a 4-D image.

Finally, such a study of synthetic aperture radar imaging for the copper mineralization geometry cannot be the isolated or ignored quantum mechanism. As we can see, the quantum mechanics are allied with the beginning of the electron signal radiated from the antenna till the image processing that involves automatic detection of open-pit mining 2-D to the 4-D images.

This chapter is devoted to delivering a new technique for automatic detection of open-pit mining in SAR data such as the sequences of TerraSAR-X. To this end,

Figure 11.23. 4-D of copper mineralization depth deformation within surrounding environment.

hologram quantum interferometry is implemented based on the PSO. In this regard, a Hamming graph based on the quantum walk of PSO search is used to convert the 2-D TerraSAR-X data into four-dimensional images. The chapter demonstrates that 4-D copper mineralization images reveal the interference between open-pit geometry deformation and the surrounding geological features such as faults and lineaments. The 4-D copper mineralization geometry is considered one of the proofs of Einstein's relativity theory. Needless to say that the consequences exhibit that the 4-D of copper mineralization extended morphological characteristic detection such as the depth of a copper mine and surrounding infrastructures. It can be stated that the integration of PSO with the 4-D phase unwrapping of TanDEM-X satellite data is known as Marghany's 4-D phase unwrapping technique, which is a high-quality promise technique for 4-D reconstruction of copper mineralization and other objects in SAR complex data.

References

[1] Bornhorst TJ, Paces JB, Grant NK, Obradovich JD, Huber NK. Age of native copper mineralization, Keweenaw Peninsula, Michigan. Economic Geology. 1988 May 1; 83(3): 619–25.

[2] Reed ST, Martens DC. Copper and zinc. Methods of Soil Analysis: Part 3 Chemical Methods. 1996 Jan 1; 5: 703–22.

[3] Richter N, Staenz K, Kaufmann H. Spectral unmixing of airborne hyperspectral data for baseline mapping of mine tailings areas. International Journal of Remote Sensing. 2008 Jul 1; 29(13): 3937–56.

[4] Marghany M. Particle swarm optimization for geological feature detection from PALSAR data. In Naypyitaw, Myanmar: 35th Asian Conference of Remote Sensing 2014.

[5] Lopes A, Touzi R, Nezry E. Adaptive speckle filters and scene heterogeneity. IEEE Transactions on Geoscience and Remote Sensing. 1990 Nov; 28(6): 992–1000.

[6] Touzi R. A review of speckle filtering in the context of estimation theory. IEEE Transactions on Geoscience and Remote Sensing. 2002 Nov; 40(11): 2392–404.

[7] Marghany M. Multi-objective evolutionary algorithm for oil spill detection from COSMO-SkeyMed satellite. pp. 355–371. In International Conference on Computational Science and Its Applications 2014 Jun 30. Springer, Cham.

[8] Yu Y, Acton ST. Speckle reducing anisotropic diffusion. IEEE Transactions on Image Processing. 2002 Dec 16; 11(11): 1260–70.

[9] Yisu J, Knowles J, Hongmei L, Yizeng L, Kell DB. The landscape adaptive particle swarm optimizer. Applied Soft Computing. 2008 Jan 1; 8(1): 295–304.

[10] Smets B, Samsonov SV, d'Oreye N. Ground deformation associated with post-mining activity at the french-german border revealed by multidimensional time series analysis of SAR data acquired in various orbital geometries. pp. H24C-02. In AGU Fall Meeting Abstracts 2012 Dec (Vol. 2012).

[11] Samsonov S, d'Oreye N, Smets B. Ground deformation associated with post-mining activity at the French–German border revealed by novel InSAR time series method. International Journal of Applied Earth Observation and Geoinformation. 2013 Aug 1; 23: 142–54.

[12] Paradella WR, Ferretti A, Mura JC, Colombo D, Gama FF, Tamburini A, Santos AR, Novali F, Galo M, Camargo PO, Silva AQ. Mapping surface deformation in open pit iron mines of Carajás Province (Amazon Region) using an integrated SAR analysis. Engineering Geology. 2015 Jul 2; 193: 61–78.

[13] Wang S, Lu X, Chen Z, Zhang G, Ma T, Jia P, Li B. Evaluating the feasibility of illegal open-pit mining identification using insar coherence. Remote Sensing. 2020 Jan; 12(3): 367.

[14] Marghany M. DEM reconstruction of coastal geomorphology from DINSAR. pp. 435–46. *In*: International Conference on Computational Science and Its Applications. Berlin, Heidelberg: Springer; 2012d.

[15] Marghany M. Three-dimensional lineament visualization using fuzzy B-spline algorithm from multispectral satellite data. pp. 213–32. *In*: Escalante-Ramirez B (ed.). Remote Sensing Advanced Techniques and Platforms. Croatia: InTech Open Access Publisher; 2012.

[16] Marghany M. DInSAR technique for three-dimensional coastal spit simulation from Radarsat-1 fine mode data. Acta Geophys 2013; 61: 478–93.

[17] Marghany M. Four-dimensional of copper mineralization using TanDEM-X satellite data. Geoscience Bulletin. 2019 May 31; 1(1).

[18] Zink M, Bachmann M, Brautigam B, Fritz T, Hajnsek I, Moreira A, Wessel B, Krieger G. TanDEM-X: The new global DEM takes shape. IEEE Geoscience and Remote Sensing Magazine. 2014 Jun 23; 2(2): 8–23.

[19] Albino F, Smets B, d'Oreye N, Kervyn F. High-resolution TanDEM-X DEM: An accurate method to estimate lava flow volumes at Nyamulagira Volcano (DR Congo). Journal of Geophysical Research: Solid Earth. 2015 Jun; 120(6): 4189–207.

[20] Rossi C, Gonzalez FR, Fritz T, Yague-Martinez N, Eineder M. TanDEM-X calibrated raw DEM generation. ISPRS Journal of Photogrammetry and Remote Sensing. 2012 Sep 1; 73: 12–20.

[21] Marghany M. Advanced Remote Sensing Technology for Tsunami Modelling and Forecasting. CRC Press; 2018 Jul 4.

[22] Shaked NT, Rosen J. Modified Fresnel computer-generated hologram directly recorded by multiple-viewpoint projections. Applied Optics. 2008 Jul 1; 47(19): D21–7.

[23] Marghany M. Four-dimensional water detection in mars using spline algorithm. International Journal Hydrology. 2018; 2(5): 607–11.

[24] Abdul-Rahman H, Gdeisat M, Burton D, Lalor M. Fast three-dimensional phase-unwrapping algorithm based on sorting by reliability following a non-continuous path. pp. 32–40. In Optical Measurement Systems for Industrial Inspection IV 2005 Jun 13 (Vol. 5856). International Society for Optics and Photonics.

[25] Song L, Chang Y, Xi J, Guo Q, Zhu X, Li X. Phase unwrapping method based on multiple fringe patterns without use of equivalent wavelengths. Optics Communications. 2015 Nov 15; 355: 213–24.

[26] Ghiglia DC, Pritt MD, Unwrapping TD. Theory, Algorithms, and Software. New York, USA: A Wiley-Interscience Publication. 1998.

[27] Hudaib AA, Hwaitat AK. Movement particle swarm optimization algorithm. Modern Applied Science. 2017; 12(1): p1–17.

[28] Marghany M. Hologram interferometric SAR and optical data for fourth-dimensional urban slum reconstruction. CD of 35th Asian Conference on Remote Sensing (ACRS 2014), Nay Pyi Taw, Myanmar 27–31, October 2014, http://www.a-a-r-s.org/acrs/administrator/components/com.../ OS-303%20.pdf [Access on April 2 2021].

[29] Marghany M. Fourth-dimensional optical hologram interferometry of RapidEye for Japan's tsunami effects. CD of 36th Asian Conference on Remote Sensing (ACRS 2015), Manila, Philippines, 24–28 October 2015, http://www.a-a-r-s.org/acrs/index.php/acrs/acrs-overview/ proceedings-1?view=publication&task=show&id=1691 [Access on April 2 2021].

[30] Herráez MA, Burton DR, Lalor MJ, Gdeisat MA. Fast two-dimensional phase-unwrapping algorithm based on sorting by reliability following a noncontinuous path. Applied Optics. 2002 Dec 10; 41(35): 7437–44.

[31] Edery A. Casimir energy of a relativistic perfect fluid confined to a D-dimensional hypercube. Journal of Mathematical Physics. 2003 Feb; 44(2): 599–610.

[32] Miller MA. Regge calculus as a fourth-order method in numerical relativity. Classical and Quantum Gravity. 1995 Dec; 12(12): 3037.

[33] Moghaddam FF, Moghaddam RF, Cheriet M. Curved space optimization: a random search based on general relativity theory. arXiv preprint arXiv:1208.2214. 2012 Aug 10.

[34] Rucker R. Geometry, relativity and the fourth dimension. Courier Corporation; 2012 Jun 8.

Index

About the Author

Prof. Maged Marghany is currently the director of Global Geoinformation, Sdn. Bhd. In 2020 he was ranked amongst the top 2 percent of scientists in a global list compiled by the prestigious Stanford University. He also ranked as the first oil spill scientist in a global list of over last 50 years compiled by the prestigious Universidade Estadual de Feira de Santana, Universidade Federal da Bahia, and Universidade Federal de Pernambuco; Brazil.

He is the author of 6 titles including: Advanced Remote Sensing Technology for Tsunami Modelling and Forecasting which is published by Routledge Taylor and Francis Group, CRC and Synthetic Aperture Radar Imaging Mechanism for Oil Spills, which is published by Elsevier, His research specializes in microwave remote sensing and remote sensing for mineralogy detection and mapping. Previously, he worked as a professor of remote sensing in Indonesian and Malaysian universities. Maged has earned many degrees including a post-doctoral in radar remote sensing from the International Institute for Aerospace Survey and Earth Sciences, a PhD in environmental remote sensing from the Universiti Putra Malaysia, a Master of Science in Physical oceanography from the University Pertanian Malaysia, general and special Diploma of Education and a Bachelor of Science in physical oceanography from the University of Alexandria in Egypt. Maged has published well over 250 papers in international conferences and journals and is active in International Geoinformatic, and the International Society for Photogrammetry and Remote Sensing (ISPRS).

For Product Safety Concerns and Information please contact our EU
representative GPSR@taylorandfrancis.com
Taylor & Francis Verlag GmbH, Kaufingerstraße 24, 80331 München, Germany